This book describes the extremely powerful technique of molecular dynamics simulation, which involves solving the classical many-body problem in contexts relevant to the study of matter at the atomic level. The method allows the prediction of the static and dynamic properties of substances directly from the underlying interactions between the molecules.

Because no aternative approach is capable of handling such a broad range of problems at the required level of detail, molecular dynamics methods have proved themselves indispensable in both pure and applied research. The author adopts a dual approach: the text is partly tutorial, but it also contains a large number of computer programs for practical use. Thus, the book will serve as an introduction to the subject for beginners, and also as a reference manual for more experienced practitioners. The material covers a wide range of practical methods and real applications, and is organized as a series of case studies, taking the reader through all the steps involved, from how to formulate the problem, how to develop the necessary software, and, finally, how to use the programs in making actual measurements.

This volume will be of value to advanced students and researchers in physics, chemistry, polymer science, and materials science in universities and industrial laboratories.

THE ART OF MOLECULAR DYNAMICS SIMULATION

THE ART OF MOLECULAR DYNAMICS SIMULATION

D. C. RAPAPORT

Bar-Ilan University

CAMBRIDGE
UNIVERSITY PRESS

Published by the Press Syndicate of the University of Cambridge
The Pitt Building, Trumpington Street, Cambridge CB2 1RP
40 West 20th Street, New York, NY 10011-4211, USA
10 Stamford Road, Oakleigh, Melbourne 3166, Australia

First published 1995

Printed in Great Britain at the University Press, Cambridge

A catalogue record of this book is available from the British Library

Library of Congress cataloguing in publication data

Rapaport, D. C.
The art of molecular dynamics simulation / D. C. Rapaport.
p. cm.
Includes bibliographical references and index.
ISBN 0-521-44561-2
1. Matter – Properties.   2. Molecular dynamics – Computer simulation.
I. Title.
QC173.3.R36   1995
539′.6–dc20   95-10724   CIP

ISBN 0 521 44561 2 hardback

Contents

Preface		*page* xi
About the software		xiv
1	Introduction	1
1.1	Historical background	1
1.2	Computer simulation	2
1.3	Molecular dynamics	4
1.4	Organization	9
1.5	Further reading	11
2	Basic molecular dynamics	12
2.1	Introduction	12
2.2	Soft-disk fluid	12
2.3	Methodology	19
2.4	Programming	20
2.5	Results	32
2.6	Further work	41
3	Simulating simple systems	42
3.1	Introduction	42
3.2	Equations of motion	42
3.3	Potential functions	44
3.4	Interaction computations	47
3.5	Integration methods	56
3.6	Initial state	63
3.7	Performance measurements	66
3.8	Trajectory sensitivity	70
3.9	Further work	73
3.10	Additional material	74

4	Equilibrium properties of simple fluids	78
4.1	Introduction	78
4.2	Thermodynamic measurements	79
4.3	Structure	85
4.4	Further work	99
4.5	Additional material	101
5	Dynamical properties of simple fluids	114
5.1	Introduction	114
5.2	Transport coefficients	114
5.3	Measuring transport coefficients	118
5.4	Space–time correlation functions	128
5.5	Measurements	135
5.6	Further work	142
5.7	Additional material	142
6	Alternative ensembles	146
6.1	Introduction	146
6.2	Feedback methods	147
6.3	Constraint methods	158
6.4	Further work	166
7	Nonequilibrium dynamics	168
7.1	Introduction	168
7.2	Homogeneous and inhomogeneous systems	168
7.3	Direct measurement	169
7.4	Modified dynamics	179
7.5	Further work	189
8	Rigid molecules	191
8.1	Introduction	191
8.2	Dynamics	192
8.3	Molecular construction	205
8.4	Measurements	211
8.5	Further work	220
9	Flexible molecules	222
9.1	Introduction	222
9.2	Description of molecule	222
9.3	Implementation details	224
9.4	Properties	228
9.5	Further work	232
10	Geometrically constrained molecules	234
10.1	Introduction	234
10.2	Geometric constraints	234

10.3	Solving the constraint problem	237
10.4	Internal forces	245
10.5	Implementation details	253
10.6	Measurements	258
10.7	Further work	261
11	Other interactions	263
11.1	Introduction	263
11.2	Long-range forces	263
11.3	Three-body potentials	277
11.4	Further work	284
12	Step potentials	285
12.1	Introduction	285
12.2	Computational approach	286
12.3	Event management	298
12.4	Results	305
12.5	Generalizations	307
12.6	Further work	314
13	Time-dependent phenomena	316
13.1	Introduction	316
13.2	Open systems	316
13.3	Thermal convection	318
13.4	Obstructed flow	323
13.5	Further work	329
14	Algorithms for supercomputers	330
14.1	Introduction	330
14.2	The quest for performance	330
14.3	Techniques for distributed processing	332
14.4	Distributed MD simulation	334
14.5	Techniques for vector processing	349
14.6	Further work	357
14.7	Additional material	357
15	The future	360
15.1	Role of simulation	360
15.2	Limits of growth	361
15.3	Interactive simulation	362
15.4	Coda	363
Appendices		364
A1	Allocating arrays	364
A2	Organizing input data	365
A3	Managing extensive computations	368

A4   Utility functions   372
A5   Header files   378
A6   Variables   378
Bibliography   383
Function index   390
Subject index   393

Preface

Molecular dynamics simulation provides the methodology for detailed microscopic modeling on the molecular scale. After all, the nature of matter is to be found in the structure and motion of its constituent building blocks, and the dynamics is contained in the solution to the N-body problem. Given that the classical N-body problem lacks a general analytical solution, the only path open is the numerical one. Scientists engaged in studying matter at this level require computational tools to allow them to follow the movement of individual molecules and it is this need that the molecular dynamics approach aims to fulfill.

The all-important question that arises repeatedly in numerous contexts is the relation between the bulk properties of matter – be it in the liquid, solid, or gaseous state – and the underlying interactions among the constituent atoms or molecules. Rather than attempting to deduce microscopic behavior directly from experiment, the molecular dynamics method – MD for short – follows the constructive approach in that it tries to reproduce the behavior using model systems. The ever-increasing power of computers makes it possible to pose questions of greater complexity, with a realistic expectation of obtaining meaningful answers; the inescapable conclusion is that MD will – if it hasn't already – become an indispensable part of the theorist's toolbox. Applications of MD are to be found in physics, chemistry, biochemistry, materials science, and in branches of engineering.

This is a recipe book. More precisely, it is a combination of an introduction to MD for the beginner, and a cookbook and reference manual for the more experienced practitioner. The hope is that through the use of a series of case studies, in which real problems are studied, both goals can be achieved. The book can be read from cover to cover to explore the principles and capabilities of MD, or it can be used in cookbook style – with a certain amount of cross-referencing – to obtain the recipe for a particular kind of computation. Some familiarity with classical and statistical mechanics, numerical methods and computer programming is assumed.

The case studies take the reader through all the stages from initial problem statement to the presentation of the results of the calculation. The link between these endpoints is the computer program – the recipe. The results of the simulations are 'experimental' observations, in the sense that the simulation is an experiment conducted on an actual, albeit highly idealized, substance. Some of these observations amount to mere measurement, while others can include the discovery of qualitatively novel effects; the custom of referring to MD simulation as computer experimentation is most certainly justified.

Computer programs are an important part of any MD project and feature prominently among the recipes. The view that programs are best kept out of sight along with the plumbing is seriously outdated, and program listings are integrated into the text, with the same status as mathematical equations. After all, a computer program is merely the statement of an algorithm (supplemented by a myriad details to assist the computer in performing its task), and an algorithm is a mathematical procedure. Without the details of the programs the recipe-oriented goal would not have been met: there are many vital, but often subtle, details that only emerge when the program is actually written, so that the program text is an essential part of any recipe and is meant to be read.

Given the near-ubiquity of MD, the choice of material had to be restricted to avoid a volume of encyclopedic size. The focus is on the simplest of models, since these form the basis of almost all later developments. Even what constitutes a simple model is open to debate, and here a modest bias on the part of the (physicist) author may be discerned. The emphasis is on showing that MD can reproduce known physical phenomena at a qualitative and semiquantitative level, but without fine-tuning potential functions, molecular structures, or other parameters, for precise quantitative agreement with experiment. Exercises such as demonstrating the solid–fluid phase transition in a system of soft-disk atoms, observing the local ordering in a simple model for water, and following the gyrations of a highly idealized polymer chain, are all far more rewarding experiences for the beginner than detailed computations of specific heats or viscosities across the entire state space of the system. Quantitative detail is not neglected, however, although here some aspects will obviously appeal to more limited segments of the audience.

The model systems to be introduced in these pages can be readily extended and adapted to problems of current interest; suggestions for further work of this kind accompany the case studies, and can serve as exercises (or even research projects) in courses devoted to simulation. The same holds true for the computational techniques. We cover a variety of methods, but not all combinations of methods and problems. In some cases all that is required is a simple modification or combination of the material covered, but in other cases more extensive efforts are called for – the literature continues to report such methodological developments. While MD can hardly be regarded as a new technique, neither can it be regarded as a fully matured method, and thus there are often several ways of approaching a particular problem, with little agreement on which is to be preferred. It is not our intent to pass judgment, and examples based on alternative methods are included.

The practical side of MD is no less important than the theoretical. A true appreciation of the capabilities and shortcomings of the various methods, an understanding of the assumptions used in the models, and a feeling for what kinds of problem are realistic candidates for MD treatment can only be obtained from experience. This is something that even users of commercial and other packaged software should be aware of. The bottom line is that the reader should be prepared to use this book like any other recipe book: off to the kitchen and start cooking!

January, 1995   Dennis C. Rapaport

About the software

Software availibility

Readers interested in downloading the software described in this book in
computer-readable form for personal, noncommercial use, should contact
the Cambridge University Press World Wide Web site (using suitable
Web-browsing software) at http://www.cup.cam.ac.uk. The home page
for this book is http://www.cup.cam.uk/onlinepubs/ArtMolecular/
ArtMoleculartop.html, where the software and other related material
can be found. The author's e-mail address – for error reports and
suggestions – is rapaport@phys8.ph.biu.ac.il.

xiv

1

Introduction

1.1 Historical background

The origins of molecular dynamics (MD) are rooted in the atomism of antiquity. The ingredients, while of more recent vintage, are not exactly new. The theoretical underpinnings amount to little more than Newton's laws of motion. The significance of the solution to the many-body problem was appreciated by Laplace [del51]: 'Given for one instant an intelligence which could comprehend all the forces by which nature is animated and the respective situation of the beings who compose it – an intelligence sufficiently vast to submit these data to analysis – it would embrace in the same formula the movements of the greatest bodies of the universe and those of the lightest atom; for it, nothing would be uncertain and the future, as the past, would be present to its eyes'. And the concept of the computer, without which there would be no MD, dates back at least as far as Babbage, even though the more spectacular hardware developments continue to this day. Thus MD is a methodology whose appearance was a foregone conclusion, and indeed not many years passed after digital computers first appeared before the first cautious MD steps were taken [ald57, gib60, rah64].

The N-body problem originated in the dynamics of the solar system, and the general problem turns out to be insoluble for three or more bodies. Once the atomic nature of matter became firmly established, quantum mechanics took charge of the microscopic world, and the situation became even more complicated because even the constituent particles seemed endowed with a rather ill-defined existence. But a great deal of the behavior of matter in its various states can still be understood in classical (meaning nonquantum) terms, and so it is that the classical N-body problem is also central to understanding matter at the microscopic

1

level. And it is the task of the numerical solution of this problem that MD addresses.

For systems in thermal equilibrium, theory, in the form of statistical mechanics, has met with a considerable measure of success, particularly from the conceptual point of view. Statistical mechanics provides a formal description – based on the partition function – of a system in equilibrium; however, with a few notable exceptions, there are no quantitative answers unless severe approximations are introduced, and even then it is necessary to assume large (essentially infinite) systems. Once out of equilibrium, theory has very little to say. Simulations of various kinds, including MD, help fill the gaps on the equilibrium side, but in the more general case it is only by means of simulation – principally MD – that progress is possible.

From the outset, the role of computers in scientific research has been a central one, both in experiment and in theory. For the theoretician, the computer has provided a new paradigm of understanding. Rather than attempting to obtain simplified closed-form expressions that describe behavior by resorting to (often uncontrolled) approximation, the computer is now able to examine the original system directly. While there are no analytic formulae to summarize the results neatly, all aspects of the behavior are open for inspection.

1.2 Computer simulation

Science requires both observation and comprehension. Without observation there are no facts to be comprehended; without comprehension science is mere documentation. The basis for comprehension is theory, and the language of theoretical science is mathematics. Theory is constructed on a foundation of hypothesis; the fewer the hypotheses needed to explain existing observations and predict new phenomena, the more 'elegant' the theory – Occam's razor.

The question arises as to how simulation is related to physical theory. University education abounds with elegant theoretical manipulation and is a repository for highly idealized problems that are amenable to closed-form solution. Despite the almost 'unreasonable applicability' of mathematics in science [wig60], the fact is that there is usually a chasm between the statement of a theory and the ability to extract quantitative information useful in interpreting experiment. In the real world, exact solutions are the notable exception. Theory therefore relies heavily on approximation, both analytical and numerical, but this is often uncontrolled and so reliability may be difficult to establish. Thus it might be said that simulation rests on the basic theoretical foundations, but tries to avoid much of the approximation normally associated with theory, replacing it by a more elaborate calculational effort. Where theory and simulation differ is in regard to cost. Theory requires few resources beyond the cerebral and is therefore 'cheap'; simulation needs the hardware and, despite plummeting prices, a computer system for tackling problems at the forefront of any field can still prove costly.

Simulation also draws from experiment. Experimental practice rests on a long (occasionally blemished) tradition; computer simulation, because of its novelty, is still somewhat more haphazard, but methodologies are gradually evolving. The output of any simulation should be treated by the same statistical methods use in the analysis of experiments. In addition to estimating the reliability of the results (on the assumption that the measurements have been made correctly) there is also the issue of adequate sampling. This is particularly important when attempting to observe 'rare' events: quantitative studies of such events require that the entire occurrence be reproduced as many times as necessary to assure adequate sampling – if computer resources cannot accommodate this requirement it is presumptuous to expect reliable results.

What distinguishes computer simulation in general from other forms of computation, if such a distinction can be made, is the manner in which the computer is used: rather than merely performing a calculation, the computer becomes the virtual laboratory in which a system is studied – a numerical experiment. The analogy can be carried even further; the results emerging from a simulation may be entirely unexpected, in that they may not be at all apparent from the original formulation of the model. A wide variety of modeling techniques have been developed over the years, and those relevant for work at the molecular level include, in addition to MD, classical Monte Carlo [all87], quantum-based techniques involving path-integral [ber86c, gil90] and Monte Carlo methods [sch92], and MD combined with electron density-function theory [rem90, tuc94], as well as discrete approaches such as cellular automata and the lattice–Boltzmann method [doo91].

Although the goal of science is understanding, it is not always obvious what constitutes 'understanding'. In the simulational context, understanding is achieved once a plausible model is able to reproduce and predict experimental observation. Subsequent study may lead to improvements in the model, or to its replacement, in order to explain further experiments, but this is no different from the way in which science is practiced in the broader context. Clearly, there is no inherent virtue in an excessively complex model if there is no way of establishing that all its features are essential for the desired results (Occam again). The practical consequence of this policy is that, despite any temptation to do otherwise, features should be added gradually. This helps with quality control in the notoriously treacherous world of computer programming: since the outcome of a simulation often cannot be predicted with enough confidence to allow full validation of the computation, the incremental approach becomes a practical necessity.

Simulation plays an important role in education. It takes little imagination to see how interactive computer demonstrations of natural phenomena can enrich any scientific presentation. Whether as an adjunct to experiment, a means of enhancing theoretical discussion, or a tool for creating hypothetical worlds, simulation is without peer. Especially in a conceptually difficult field such as physics, simulation can be used to help overcome some of the more counter-intuitive concepts encountered even at a relatively elementary level. As to the role of MD, it can bring to life the entire invisible universe of the atom, an experience no less rewarding for the experienced scientist than for the utter tyro. But, as with education in general, simulation must be kept honest, because seeing is believing, and animated displays can be very convincing irrespective of their veracity.

1.3 Molecular dynamics

1.3.1 Foundations

The theoretical basis for MD embodies many of the important results produced by the great names of analytical mechanics – Euler, Hamilton, Lagrange, Newton. Their contributions are now to be found in introductory mechanics texts (such as [gol80]). Some of these results contain fundamental observations about the apparent workings of nature; others are elegant reformulations that spawn further theoretical development. The simplest form of MD, that of structureless particles, involves little more than Newton's second law. Rigid molecules require the use of the Euler equations, best expressed in terms of Hamilton's quaternions. Molecules with internal degrees of freedom, but that are also subject to structural constraints, involve the Lagrange method for incorporating geometric constraints into the dynamical equations. Normal equilibrium MD corresponds to the microcanonical ensemble of statistical mechanics, but in certain cases properties at constant temperature (and sometimes pressure) are required; there are ways of modifying the equations of motion to produce such systems, but of course the individual trajectories no longer represent the solution of Newton's equations.

The equations of motion can only be solved numerically. Because of the nature of the interatomic interaction, exemplified by the Lennard-Jones potential with a strongly repulsive core, atomic trajectories are unstable in the sense that an infinitesimal perturbation will grow at an exponential rate (see Chapter 3), and it is fruitless to seek more than moderate accuracy in the trajectories, even over limited periods of time. Thus a comparatively low-order numerical integration method often suffices; whether or not this is adequate emerges from the results, but the reproducibility of MD measurements speaks for itself. Where softer interactions are involved, such as harmonic springs or torsional interactions, either or both of which are often used for modeling molecules with internal degrees of freedom, a higher-order integrator, as well as a smaller timestep than before, may be more appropriate to accommodate the fast internal motion. The numerical treatment of constraints introduces an additional consideration, namely that the constraints themselves must be preserved to much higher accuracy than is provided by the integration method, and methods exist that address this problem. All these issues, and more, are covered in later chapters.

While MD is utterly dependent on the now ubiquitous computer, an invention of the twentieth century, it pays little heed to the two greatest developments that occurred in physics in the very same century – relativity and quantum mechanics. Special relativity proscribes information transfer at speeds greater than that of light; MD simulation assumes forces whose nature implies an infinite speed of propagation. Quantum mechanics has at its base the uncertainty principle; MD requires (and provides) complete information about position and momentum at all times. In practice, the phenomena studied by MD simulation are those where relativistic effects are not observed and quantum effects can, if necessary, be incorporated as semi-classical corrections (quantum theory shows how this should be done) [mai81]. But, strictly speaking, MD deals with a world that, while intuitively appealing to late-nineteenth-century science (not to mention antiquity), has little concern for anything that is 'nonclassical'. This fact has in no way diminished the power and effectiveness of the method.

1.3.2 Relation to statistical mechanics

Statistical mechanics (for example [mcq76]) deals with ensemble averages. For the canonical ensemble, in which the temperature T and number of particles N are fixed, the equilibrium average of some quantity G is expressed in terms of phase-space integrals involving the potential energy $U(r_N)$:

$$\langle G \rangle = \frac{\int G(r_N)e^{-\beta U(r_N)}dr_N}{\int e^{-\beta U(r_N)}dr_N} \tag{1.1}$$

where r_N denotes the full set of coordinates, $\beta = 1/k_B T$, and k_B is the Boltzmann constant. This average corresponds to a series of measurements over an ensemble of independent systems.

The ergodic hypothesis relates the ensemble average to measurements carried out for a single equilibrium system during the course of its natural evolution – both kinds of measurement should produce the same result. Molecular dynamics simulation follows the dynamics of a single system and produces averages of the form

$$\langle G \rangle = \frac{1}{M} \sum_{\mu=1}^{M} G_\mu(r_N) \tag{1.2}$$

over a series of M measurements made as the system evolves. Assuming that the sampling is sufficiently thorough to capture the typical behavior, the two kinds of averaging will be identical. The observation that MD corresponds to the microcanonical (constant energy) ensemble, rather than to the canonical (constant-temperature) ensemble, will be addressed when it appears likely to cause problems.

1.3.3 Relation to other classical simulation methods

The basic Monte Carlo method [all87] begins by replacing the phase-space integrals in (1.1) by sums over states:

$$\langle G \rangle = \frac{\sum_s G(s)e^{-\beta U(s)}}{\sum_s e^{-\beta U(s)}} \tag{1.3}$$

Then, by a judicious weighting of the states included in the sum, which for the general case results in

$$\langle G \rangle = \frac{\sum_s W(s)^{-1} G(s)e^{-\beta U(s)}}{\sum_s W(s)^{-1} e^{-\beta U(s)}} \tag{1.4}$$

where $W(s)$ is the probability with which states are chosen, (1.4) can be reduced to a simple average over the S states examined, namely,

$$\langle G \rangle = \frac{1}{S} \sum_{s=1}^{S} G(s) \tag{1.5}$$

Clearly, we require $W(s) = \exp(-\beta U(s))$ for this to be true, and much of the art of Monte Carlo is to ensure that states are actually produced with this probability; the approach is called importance sampling. The Monte Carlo method considers only configuration space, having eliminated the momentum part of phase space. Since there are no dynamics it can only be used to study systems in equilibrium, although if dynamical processes are represented in terms of collision cross-sections it becomes possible to study the consequences of the process, even if not the detailed dynamics [bir94].

Molecular dynamics operates in the continuum, in contrast to lattice-based methods [doo91], such as cellular automata, which are spatially discrete. While these are very effective from a computational point of view, they suffer from certain design problems such as the lack of a range of particle velocities, or unwanted effects due to lattice symmetry, and are also not easily extended. The MD approach is computationally demanding, but since it attempts to mimic nature it has few inherent limitations. One further continuum-dynamical method, known as Brownian dynamics [erm80], is based on the Langevin equation; the forces are no longer computed explicitly but are replaced by stochastic quantities that reflect the fluctuating local environment experienced by the molecules.

1.3.4 Applications and achievements

Given the modeling capability of MD and the variety of techniques that have emerged, what kinds of problem can be studied? Certain applications can be eliminated owing to the classical nature of MD. There are also hardware imposed limitations on the amount of computation that can be performed over a given period of time – be it an hour or a month – thus restricting the number of molecules of a given complexity that can be handled, as well as storage limitations having similar consequences.

The phenomena that can be explored must occur on length and time scales that are encompassed by the computation. Some classes of phenomena may require repeated runs based on different sets of initial conditions to sample adequately the kinds of behavior that can develop, adding to the computational demands. Small system size enhances the fluctuations and sets a limit on the measurement accuracy; finite-size effects – even the shape of the simulation region – can also influence certain results. Rare events present additional problems of observation and measurement.

Liquids represent the state of matter most frequently studied by MD methods. This is due to historical reasons, since both solids and gases have well-developed theoretical foundations, but there is no general theory of liquids. For solids, theory begins by assuming that the atomic constituents undergo small oscillations about fixed lattice positions; for gases, independent atoms are assumed and interactions are introduced as weak perturbations. In the case of liquids, however, the interactions are as important as in the solid state, but there is no underlying ordered structure to begin with.

A somewhat random and far from complete assortment of the kinds of use to which MD simulation is put follows. Fundamental studies: equilibration, tests of molecular chaos, kinetic theory, diffusion, transport properties, size-dependence, tests of models and potential functions. Phase transitions: first- and second-order, phase coexistence, order parameters, critical phenomena. Collective behavior: decay of space- and time-correlation functions, coupling of translation and rotation, vibration, spectroscopic measurements, orientational order, dielectric properties. Complex fluids: structure and dynamics of glasses, molecular liquids, pure water and aqueous solutions, liquid crystals, ionic liquids, fluid interfaces, films and monolayers. Polymers: chains, rings and branched molecules, membranes, equilibrium conformation, relaxation and transport processes. Solids: defect formation and migration, fracture, grain boundaries, structural transformations, radiation damage, elastic and plastic mechanical properties, friction, shock waves, molecular crystals, epitaxial growth. Biomolecules: structure and dynamics of proteins, micelles, docking of molecules. Fluid dynamics: laminar flow, boundary layers, rheology of non-Newtonian fluids, unstable flow. And there is much more.

The elements involved in an MD study, the way the problem is formulated, and the relation to the real world, can be used to classify MD problems into various categories. Examples of this classification include whether the interactions are short- or long-ranged; whether the system is thermally and mechanically isolated or open to outside influence; whether, if in equilibrium, normal dynamical laws are used or the equations of motion are modified to produce a particular statistical-

mechanical ensemble; whether the constituent particles are simple structureless atoms or more complex molecules, and, if the latter, whether
the molecules are rigid or flexible; whether simple interactions are represented by continuous potential functions or by step potentials; whether
interactions involve just pairs of particles or multi-particle contributions
as well; and so on and so on.

Despite the successes, many challenges remain. Multiple phases introduce the issue of interfaces that often have a thickness comparable
to the typical simulated region size. Inhomogeneities such as density or
temperature gradients can be difficult to maintain in small systems, given
the magnitude of the inherent fluctuations. Slow relaxation processes,
such as those characterizing the glassy state, diffusion that is hindered
by structure as in polymer melts, and the very gradual appearance of
spontaneously forming spatial organization, are all examples of problems
involving temporal scales many orders of magnitude larger than those
associated with the underlying molecular motion.

1.4 Organization

Case studies are used throughout. The typical case study includes a
summary of the theoretical background used for formulating the computational approach. The computation is described either by means of
a complete program listing or as a series of additions and modifications
to an earlier program. Essential but often neglected details such as
the initial conditions, organization of the input and output, accuracy,
convergence, and efficiency, are also addressed. Results obtained from
running each program are shown; these sometimes reproduce published
results, although no special effort is made to achieve a similar level of
accuracy since our goal is one of demonstration, not of compiling a
collection of definitive measurements. Suggested extensions and assorted
other projects are included as exercises for the reader.

We begin with the simplest possible example to demonstrate that
MD actually works. Later chapters extend the basic model in various
directions, improve the computational methods, deal with various kinds
of measurement, and introduce new models for more complex problems.
The programs themselves are constructed incrementally, with most case
studies building on programs introduced earlier. In order to avoid a
combinatorial explosion, the directions explored in each chapter tend to
be relatively independent, but in more ambitious MD applications it is
quite likely that combinations of the various techniques will be needed.

Some care is necessary here, because what appears obvious and trivial for simple atoms may, for example, require particular attention for molecules subject to constraints – each case must be treated individually.

Chapter 2 introduces the MD approach using the simplest possible example and demonstrates how the system behaves in practice; general issues of programming style and organization that are used throughout the book are also introduced here. In Chapter 3 we discuss the methodology for simulating monatomic systems, the algorithms used, and the considerations involved in efficient and accurate computation. Chapter 4 focuses on measuring the thermodynamic and structural properties of systems in equilibrium; some of these properties correspond to what can be measured in the laboratory, while others provide a microscopic perspective unique to simulation. The dynamical properties of equilibrium systems are the subject of Chapter 5, including transport coefficients and the correlation functions that characterize space- and time-dependent processes.

More complex systems and environments form the subject of subsequent chapters. Modifications of the dynamics to allow systems to be studied under constant temperature and pressure conditions, as opposed to the constant energy and volume implicit in the basic MD approach, are covered in Chapter 6. In Chapter 7 we discuss further methods for measuring transport properties, both by modeling the relevant process directly, and by using a modified form of the dynamics designed for systems not in thermal equilibrium. The dynamics of rigid molecules forms the subject of Chapter 8, and after describing the general problem a model for water is treated in some detail. Flexible molecules are discussed in Chapter 9. Molecules possessing internal degrees of freedom but also subject to geometric constraints that provide a certain amount of rigidity are analyzed in Chapter 10, together with a model used for modeling alkane chains. Chapter 11 addresses two further classes of system, one in which long-range forces dominate the behavior and require special treatment, the other involving three-body interactions.

Chapter 12 describes an alternative approach to MD based on step potentials rather than on the continuous potentials of earlier chapters; this calls for an entirely different computational technique. In Chapter 13 we focus on the study of time-dependent behavior and demonstrate the ability of MD to reproduce phenomena normally associated with macroscopic hydrodynamics. The special considerations involved in implementing MD computations on parallel and vector supercomputers form the subject of Chapter 14. Appendices cover miscellaneous topics related to programming and data, describe several utility functions used in the case studies, and provide a concise alphabetical summary of the variables used in the programs. A separate index of program functions is included.

1.5 Further reading

A great deal of information about MD methodology and applications is scattered throughout the scientific literature, and references to material relevant to the subjects covered here will appear in the appropriate places. Three volumes of conference proceedings include pedagogical expositions of various aspects of MD simulation [cic86a, cat90, all93b], and a monograph on liquid simulation covers both MD and Monte Carlo techniques [all87]. Another book devoted in part to MD is [hoo91]. Three evenly-spaced reviews of the role of simulation in statistical mechanics are [bee66, woo76, abr86]. Two extensive literature surveys on liquid simulation [lev84, lev92] and a collection of reprints [cic87] are also available.

2
Basic molecular dynamics

2.1 Introduction

This chapter provides the introductory appetizer and aims to leave the reader new to MD with a feeling for what the subject is all about. Later chapters will address the techniques in detail; here the goal is to demonstrate a working example with a minimum of fuss and so convince the beginner that MD is not only straightforward but also that it works successfully. Of course, the techniques discussed here are not particularly efficient from a computational point of view and the model itself is about the simplest there is. Such matters will be rectified later. The general program organization and stylistic conventions used in case studies throughout the book are also introduced here.

2.2 Soft-disk fluid

2.2.1 Equations of motion

The most rudimentary microscopic model for a substance capable of existing in any of the three most familiar states of matter – solid, liquid, and gas – is based on spherical particles that interact with one another; in the interest of brevity such particles will be referred to as atoms (albeit without hint of their quantum origins). The interactions, again at the simplest level, occur between pairs of atoms and are responsible for providing the two principal features of an interatomic force. The first is a resistance to compression, hence the interaction repels at close range. The second is to bind the atoms together in the solid and liquid states, and for this the atoms must attract each other over a range of separations. Potential functions exhibiting these characteristics can adopt a variety of forms and, when chosen carefully, actually provide useful models for real substances.

The best known of these potentials, originally proposed for liquid argon, is the Lennard-Jones (LJ) potential [mcq76, mai81]; for a pair of atoms i and j located at r_i and r_j the potential energy is

$$u(r_{ij}) = 4\epsilon \left[\left(\frac{\sigma}{r_{ij}} \right)^{12} - \left(\frac{\sigma}{r_{ij}} \right)^6 \right], \quad r_{ij} \le r_c \tag{2.1}$$

where $r_{ij} = r_i - r_j$ and $r \equiv |r|$. The parameter ϵ governs the strength of the interaction and σ defines a length scale; the interaction repels at close range, then attracts, and is eventually cut off at some limiting separation r_c. While the strongly repulsive core arising from (in the language of quantum mechanics) the nonbonded overlap between the electron clouds has a rather arbitrary form (and other powers and functional forms are sometimes used), the attractive tail actually represents the van der Waals interaction due to electron correlations. The interactions involve individual pairs of atoms: each pair is treated independently, with other atoms in the neighborhood having no effect on the force between them.

We will simplify the interaction even further by ignoring the attractive tail and changing (2.1) to

$$u(r_{ij}) = 4\epsilon \left[\left(\frac{\sigma}{r_{ij}} \right)^{12} - \left(\frac{\sigma}{r_{ij}} \right)^6 \right] + \epsilon, \quad r_{ij} \le r_c = 2^{1/6}\sigma \tag{2.2}$$

with r_c chosen so that $u(r_c) = 0$. A model fluid constructed using this potential is little more than a collection of colliding balls that are both soft (though the softness is limited) and smooth. All that holds the system together is the container within which the atoms (or balls) are confined. While the kinds of system that can be represented quantitatively by this highly simplified model are limited, typically gases at low density, it does nevertheless have much in common with more detailed models, and has a clear advantage in terms of computational simplicity. If certain kinds of behavior can be shown to be insensitive to specific features of the model, in this instance the attractive tail of the potential, then it is clearly preferable to eliminate them from the computation in order to reduce the amount of work, and for this reason the soft-sphere system will reappear in many of the case studies.

The force corresponding to $u(r)$ is $f = -\nabla u(r)$, so the force that atom

j exerts on atom i is

$$f_{ij} = \left(\frac{48\epsilon}{\sigma^2}\right) \left[\left(\frac{\sigma}{r_{ij}}\right)^{14} - \frac{1}{2}\left(\frac{\sigma}{r_{ij}}\right)^{8}\right] r_{ij} \tag{2.3}$$

provided $r_{ij} \leq r_c$, and zero otherwise. As r increases towards r_c the force drops to zero, so that there is no discontinuity at r_c; ∇f and higher derivatives are discontinuous, though this has no real impact on the numerical solution. The equations of motion follow from Newton's second law:

$$m\ddot{r}_i = F_i = \sum_{\substack{j=1 \\ (j \neq i)}}^{N_a} f_{ij} \tag{2.4}$$

where the sum is over all N_a atoms, excluding i itself, and m is the atomic mass. It is these equations which must be numerically integrated. The fact that $f_{ji} = -f_{ij}$ (Newton's third law) means that each atom pair need only be examined once. The amount of work is proportional to N_a^2, so that for models in which r_c is small compared to the size of the container it would obviously be a good idea to determine those atom pairs for which $r_{ij} \leq r_c$ and use this information to reduce the computational effort; we will indeed adopt such an approach later on. In the present example, which focuses on just the smallest of systems, we continue with this all-pairs approach. Note that for the potential function (2.2) it is never necessary to evaluate r_{ij}; only r_{ij}^2 is needed, so that the (sometimes costly) square root computation is avoided.

At this point we introduce a set of dimensionless, or reduced, MD units in which all physical quantities will be expressed. There are several reasons for doing this, not the least being the ability to work with numerical values that are not too distant from unity, instead of the extremely small values normally associated with the atomic scale. Another benefit of dimensionless units is that the equations of motion are simplified because some, if not all, of the parameters defining the model are absorbed into the units. The most familiar reason for using such units is related to the general notion of scaling, namely, that a single model can describe a whole class of problems, and once the properties have been measured in dimensionless units they can easily be scaled to the appropriate physical units for each problem of interest. From a strictly practical point of view, the switch to such units removes any risk of encountering values lying outside the range that is representable by the computer hardware.

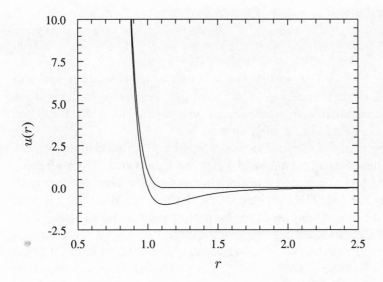

Fig. 2.1 LJ and soft-sphere interaction energy (dimensionless MD units).

For MD studies using potentials based on the LJ form (2.1) the most suitable dimensionless units are based on the choice of σ, m, and ϵ as the units of length, mass, and energy, respectively, and making the replacements $r \to r\sigma$ for length, $e \to e\epsilon$ for energy, and $t \to t\sqrt{m\sigma^2/\epsilon}$ for time. The resulting form of the equation of motion, now in MD units, is

$$\ddot{r}_i = 48 \sum_{j(\neq i)} \left(r_{ij}^{-14} - \tfrac{1}{2} r_{ij}^{-8} \right) r_{ij} \qquad (2.5)$$

The dimensionless kinetic and potential energies, per atom, are

$$E_k = \frac{1}{2N_a} \sum_{i=1}^{N_a} v_i^2 \qquad (2.6)$$

$$E_u = \frac{4}{N_a} \sum_{1 \leq i < j \leq N_a} \left(r_{ij}^{-12} - r_{ij}^{-6} \right) \qquad (2.7)$$

The functional forms of the LJ and soft-sphere potentials, in MD units, are shown in Figure 2.1.

The unit of temperature is ϵ/k_B, and since each translational degree of freedom contributes $k_B T/2$ to the kinetic energy, the temperature of a d-dimensional ($d = 2$ or 3) system is

$$T = \frac{1}{dN_a} \sum_i v_i^2 \qquad (2.8)$$

We have set $k_B = 1$, so that the MD unit of temperature is now also defined. Strictly speaking, of the total dN_a degrees of freedom, d are eliminated because of momentum conservation, but if N_a is not too small this detail can be safely ignored.

If the model is intended to model liquid argon, the relations between the dimensionless MD units and real physical units are as follows [rah64]. Lengths are expressed in terms of $\sigma = 3.4$ Å. The energy units are specified by $\epsilon/k_B = 120$ K, implying that $\epsilon = 120 \times 1.3806 \times 10^{-16}$ erg/atom (several kinds of units are in use for energy, and conversion among them is based on standard relations that include 1.3806×10^{-16} erg/atom $= 1.987 \times 10^{-3}$ kcal/mole $= 8.314$ J/mole). Given the mass of an argon atom $m = 39.95 \times 1.6747 \times 10^{-24}$ g, the MD time unit corresponds to 2.161×10^{-12} s; thus a typical timestep size of $\Delta t = 0.005$ used in the numerical integration of the equations of motion corresponds to approximately 10^{-14} s. Finally, if N_a atoms occupy a cubic region of edge length L, then a typical liquid density of 0.942 g/cm^3 implies that $L = 4.142 N_a^{1/3}$ Å, which in reduced units amounts to $L = 1.218 N_a^{1/3}$.

2.2.2 Boundary conditions

Finite and infinite systems are very different, and the question of how large a relatively small system must be to yield results that resemble the behavior of the infinite system faithfully lacks a unique answer. The simulation takes place in a container of some sort, and it is tempting to regard the container walls as rigid boundaries against which atoms collide while trying to escape from the simulation region. In systems of macroscopic size, only a very small fraction of the atoms is close enough to a wall to experience any deviation from the environment prevailing in the interior. Consider, for example, a three-dimensional system with $N_a = 10^{21}$ at liquid density. Since the number of atoms near the walls is of order $N_a^{2/3}$, this amounts to 10^{14} atoms – a mere one in 10^7. But for a more typical MD value of $N_a = 1000$, roughly 500 atoms are immediately adjacent to the walls, leaving very few interior atoms; if the first two layers are excluded a mere 216 atoms remain. Thus the simulation will fail to capture the typical state of an interior atom and the measurements will reflect this fact. Unless the goal is the study of behavior near real

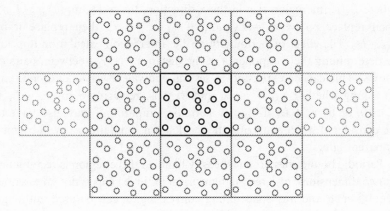

Fig. 2.2 The meaning of periodic boundary conditions (the two-dimensional case is shown).

walls, a problem that is actually of considerable importance, walls are best eliminated.

A system that is bounded but free of physical walls can be constructed by resorting to periodic boundary conditions, shown schematically in Figure 2.2. The introduction of periodic boundaries is equivalent to considering an infinite space-filling array of identical copies of the simulation region. There are two consequences of this periodicity. The first is that an atom that leaves the simulation region through a particular bounding face immediately reenters the region through the opposite face. The second is that atoms lying within a distance r_c of a boundary interact with atoms in an adjacent copy of the system, or, equivalently, with atoms near the opposite boundary – a wraparound effect. Another way of regarding periodic boundaries is to think of mapping the region (topologically, not spatially) onto the equivalent of a torus in four dimensions (a two-dimensional system is mapped onto a torus); then it is obvious that there are no physical boundaries. In this way it is possible to model systems that are effectively bounded but that are nevertheless spatially homogeneous insofar as boundaries are concerned.

The wraparound effect of the periodic boundaries must be taken into account in both the integration of the equations of motion and the interaction computations. After each integration step the coordinates must be examined, and if an atom is found to have moved outside the region its coordinates must be adjusted to bring it back inside. If, for example, the x-coordinate is defined to lie between $-L_x/2$ and $L_x/2$, where L_x is the region size in the x-direction, the tests are: if $r_{xi} \geq L_x/2$ then replace r_{xi} by $r_{xi} - L_x$, and if $r_{xi} < -L_x/2$ then replace it by $r_{xi} + L_x$. The effect of periodicity on the interaction calculation appears in determining the components of the distance r ($\equiv r_{ij}$) between pairs of atoms. This can be expressed in several equivalent ways, all amounting to asking whether $|r_x| > L_x/2$, and, if it is, replacing r_x by $r_x - L_x$ if $r_x > 0$, or by $r_x + L_x$ if $r_x < 0$. Periodic wraparound may also have to be considered when analyzing the results of a simulation, as will become apparent later.

Periodic boundaries are most easily handled if the region is rectangular in two dimensions, or a rectangular prism in three. This is not an essential requirement, and any space-filling convex region can be used, although the boundary computations will not be as simple as those just illustrated. The motivation for choosing alternative region shapes is to enlarge the volume to surface ratio, and thus increase the maximum distance between atoms before periodic ambiguity appears (it is obviously meaningless to speak of interatomic distances that exceed half the region size), the most desirable shape in three dimensions – though not space-filling – being the sphere. In two dimensions a hexagon might be used, while in three the truncated octahedron [ada80] is one such candidate. Another reason for choosing more complex region shapes is to allow the modeling of crystalline structures with non-orthogonal axes (for example a trigonal unit cell); there, too, an alternative region shape (such as a sheared cube) might be worth considering.

Although not an issue in this particular case, the use of periodic boundaries limits the interaction range to no more than half the smallest region dimension – in practice the range is generally much less. Long-range forces require an entirely different approach that will be described in Chapter 11. (Problems can also arise if there are strong correlations between atoms separated by distances approaching the region size, because periodic wraparound can then lead to spurious effects. One example is the vibration of an atom producing what are essentially sound waves; the disturbance, if not sufficiently attenuated, can propagate around the system and eventually return to affect the atom itself.)

Even with periodic boundaries finite-size effects are still present, so how big does the system have to be before they can be neglected? The answer depends on the kind of system and the properties of interest. As a minimal requirement, the size should exceed the range of any significant correlations, but there may be more subtle effects even in larger systems. Only detailed numerical study can hope to resolve this question.

2.2.3 *Initial state*

In order for MD to serve a useful purpose it must be capable of sampling a representative region of the total phase space of the system. An obvious corollary of this requirement is that the results of a simulation of adequate duration are insensitive to the initial state, so that any convenient initial state is allowed. A particularly simple choice is to start with the atoms at the sites of a regular lattice – such as the square or simple cubic lattice – spaced to give the desired density. The initial velocities are assigned random directions and a fixed magnitude based on temperature; they are also adjusted to ensure that the center of mass of the system is at rest and so they eliminate any overall flow. The speed of equilibration to a state in which there is no memory of this arbitrarily selected initial configuration is normally quite rapid, so that more careful attempts at constructing a 'typical' state are of little benefit.

2.3 Methodology

2.3.1 *Integration*

Integration of the equations of motion uses the simplest of numerical techniques, the leapfrog method. The origin of the method will be discussed in Chapter 3; for the present it is sufficient to state that, despite its low order, the method has excellent energy conservation properties and is widely used.

If $h = \Delta t$ denotes the size of the timestep used for the numerical integration, then the integration formulae applied to each component of an atom's coordinates and velocities are

$$v_{xi}(t + h/2) = v_{xi}(t - h/2) + ha_{xi}(t) \tag{2.9}$$
$$r_{xi}(t + h) = r_{xi}(t) + hv_{xi}(t + h/2) \tag{2.10}$$

The name 'leapfrog' stems from the fact that coordinates and velocities are evaluated at different times; if a velocity estimate is required to correspond to the time at which coordinates are evaluated, then

$$v_{xi}(t) = v_{xi}(t - h/2) + (h/2)a_{xi}(t) \tag{2.11}$$

can be used. The local errors introduced at each timestep due to the truncation of what should really be infinite series in h are of order $O(h^4)$ for the coordinates and $O(h^2)$ for velocities.

2.3.2 Measurements

The most accessible properties of systems in equilibrium are those introduced in elementary thermodynamics, namely, energy and pressure, each expressed in terms of the independent temperature and density variables T and ρ. Measuring such quantities during an MD simulation is relatively simple, and provides the link between the world of thermodynamics – which predates the recognition of the atomic nature of matter – and the detailed behavior at the microscopic level. However, it is energy rather than temperature that is constant in our MD simulation, so the thermodynamic results are expressed in terms of the average $\langle T \rangle$ rather than T itself.

In this case study, energy and pressure are the only properties measured. Pressure is defined in terms of the virial expression [han86b] (with $k_B = 1$):

$$PV = N_a T + \frac{1}{d}\left\langle \sum_{i=1}^{N_a} r_i \cdot F_i \right\rangle \tag{2.12}$$

(in two dimensions the region volume V is replaced by the area). For pair potentials (2.12) can be written as a sum over interacting atom pairs, namely,

$$PV = N_a T + \frac{1}{d}\left\langle \sum_{i<j} r_{ij} \cdot f_{ij} \right\rangle \tag{2.13}$$

and for the force (2.3) this is

$$PV = \frac{1}{d}\left\langle \sum_i v_i^2 + 48 \sum_{i<j}\left(r_{ij}^{-12} - \tfrac{1}{2}r_{ij}^{-6}\right)\right\rangle \tag{2.14}$$

While the total energy (per atom) $E = E_k + E_u$ is conserved, apart from any numerical integration error, quantities such P and T $(= 2E_k/d)$ fluctuate, and averages must be computed over sequences of timesteps; such averaging will be included in the program and used for estimating the mean values as well as the statistical measurement errors.

2.4 Programming

2.4.1 Style and conventions

In this section we present the full listing of the program used in the case study. Not only is the program the tool for getting the job done, it also incorporates a definitive statement of all the computational details.

But before addressing these details a few general remarks on matters of organization and programming style are in order; further discussion of this subject appears later in the section. Style, to a considerable degree, is a matter of personal taste; the widely used C language chosen for this work offers a certain amount of flexibility in this respect, a boon for some, but a bane for others.

A similar form of organization is used for most programs in the book. Parts of the program discussed in this chapter may seem to be expressed in a more general form than is absolutely necessary; this is to provide a basis for extending the program to handle later case studies. We use a subset of the C language that can be readily translated to Fortran and assume that the reader has at least a passing (and easily acquired) familiarity with C. To simplify such translation (Fortran is still the choice of many), the data structuring capability of C is not used, except in the occasional utility function, but even then only in ways that are hidden from the simulation proper. C requires that all variables be defined prior to use; all the definitions will be included, but because the material is presented in a 'functional' manner, rather than as a serial listing of the program text, variables may first appear in the recipe before they are formally defined (this is of course not the case in the program sources). Local variables used within functions are not preserved between calls.

We adopt the convention that all variable names begin with a lower-case letter; multi-word names use intermediate capitals to clarify meaning. Function names begin with an upper-case letter. Fixed program parameters that are specified using the #define statement are fully capitalized.

The format of a C program is also subject to taste. The layout used here is one designed to economize on vertical space, while at the same time using indentation and the positioning of block-delimiting braces to emphasize the logical structure. The line numbers are of course not part of the program, and are included merely as visual markers.

2.4.2 Program organization

The main program of this elementary MD exercise, which forms the basis of most of the subsequent case studies as well, is as follows:

```
main (int argc, char **argv) {
  GetNameList (argc, argv);
  PrintNameList (stdout);
```

```
      SetParams ();
5     SetupJob ();
      moreCycles = 1;
      while (moreCycles) {
        SingleStep ();
        if (stepCount >= stepLimit) moreCycles = 0;
10    }
      }
```

After the initialization phase (GetNameList, SetParams, SetupJob), in the course of which parameters and other data are read in or initialized, and storage arrays allocated, the program enters a loop. Each loop cycle advances the system by a single timestep. The loop terminates when moreCycles is set to zero; here this occurs after a preset number of timesteps, but in a more general context moreCycles can be zeroed once the total processing time exceeds a preset limit, or even by means of an interrupt generated by the user from outside the program when she feels the run has produced enough results (there are also more drastic ways of terminating a program). As a reminder to lapsed C users, main is where the program begins, argc is the number of arguments passed to the program from the command line (as in Unix), and the array argv provides access to the text of each of these arguments; the command line contents are used in subsequent case studies.

The function that handles the processing for a single timestep, including calls to functions that deal with the force evaluation, integration of the equations of motion, adjustments required by periodic boundaries, and the measurements, is

```
      SingleStep () {
        stepCount = stepCount + 1;
        timeNow = stepCount * deltaT;
        ComputeForces ();
5       LeapfrogStep ();
        ApplyBoundaryCond ();
        EvalProps ();
        AccumProps (1);
        if (stepCount % stepAvg == 0) {
10        AccumProps (2);
          PrintSummary (stdout);
          AccumProps (0);
        }
      }
```

All the work needed for initializing the computation is concentrated in the following function:

```
SetupJob () {
  AllocArrays ();
  InitCoords ();
  InitVels ();
5 AccumProps (0);
  stepCount = 0;
}
```

Having dealt with the top-level functions of the program it is appropriate to insert a few comments on the program structure adopted in these recipes. The order of presentation of this introductory case study reflects the organization of the program itself: the organization is modular, with separate functions being responsible for distinct portions of the computation. In this case study the emphasis on organization may appear overdone, given the simplicity of the problem, but, as indicated earlier, our aim is to provide a more general framework that will be utilized later. On the other hand, in order to avoid the risk of tedium, we have not carried this functional decomposition to the extremes practiced in professional software development.

The meaning of most program variables should be apparent from the names, with the same being true for functions. Where the meanings are not obvious, or when additional remarks are called for, the text will include further details. An alphabetically ordered summary of the globally defined variables appears in the Appendix. Other questions ought to be resolved by examining functions that appear later on.

There are many program elements that are common to MD simulations of various kinds. Some of these already appear in this initial case study, others will be introduced later on. Examples include

(a) parameter input with completeness and consistency checks;
(b) runtime array allocation, with array sizes determined by the actual system size;
(c) initialization of variables;
(d) the main loop which cycles through the force computations and trajectory integration, and performs data collection at specified intervals;
(e) the processing and statistical analysis of various kinds of measurement;
(f) storage of accumulated results and condensed configurational snapshots for later analysis;
(g) run termination based on various criteria;

(h) provision for checkpointing (or saving) the current computational
     state of a long simulation run, both as a safety measure, and to
     permit the run to be interrupted and continued at some later time.

2.4.3 Computational functions

The next function is responsible for the interaction computations, and is
immediately followed by a discussion of how vector and other array-like
quantities are organized. This version handles both two- and three-
dimensional systems, depending on the value of the (spatial dimensionality) parameter NDIM:

```
   ComputeForces () {
     real dr[NDIM + 1], f, fcVal, rrCut, rr, rri, rri3;
     int j1, j2, k, n;
     rrCut = Sqr (rCut);
5    for (n = 1; n <= nAtom; n ++) {
       for (k = 1; k <= NDIM; k ++) ra[k][n] = 0.;
     }
     uSum = 0.;    virSum = 0.;
     for (j1 = 1; j1 <= nAtom - 1; j1 ++) {
10     for (j2 = j1 + 1; j2 <= nAtom; j2 ++) {
         rr = 0.;
         for (k = 1; k <= NDIM; k ++) {
           dr[k] = r[k][j1] - r[k][j2];
           if (fabs (dr[k]) > regionH[k])
15           dr[k] = dr[k] - SignR (region[k], dr[k]);
           rr = rr + Sqr (dr[k]);
         }
         if (rr < rrCut) {
           rri = 1. / rr; rri3 = rri * rri * rri;
20         fcVal = 48. * rri3 * (rri3 - 0.5) * rri;
           for (k = 1; k <= NDIM; k ++) {
             f = fcVal * dr[k];
             ra[k][j1] = ra[k][j1] + f;
             ra[k][j2] = ra[k][j2] - f;
25         }
           uSum = uSum + 4. * rri3 * (rri3 - 1.) + 1.;
           virSum = virSum + fcVal * rr;
       } } }
   }
```

The above function computes the forces – identical to the accelerations
in the MD units defined earlier – as well as the potential energy and
the contribution of the interactions to the virial. Periodic boundaries
are included; the function SignR (see Appendix) transfers the sign of

its second argument to that of the first. It is worth reiterating that this approach to the force computation involves all $N_a(N_a - 1)/2$ pairs of atoms, and is not the way to carry out serious simulations of this type; however, a small performance improvement might be achieved here by testing the magnitudes of the individual dr[k] values as they are computed to see if they exceed rCut, and bypassing the atom pair as soon as this happens.

In three dimensions the atom coordinates r[k][n] are defined for n=1, ..., nAtom, and for k=1,2,3, corresponding to the x-, y-, and z-components (the last is dropped in two dimensions); ra[k][n] denotes acceleration. In adopting this method of indexing, which is certainly not the conventional C style, we have attempted to adhere to the original algebraic formulation of the problem, as well as make the program more acceptable to Fortran users. To elaborate further: C indexes arrays beginning from zero, so that the final element has an index one less than the array size; Fortran, on the other hand, begins at unity, conforming to the normal way of counting objects. There is a minor aesthetic inconvenience to using this approach in C, namely, that one-dimensional arrays whose sizes are specified at compilation time must be one element larger than necessary to allow for the unused zeroth element; the vector quantity dr appearing in ComputeForces provides an example of this. It will be shown shortly that most arrays (such as r[k][n]) are allocated dynamically at runtime, and this can be done in a manner that does not waste any storage space.

One other detail where this listing differs from conventional C is in the type of floating-point variable defined, since real is not a standard variable type in C – the name is borrowed from Fortran. To allow flexibility, real can be set to correspond to either single or double floating-point precision, known respectively as float and double in C; single precision saves storage whereas double precision provides accuracy, but as for relative computation speed, this depends on the particular processor hardware (either precision may be faster, sometimes by a substantial amount).

There are other stylistic issues implicit in this listing. The coordinates r[k][n], for example, are represented as two-dimensional arrays, with a particular order specified for the two indices. This decision is arbitrary to a certain extent; coordinates can equally well be represented (for two-dimensional simulations) as distinct vectors rx[n] and ry[n], or as a single one-dimensional array in which r[2*n-1] and r[2*n] are the x- and y-components for a given atom. Furthermore, instead of r[k][n] we could have used r[n][k], as a programmer familiar with Fortran storage conventions might have done, but that is definitely not advisable in C where multi-dimensional arrays are stored in exactly the opposite order from Fortran (C also employs an additional array of pointers, a matter we will not discuss here). From an efficiency point of view the single one-dimensional array usually produces the best performance, but the program tends to be less readable. Introducing a separate array for each component prevents the use of loops over the spatial dimension; short loops of this kind are generally inefficient, but they have the advantage of conciseness, and any reasonably smart compiler will eliminate them by unrolling (fully expanding) the loop as part of the optimization process. In a related attempt to aid readability we have avoided some of the operations in C that permit writing extremely concise code (often bordering on obfuscation); while the experienced C user will perceive their absence, the efficiency of the compiled program is unlikely to be affected.

Sometimes, style effects efficiency in a very serious manner. Patterns of memory access can have a strong impact on computation speed, especially in modern computers with mapped and interleaved memory and multi-level caches. Awareness of the issues involved, which may demand some familiarity with the specific processor architecture, can suggest how to arrange data and organize the loop structure of a program; this is a rather specialized subject – ignored by most – and will not be covered here.

Returning to the listing of SingleStep, the next function we encounter handles the task of leapfrog integration:

```
LeapfrogStep () {
  int k, n;
  for (n = 1; n <= nAtom; n ++) {
    for (k = 1; k <= NDIM; k ++) {
      rv[k][n] = rv[k][n] + deltaT * ra[k][n];
      r[k][n] = r[k][n] + deltaT * rv[k][n];
} }
}
```

This is followed by a call to the function responsible for taking care of periodic wraparound:

```
ApplyBoundaryCond () {
  int k, n;
  for (n = 1; n <= nAtom; n ++) {
    for (k = 1; k <= NDIM; k ++) {
      if (r[k][n] >= regionH[k])
```

```
            r[k][n] = r[k][n] - region[k];
          else if (r[k][n] < - regionH[k])
            r[k][n] = r[k][n] + region[k];
      } }
10 }
```

2.4.4 Initial state

Preparation of the initial state uses the following two functions, one for the atomic coordinates, the other for velocities. The versions shown here are for a two-dimensional simulation; the three-dimensional case will be covered in Chapter 3, but to make spatial dimensionality changes easier (here and elsewhere) we use the value NDIM in specifying array sizes and loop counts.

The number of atoms in the system is expressed in terms of the size of the array of unit cells in which the atoms are initially arranged, the relevant values appear in initUcell; here the simple square lattice (with the option of unequal edge lengths) is used, so that each unit cell contains just one atom, and the system is centered about the origin. The velocities are set to a fixed value vMag that depends on the temperature (see below), and after assignment of random velocity directions the velocities are adjusted to ensure that the center of mass is stationary. The function RandR (see Appendix) serves as a source of uniformly distributed random values in the range $(0, 1)$, with randSeed the seed value used by the generator:

```
    InitCoords () {
      real c[NDIM + 1], gap[NDIM + 1];
      int k, n, nX, nY;
      for (k = 1; k <= NDIM; k ++)
5       gap[k] = region[k] / initUcell[k];
      n = 0;
      for (nY = 1; nY <= initUcell[2]; nY ++) {
        c[2] = (nY - 0.5) * gap[2] - regionH[2];
        for (nX = 1; nX <= initUcell[1]; nX ++) {
10        c[1] = (nX - 0.5) * gap[1] - regionH[1];
          n = n + 1;
          for (k = 1; k <= NDIM; k ++) r[k][n] = c[k];
      } }
    }

    InitVels () {
      real vSum[NDIM + 1], ang;
      int k, n;
```

```
     for (k = 1; k <= NDIM; k ++) vSum[k] = 0.;
5    for (n = 1; n <= nAtom; n ++) {
       ang = 2. * pi * RandR (&randSeed);
       rv[1][n] = vMag * cos (ang);
       rv[2][n] = vMag * sin (ang);
       for (k = 1; k <= NDIM; k ++)
10         vSum[k] = vSum[k] + rv[k][n];
     }
     for (k = 1; k <= NDIM; k ++) vSum[k] = vSum[k] / nAtom;
     for (n = 1; n <= nAtom; n ++) {
       for (k = 1; k <= NDIM; k ++)
15         rv[k][n] = rv[k][n] - vSum[k];
     }
   }
```

2.4.5 *Variables*

It is debatable which should be discussed first, the program or the variables on which it operates. Here we have picked the former in order to provide some motivation for a discussion of the latter.

The scheme we have chosen is that all variables needed by more than one function are defined globally. This implies that they are accessible to all functions, and is not an approach recommended for large software projects because it is difficult to keep track of (and control) which variables are used where. The alternative is to make extensive use of argument lists, perhaps using structures to organize the data transferred between functions; while offering a means of regulating access to variables, it makes the program longer and more tedious to read, so we forgo the practice.

Having settled this issue, what are the global variables used by the program? The list of declarations (ordered roughly alphabetically, with arrays first) follows:

```
#define NDIM 2
real **r, **rv, **ra, region[NDIM + 1], regionH[NDIM + 1],
    deltaT, density, kinEnergy, pi, potEnergy, pressure,
    rCut, sKinEnergy, sPressure, sTotEnergy, ssKinEnergy,
5   ssPressure, ssTotEnergy, temperature, timeNow,
    totEnergy, uSum, virSum, vMag, vSum, vvSum;
int initUcell[NDIM + 1], moreCycles, nAtom, randSeed,
    runId, stepAvg, stepCount, stepEquil, stepLimit;
```

Most of the names should be self-explanatory. Three of the variables, r, rv, and ra – corresponding to the coordinates, velocities, and accelerations – are actually pointers to two-dimensional arrays that are allocated dynamically at the start of the run and sized according to the values of nAtom and NDIM. Writing **r is equivalent to r[...][...] with specific array sizes, except that in the former case the array sizes are established when the program is run rather than at compilation time. The advantage of dynamic allocation (in addition to bypassing any size limitations that some compilers might impose on arrays whose limits are included in the program source) is that it enhances program flexibility by eliminating any arbitrary built-in size assumptions, a benefit not available in normal Fortran.

The array region contains the edges lengths of the simulation region and, because of their frequent use, the half-lengths are stored in regionH. The variables with names prefixed by a single s are sums accumulated in order to evaluate averages, and those with ss are sums of squares used in evaluating standard deviations. The other quantities, as well as a list of those variables supplied as input to the program, will be covered by the remaining functions below.

All dynamic array allocations are carried out inside a single function AllocArrays; for this example it has the form

```
   AllocArrays () {
     r  = AllocMatR (NDIM, nAtom);
     rv = AllocMatR (NDIM, nAtom);
     ra = AllocMatR (NDIM, nAtom);
5  }
```

Each call to AllocMatR allocates a real two-dimensional array of size NDIM by nAtom, with both indices starting at unity. This and other general utility functions are described in the Appendix.

Other variables required for the simulation, excluding those that will be included in the input data, are set by the function SetParams:

```
   SetParams () {
     int k;
     pi = 4. * atan (1.);
     rCut = pow (2., 1./6.);
5    for (k = 1; k <= NDIM; k ++) {
       region[k] = initUcell[k] / sqrt (density);
       regionH[k] = 0.5 * region[k];
     }
     nAtom = initUcell[1] * initUcell[2];
10   vMag = sqrt (NDIM * (1. - 1. / nAtom) * temperature);
   }
```

Here we have evaluated nAtom and region assuming just one atom per unit cell, and made allowance for momentum conservation when computing vMag from the temperature.

2.4.6 Measurements

In this introductory case study the emphasis is on demonstrating a minimal working program. The measurements of basic thermodynamic properties of the system that are included are covered by the following functions. Note that the use of the leapfrog method requires that the velocity components be shifted back half a timestep when evaluating kinetic energy. The quantity vSum is used to accumulate the sum of all the components of the total momentum (or velocity in reduced MD units); the fact that this should remain exactly zero is provided as a simple (partial) check on the correctness of the calculation.

The first of the functions computes the adjusted velocity and velocity-squared sums and the instantaneous energy and pressure values:

```
   EvalProps () {
     real v, vv;
     int k, n;
     vSum = vvSum = 0.;
5    for (n = 1; n <= nAtom; n ++) {
       vv = 0.;
       for (k = 1; k <= NDIM; k ++) {
         v = rv[k][n] - 0.5 * ra[k][n] * deltaT;
         vSum = vSum + v;     vv = vv + Sqr (v);
10     }
       vvSum = vvSum + vv;
     }
     kinEnergy = 0.5 * vvSum / nAtom;
     potEnergy = uSum / nAtom;
15   totEnergy = kinEnergy + potEnergy;
     pressure = density * (vvSum + virSum) / (nAtom * NDIM);
   }
```

The second function collects the results of the measurements, and evaluates means and standard deviations upon request. (We assume that the arguments of the square-root function sqrt are non-negative; for quantities that barely fluctuate, rounding errors can sometimes lead to very small but negative values, in which case the arguments should be set to zero.)

```
   AccumProps (int icode) {
     if (icode == 0) {
       sTotEnergy = ssTotEnergy = 0.;
```

```
      sKinEnergy = ssKinEnergy = 0.;
5     sPressure = ssPressure = 0.;
   } else if (icode == 1) {
      sTotEnergy = sTotEnergy + totEnergy;
      ssTotEnergy = ssTotEnergy + Sqr (totEnergy);
      sKinEnergy = sKinEnergy + kinEnergy;
10    ssKinEnergy = ssKinEnergy + Sqr (kinEnergy);
      sPressure = sPressure + pressure;
      ssPressure = ssPressure + Sqr (pressure);
   } else if (icode == 2) {
      sTotEnergy = sTotEnergy / stepAvg;
15    ssTotEnergy = sqrt (ssTotEnergy / stepAvg -
         Sqr (sTotEnergy));
      sKinEnergy = sKinEnergy / stepAvg;
      ssKinEnergy = sqrt (ssKinEnergy / stepAvg -
         Sqr (sKinEnergy));
20    sPressure = sPressure / stepAvg;
      ssPressure = sqrt (ssPressure / stepAvg -
         Sqr (sPressure));
   }
}
```

Depending on the value of the argument icode (0, 1, or 2), AccumProps will initialize the accumulated sums (sTotEnergy, ...), accumulate the current values, or produce the final averaged estimates (the same variables are used for these final values).

2.4.7 *Input and output*

The function GetNameList, called from main, reads all the data required to specify the simulation from an input file. It uses a Fortran-style (almost) 'namelist' to group all the data conveniently and automate the input task. It also checks that all requested data items have been provided, and that there are no extraneous data. For this case study the list of variables is specified in the following way:

```
   NameList nameList[] = {
      INAME (runId),
      INAME (initUcell),
      RNAME (density),
5     RNAME (temperature),
      RNAME (deltaT),
      INAME (randSeed),
      INAME (stepAvg),
      INAME (stepEquil),
10    INAME (stepLimit),
   };
```

The C macros RNAME and INAME are used to signify real and integer variables, and if a variable happens to denote an array the processing ensures that the correct number of elements are obtained. The function PrintNameList, also called by main, outputs an annotated copy of the data. For the curious, full details of these functions appear in the Appendix.

The input data is read from a file whose default name is sd.data, the letters sd standing for soft disks (or ss for soft spheres in three dimensions). This name is specified by the variable

```
char *progId = "sd";
```

If the command used to run the program includes a numerical argument, for example

```
mdsoft 69
```

then the name of the data file is sd69.data instead. Such identification numbers are used to distinguish between different runs of the same program; output files, when they are introduced later, will also have this identifier as part of their names. The number itself is stored in runId.

Output from the run is produced by the following function:

```
PrintSummary (FILE *fp) {
  fprintf (fp,
     "%5d %8.4f %7.4f %7.4f %7.4f %7.4f %7.4f %7.4f",
     stepCount, timeNow, vSum, sTotEnergy, ssTotEnergy,
5    sKinEnergy, ssKinEnergy, sPressure, ssPressure);
  fprintf (fp, "\n");
}
```

Data is written to a file, which in the present case is just the user's terminal because the call to PrintSummary in SingleStep used the argument stdout. By calling this function twice, with different arguments, output can be sent both to the terminal and to a file that logs all the output. The reader unfamiliar with standard C library functions will find fprintf, and numerous other functions used later, described in any text on the C language.

2.5 Results

In this section we present a few of the results that can be obtained from simulations of the two-dimensional soft-disk fluid. In view of the fact that the MD algorithm described here is far from efficient, the results will mostly be confined to short simulation runs of small systems, just to give a foretaste of what is to come. More detailed results based on more extensive computations will come later.

The input file used in the first demonstration contains the following entries:

```
runId           1
initUcell       20 20
density         0.8
temperature     1.
deltaT          0.005
randSeed        17
stepAvg         100
stepEquil       0
stepLimit       999999
```

The total number of atoms is $N_a = 400$ because the initial configuration is a 20×20 square lattice. The timestep value (deltaT) is determined by the requirement that energy be conserved by the leapfrog method (to be discussed in Chapter 3). The initial temperature is $T = 1$; temperature will fluctuate during the run, and no attempt will be made here to set the mean temperature to any particular value.

2.5.1 Conservation laws

The most obvious test that the computation must pass is that of momentum and energy conservation. While the former is intrinsic to the algorithm, and – assuming periodic boundaries – can only be violated by some error in the algorithm, the latter is sensitive to the choice of integration method and the size of Δt. A quantity that is not conserved is angular momentum; a conservation law requires the system to be invariant under some change, such as translation, but, because of the periodic boundaries, the rotational invariance needed for angular momentum conservation is not applicable. Programming errors can sometimes (but not always) be detected by the violation of a conservation law; when this occurs the effect can be gradual, intermittent, or catastrophic, depending on the cause of error.

In Table 2.1 we show an edited version of the output of the run specified above; the results listed are the sum of the velocity components, the mean energy and kinetic energy per atom, their standard deviations, and the mean pressure. Clearly, energy and momentum are conserved as expected, kinetic energy fluctuates by a limited amount, and it is also apparent that as a result of some of the initial kinetic energy

Table 2.1. *Edited run output*

step	t	$\sum v$	$\langle E \rangle$	$\sigma(E)$	$\langle E_k \rangle$	$\sigma(E_k)$	$\langle P \rangle$
100	0.5	0.0000	0.9952	0.0002	0.6592	0.0970	4.5371
200	1.0	0.0000	0.9951	0.0001	0.6490	0.0121	4.5829
300	1.5	0.0000	0.9951	0.0001	0.6397	0.0168	4.6454
400	2.0	0.0000	0.9951	0.0000	0.6477	0.0156	4.5675
500	2.5	0.0000	0.9951	0.0000	0.6596	0.0166	4.4702
1000	5.0	0.0000	0.9950	0.0000	0.6480	0.0255	4.5496
2000	10.0	0.0000	0.9951	0.0001	0.6497	0.0122	4.5371
3000	15.0	0.0000	0.9952	0.0000	0.6539	0.0117	4.5184
7000	35.0	0.0000	0.9952	0.0001	0.6370	0.0111	4.6203

being converted to potential energy the temperature of the system (here $T = E_k$) has dropped considerably below the initial setting.

2.5.2 Equilibration

Characterizing equilibrium is by no means an easy task, especially in small systems whose properties fluctuate considerably. Averaging over a series of timesteps will reduce the fluctuations, but different quantities relax to their equilibrium averages at different rates, and this must also be taken into account when trying to establish when the time is ripe to begin making measurements. Fortunately, relaxation is generally quite rapid, but one must always beware of those situations where this is not true. Equilibration can be accelerated by starting the simulation at a higher temperature and later cooling by rescaling the velocities (this is similar, but not identical, to using a larger timestep initially); too high a temperature will, however, lead to numerical instability.

One simple measure of equilibration is the rate at which the velocity distribution converges to its expected final form. Theory [mcq76] predicts the Maxwell distribution:

$$f(\boldsymbol{v}) = \rho \left(\frac{m}{2\pi k_B T} \right)^{d/2} \exp(-mv^2/2k_B T) \tag{2.15}$$

and after angular integration this becomes (in MD units)

$$f(v) \propto v^{d-1} \exp(-v^2/2T) \tag{2.16}$$

The distribution can be measured by constructing a velocity histogram

$\{h_n\}, n = 1, \ldots, N_b$, where h_n is the number of atoms with velocity magnitude between $(n-1)\Delta v$ and $n\Delta v$, $\Delta v = v_m / N_b$, and v_m is a suitable upper limit to v. The normalized histogram represents a discrete approximation to $f(v)$.

The function that carries out this computation is the following (this is the two-dimensional version; the change for three dimensions is trivial):

```
   EvalVelDist () {
     real deltaV, histSum, vv;
     int j, n;
     countVel = countVel + 1;
5    if (countVel == 1) {
       for (j = 1; j <= sizeHistVel; j ++) histVel[j] = 0.;
     }
     deltaV = rangeVel / sizeHistVel;
     for (n = 1; n <= nAtom; n ++) {
10     vv = Sqr(rv[1][n]) + Sqr (rv[2][n]);
       j = (int) (sqrt (vv) / deltaV) + 1;
       if (j > sizeHistVel) j = sizeHistVel;
       histVel[j]= histVel[j] + 1.;
     }
15   if (countVel == limitVel) {
       histSum = 0.;
       for (j = 1; j <= sizeHistVel; j ++)
         histSum = histSum + histVel[j];
       for (j = 1; j <= sizeHistVel; j ++)
20       histVel[j] = histVel[j] / histSum;
       PrintVelDist (stdout);
       countVel = 0;
     }
   }
```

Depending on the value of countVel the function will, in addition to adding the latest results to the accumulated total, either initialize the histogram counts, or carry out the final normalization. Other kinds of analysis in subsequent case studies will use functions that operate in a similar manner.

In order to use this function, storage for the histogram array must be allocated, and a number of additional variables defined and assigned values. The variables are

```
real *histVel, rangeVel;
int countVel, limitVel, sizeHistVel, stepVel;
```

and those included in the input data must be added to the array nameList:

```
INAME (limitVel),
RNAME (rangeVel),
INAME (sizeHistVel),
INAME (stepVel),
```

Allocation of the histogram array is included in `AllocArrays`:

```
histVel = AllocVecR (sizeHistVel);
```

initialization, in `SetupJob`, requires the additional statement

```
countVel = 0;
```

and the histogram function itself is called from `SingleStep` by

```
if (stepCount >= stepEquil && (stepCount - stepEquil)
    stepVel == 0) EvalVelDist ();
```

Histogram output is provided by the function

```
  PrintVelDist (FILE *fp) {
    real vBin;
    int n;
    for (n = 1; n <= sizeHistVel; n ++) {
5     vBin = (n - 0.5) * rangeVel / sizeHistVel;
      fprintf (fp, "%8.3f %8.3f\n", vBin, histVel[n]);
    }
  }
```

To demonstrate the way in which the velocity distribution evolves over time during the early portion of the simulation, we study a system with $N_a = 2500$; use of a larger system produces smoother results without needing to average over multiple runs (to simulate a system of this size quickly we resorted to the methods described in Chapter 3, although this has no effect on the results). The input data are as above, except for the following:

```
initUcell      50 50
deltaT         0.001
limitVel       4
rangeVel       3.0
sizeHistVel    50
stepVel        5
```

The results are shown in Figure 2.3; the final distribution develops rapidly and is reached within about 0.4 time units. From results of this kind it is clear that there is no need to assign an initial velocity

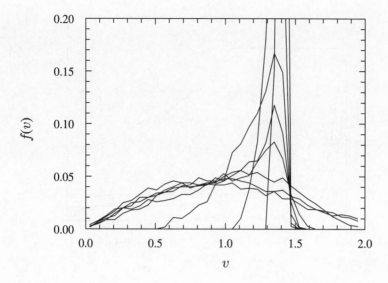

Fig. 2.3 Velocity distribution as a function of time; successively broader graphs are for t values 0.02, 0.04, 0.06, 0.08, 0.1, 0.2, 0.4, and 1.0 (the initial state – not shown – is a spike at $v = \sqrt{2}$).

distribution carefully – the system takes care of this matter on its own (for very small systems there are deviations from the theoretical distribution [ray91]).

The Boltzmann H-function occupies an important position in the development of statistical mechanics [mcq76]. It is defined as

$$H(t) = \int f(v, t) \log f(v, t) dv \tag{2.17}$$

and it can be proved that $\langle dH/dt \rangle \leq 0$, with equality only applying when $f(v)$ is the Maxwell distribution. In order to compute $H(t)$ we use the velocity histogram $\{h_n\}$ obtained previously; if we neglect constants, $H(t)$ can be approximated by

$$h(t) = \sum_n h_n \log(h_n/v_n^{d-1}) \tag{2.18}$$

An additional variable is required for this computation, namely,

```
real hFunction;
```

and the following code must be added to the summary phase of Eval-

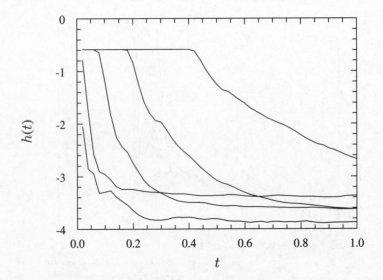

Fig. 2.4 Time-dependence of the Boltzmann H-function (neglecting constants) starting from an ordered state, with ρ ranging from 0.2 to 1.0 in steps of 0.2; convergence is faster at higher ρ.

VelDist (for the two-dimensional case):

```
   hFunction = 0.;
   for (j = 1; j <= sizeHistVel; j ++) {
     if (histVel[j] > 0.) {
       hFunction = hFunction + histVel[j] *
5        log (histVel[j] / ((j - 0.5) * deltaV));
   } }
```

For output, add the extra line to PrintVelDist:

```
   fprintf (fp, "%8.3f\n", hFunction);
```

In Figure 2.4 we show the results of this analysis for several densities, using the above system. The large-t limit of the H-function depends on T(as well as ρ), and since no attempt is made to force the system to a particular temperature the limiting values will differ. Convergence is fastest at high density, while at lower density $h(t)$ does not begin to change until atoms come within interaction range. Finite systems lack the monotonicity suggested by the theorem, but the overall trend is clear and, strictly speaking, the theorem only addresses average quantities. A computation of this kind was carried out in the early days of MD [ald58]; Boltzmann would presumably have found the results much to his taste.

Table 2.2. *Soft-disk results*

ρ	$\langle E \rangle$	$\langle E_k \rangle$	$\sigma(E_k)$	$\langle P \rangle$	$\sigma(P)$
0.8	0.9951	0.650	0.019	4.537	0.147
0.6	0.9936	0.823	0.019	1.954	0.115
0.4	0.9936	0.915	0.014	0.815	0.057

2.5.3 Thermodynamics

To provide a glimpse of what can be done, we show a few measurements made during some short test runs using the following input data:

```
initUcell      20 20
density        0.8
temperature    1.0
deltaT         0.005
stepAvg        1000
stepEquil      500
stepLimit      1500
```

Various values of density are used; any data items not explicitly shown here take the values specified previously. The output is summarized in Table 2.2.

It is unlikely that the temperature (here just $\langle E_k \rangle$) is the one wanted, and the value will certainly not be the one used to create the initial state. To obtain a particular $\langle T \rangle$ the velocities must be adjusted over a series of timesteps until the system settles down at the correct state point. The actual velocity rescaling should be based on $\langle T \rangle$, and not on instantaneous T values that are subject to considerable fluctuation. Though not apparent here, the energy itself can gradually drift upward because of the numerical error in the leapfrog method; the drift rate for a given temperature depends on Δt and is negligible for sufficiently small values. We will return to these matters in Chapter 3.

2.5.4 Trajectories

The first opportunity for using MD to provide results that are unobtainable by other means is in the study of the trajectories followed by individual atoms. Clearly, a single trajectory conveys very little information, but if the trajectories of groups of nearby atoms are examined a clear picture emerges of the different behavior in the solid, liquid, and gaseous states of matter. In the solid phase the atoms are confined to

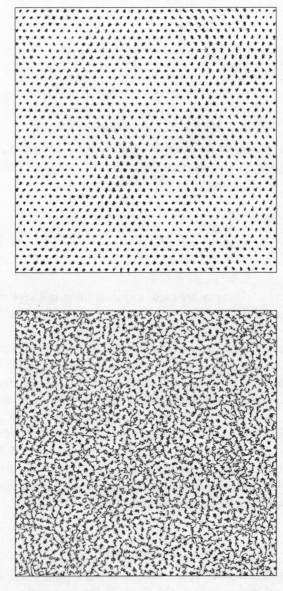

Fig. 2.5 Pictures of trajectory plots at two densities, $\rho = 0.95$ and $\rho = 0.75$, emphasizing the difference between solid and fluid phases, namely localized and diffusing trajectories.

small vibrations around the sites of a regular lattice, the gas is distinguished by trajectories that are ballistic over relatively long distances, while the liquid is characterized by generally small steps, occasional rearrangement, and no long-range positional order. The differences in the trajectories are reflected at the macroscopic level by the values of the diffusion coefficient. Diffusion is just the mean-square atomic displacement (in the MD case, after allowing for periodic wraparound), and is one example of a transport process that MD can examine directly; we will return to this in Chapter 5.

The best way to observe these features is by running an MD simulation interactively and watching the trajectories as they develop for different T and ρ. Trajectories can be shown on a computer display screen by simply drawing a line between the atomic positions every few timesteps; whenever a periodic boundary is crossed, simply interrupt the trajectory drawing and restart it from the opposite boundary. Suitable graphics functions are readily added to the program (with the details depending on the computer and software environment); all that is required, apart from setting up the display functions and arranging for atomic coordinates to be converted to screen coordinates, is the decision as to how frequently the display should be updated. Examples appear in Figure 2.5. Adding full interactivity to allow the user to change T and ρ while the simulation is running requires a little more effort, but leads to visually interesting results.

2.6 Further work

2.1 Compare the observed and theoretical velocity distributions.

2.2 Check that the correct limiting values of $H(t)$ are obtained.

2.3 Add the necessary graphics capability so that trajectories can be examined.

3

Simulating simple systems

3.1 Introduction

In this chapter we focus on a number of techniques used in MD simulation, primarily the methods for computing the interactions and integrating the equations of motion. The goal is to generate the atomic trajectories; subsequent chapters will deal with the all-important question of analyzing this raw 'experimental' data. We continue to work with the simplest atomic systems, in other words, with monatomic fluids based on the LJ potential, not only because we want to introduce the methodology gradually, but also because a lot of the actual qualitative (and even quantitative) behavior of many-body systems is already present in this simplest of models. Models of this kind are widely used in MD studies of basic many-body behavior, examples of which will be encountered in subsequent chapters.

3.2 Equations of motion

While Newton's second law suffices for the dynamics of the simple atomic fluid discussed in this chapter, later chapters will require more complex forms of the equations of motion. The Lagrangian formulation of classical mechanics provides a general basis for dealing with these more advanced problems, and we begin with a brief summary of the relevant results. There are of course other ways of approaching the subject, and we will also make passing reference to Hamilton's equations. A full treatment of the subject can be found in textbooks on classical mechanics, for example [gol80].

3.2.1 Lagrange equations of motion

The starting point is Hamilton's variational principle, which concisely summarizes most of classical mechanics into the statement that the phase-space trajectory followed by a mechanical system is the one for which the time integral $\int \mathcal{L} dt$ is an extremum, where $\mathcal{L}$ is the Lagrangian. Given a set of N independent generalized coordinates and velocities $\{q_i, \dot{q}_i\}$ that describe the state of a conservative system (one in which all forces derive from some potential energy function U), so that $\mathcal{L} = \mathcal{L}(\{q_i\}, \{\dot{q}_i\}, t)$, then $\mathcal{L}$ can be shown to satisfy the Lagrange equations

$$\frac{d}{dt}\left(\frac{\partial \mathcal{L}}{\partial \dot{q}_i}\right) - \frac{\partial \mathcal{L}}{\partial q_i} = 0, \quad i = 1, \ldots, N \tag{3.1}$$

These equations form the starting point for many of the subsequent developments. Newton's second law is a simple consequence of this result, where, if q_i denotes a component of the Cartesian coordinates for one of the atoms (and assuming identical masses m):

$$\mathcal{L} = \tfrac{1}{2} m \sum_i \dot{q}_i^2 - U(\{q_i\}) \tag{3.2}$$

so that (3.1) becomes

$$m\ddot{q}_i = -\frac{\partial U}{\partial q_i} = F_i \tag{3.3}$$

where F_i is the corresponding force component.

3.2.2 Lagrange equations with constraints

There are situations where it is desirable to define the dynamics in ways which cannot be based just on forces obtained from some potential function. For example, in the case of partially rigid molecules the lengths of interatomic bonds should be kept constant. Such restrictions on the dynamics are called constraints and their effect on the equations of motion is the appearance of extra terms that play the role of internal forces, although these terms have an entirely different origin. Here we outline the general framework; the details depend on the problem, and examples will be encountered in Chapters 6 and 10.

Hamilton's principle can be extended to systems with constraints having the general form

$$\sum_k a_{lk} \dot{q}_k + a_l = 0, \quad l = 1, \ldots, M \tag{3.4}$$

This includes the special case of holonomic constraints for which there exist relations between the coordinates of the type $g_l(\{q_k\}, t) = 0$, in which case

$$a_{lk} = \partial g_l / \partial q_k, \qquad a_l = \partial g_l / \partial t \tag{3.5}$$

The resulting Lagrange equations are

$$\frac{d}{dt}\left(\frac{\partial \mathscr{L}}{\partial \dot{q}_i}\right) - \frac{\partial \mathscr{L}}{\partial q_i} = \sum_l \lambda_l a_{li}, \quad i = 1, \ldots, N \tag{3.6}$$

where the time evolution of the M Lagrange multipliers $\{\lambda_l\}$ is evaluated along with the N coordinates: there is a total of $N + M$ equations with a similar number of unknowns. The sum on the right-hand side of (3.6) can be regarded as a generalized force, equivalent in its effect to the imposed constraints.

3.2.3 Hamilton equations of motion

An alternative formulation of the equations of motion sometimes appears in the MD literature. Replace the generalized velocities $\{\dot{q}_i\}$ in the Lagrange formulation by generalized momenta $p_i = \partial \mathscr{L} / \partial \dot{q}_i$ (if the coordinates are Cartesian, then $p_i = m\dot{q}_i$) and consider the Hamiltonian $\mathscr{H} = \mathscr{H}(\{q_i\}, \{p_i\}, t)$ defined by

$$\mathscr{H} = \sum_i \dot{q}_i p_i - \mathscr{L} \tag{3.7}$$

The two first-order equations of motion associated with each coordinate are

$$\dot{q}_i = \frac{\partial \mathscr{H}}{\partial p_i}, \qquad \dot{p}_i = -\frac{\partial \mathscr{H}}{\partial q_i} \tag{3.8}$$

If $\mathscr{H}$ has no explicit time-dependence, then $\dot{\mathscr{H}} = 0$, and $\mathscr{H}$ – the total energy – is a conserved quantity.

3.3 Potential functions

3.3.1 Origins

Modeling of matter at the microscopic level is based on a comprehensive description of the constituent particles. Although such a description must in principle be based on quantum mechanics, MD generally adopts a classical point of view, typically representing atoms or molecules as point masses interacting through forces that depend on the separation of these objects. More complex applications are likely to require extended molecular structures, in which case the forces will also depend on relative orientation. The quantum picture of interactions arising from overlapping electron clouds has been transformed into a system of masses coupled by exotic 'springs'. The justification for this antithesis of quantum mechanics is that not only does it work, but it appears to work surprisingly well; on the other hand the rigorous quantum mechanical description is still hard-pressed in dealing with even the smallest systems.

Obviously, the structural models and potential functions used in classical MD simulation should not be taken too literally, and the potentials are often referred to as effective potentials in order to clarify their status. The classical approximation to the quantum mechanical description of a molecule and its interactions is not derived directly from 'first principles', but, rather, is the result of adapting both structure and potential function to a variety of different kinds of information; these include the results of quantum mechanical energy calculations, experimental data obtained by thermodynamic and various kinds of spectroscopic means, the structure of the crystalline state, measurements of transport properties, collision studies using molecular beams, and so on [hir54, mai81]. These models undergo refinement as new comparisons between simulation and experiment become available, and whenever the evidence against a particular model becomes overwhelming, a revised or even an entirely new model must be developed. (From a strictly theoretical point of view the interactions between molecules can always be written in terms of a multipole expansion [pri84]; if most of the important behavior can be confined to the leading-order terms, then this could be used as the basis for a model potential. While such a systematic approach is appealing, it is not generally used in practice.)

It is always the simplest models that are tried first. Atoms are modeled as point particles interacting through pair potentials. Molecules are represented by atoms with orientation-dependent forces, or as extended structures, each containing several interaction sites; the molecules may be rigid, flexible, or somewhere in between, and if there are internal degrees of freedom there will be internal forces as well. The purpose of this book is not to discuss the design of molecular models; we will make use of existing models and – from a pedagogical viewpoint – the simpler the model the better. Our aim is to demonstrate the general methodology by example, not to review the enormous body of literature devoted to many different kinds of model developed for specific applications. As far

as MD is concerned the complexity of the model has little effect on the
nature of the computation, merely on the amount of work involved.

3.3.2 Example potentials

The most familiar pair interaction is the LJ potential introduced in
Chapter 2. It has been used quite successfully for liquid argon [rah64,
ver67] (although there are better potentials [bar71, mai81]), and is also
often used as a generic potential for qualitative explorations not involving
specific substances. The LJ interaction is characterized by its strongly
repulsive core and weakly attractive tail. To keep computation to a
reasonable level the interaction is truncated at a relatively short range;
at a typical cutoff distance of $r_c = 2.5$ (in MD units) the interaction
energy is just 0.016 of the well depth.

The discontinuity at r_c affects both the apparent energy conservation
and the actual atomic motion, with atoms separated by a distance close
to r_c sometimes moving repeatedly in and out of interaction range. The
discontinuity can be smeared out by changing the form of the potential
function slightly, although this must be done carefully since it is the
potential that defines the model. For example, a potential function $u(r)$
can be modified to eliminate the discontinuity both in itself and in its
first derivative (the force) by replacing it with

$$u_1(r) = u(r) - u(r_c) - du(r)/dr\big|_{r=r_c} (r - r_c) \qquad (3.9)$$

This modification applies across the entire interaction range; an example
of an alternative that confines the change to the vicinity of r_c involves the
use of a cubic polynomial that interpolates smoothly and differentiably
between the value of u at $r = r_c - \delta$ (where $\delta \ll r_c$) and the value 0 at r_c.

A slight change to the LJ interaction leads to a potential that is entirely
repulsive in nature and very short-ranged; a two-dimensional fluid with
this potential was studied in Chapter 2. The particles represented by
this potential are little more than soft spheres (in three dimensions, or
disks in two), although softness is confined to a very narrow range of
separations and the spheres rapidly tend to become hard as they are
driven together. (Another version of the 'soft-sphere' interaction retains
just the r^{-12} term of the LJ potential, again with a cutoff at which u is
discontinuous; we will not consider this variant here.) A system subject
to the original LJ potential can exist in the solid, liquid, or gaseous
states; the attractive part of the potential is used to bind the system
when in the solid and liquid states, and the repulsive part prevents

collapse. When the attractive interaction is eliminated, the behavior is determined primarily by density; at high density the soft-sphere system is packed into a crystalline state, but once melted, unlike the LJ case where there is also a liquid–gas phase transition, the liquid and gas states are thermodynamically indistinguishable.

Other functional forms can be used for interactions between atoms, and between small molecules in cases where spherical symmetry applies [mai81]. Some prove more suitable than others for particular problems. There is even an alternative to the LJ potential for use in simple cases, namely, a function in which the r^{-12} term is replaced by $A\exp(-\alpha r)$; while such a potential produces a softer central core, the repulsive part contributes over a longer range. But since the subject is MD, not the construction of potential functions, we will not pursue this subject any further. Interactions suitable for describing other kinds of molecule will be introduced in subsequent chapters.

3.4 Interaction computations

3.4.1 All pairs

This method was described in Chapter 2. It is the simplest to implement, but extremely inefficient when the interaction range r_c is small compared to the linear size of the simulation region. All pairs of atoms must be examined, because it is not known in advance which atoms actually interact owing to the continual rearrangement that characterizes the fluid state. Although testing whether atoms are separated by less than r_c is only a part of the overall interaction computation, the fact that the amount of computation needed grows as $O(N_a^2)$ rules out the method for all but the smallest values of N_a. Two techniques for reducing this growth rate to a more acceptable $O(N_a)$ level, often used in tandem, will be discussed here; to within a numerical factor this clearly represents the lower bound for the amount of work required to process all N_a atoms. A schematic summary of the methods appears in Figure 3.1.

3.4.2 Cell subdivision

Cell subdivision [sch73, hoc74] provides a means of organizing the information about atom positions into a form that avoids most of the unnecessary work and reduces the computational effort to the $O(N_a)$ level. Imagine that the simulation region is divided into a lattice of small

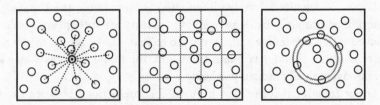

Fig. 3.1 The different approaches to computing interactions: all pairs, cell subdivision (the cell size exceeds the interaction range), and neighbor lists (the concentric circles show the interaction range and the area covered by the neighbor list for one of the atoms).

cells, and that the cell edges all exceed r_c in length. Then if atoms are assigned to cells on the basis of their current positions it is obvious that interactions are only possible between atoms that are either in the same cell or in immediately adjacent cells; if neither of these conditions are met, then the atoms must be at least r_c apart. Because of symmetry only half the neighboring cells need be considered; thus a total of 14 neighboring cells must be examined in three dimensions, and five in two dimensions (these numbers include the cell itself). The wraparound effect due to periodic boundaries is readily incorporated into this scheme. Clearly, the region size must be at least $4r_c$ for the method to be useful, but this requirement is usually met. It is not essential that the cell edges exceed r_c, but if this condition is not satisfied, further cells, not merely nearest neighbors, will have to be included [que73].

The program for the cell-based force calculation involves a form of data organization known as a linked list [knu68]. Rather than accessing data sequentially, the linked list associates a pointer p_n with each data item x_n, the purpose of which is to provide a nonsequential path through the data. Each linked list requires a separate pointer f to access the first data item, and the item terminating the list must have a special pointer value, such as zero, that cannot be mistaken for anything else. Thus $f = a$ points to x_a as the first item in the list, $p_a = b$ points to x_b as the second item, and so on, until a pointer value $p_z = 0$ is encountered, terminating the list. (This type of data organization will reappear in other contexts in subsequent chapters.)

In the cell algorithm, linked lists are used to associate atoms with the cells in which they reside at any given instant; a separate list is required for each cell. The reason for using linked lists is to economize on storage. It is not known in advance how many atoms occupy each cell – a number that can be anywhere between zero and a value determined by the highest possible packing density; the use of sequential tables that list the atoms in each cell, while guaranteeing sufficient storage so that any cell can be maximally occupied, is extremely wasteful. The linked-list approach does not have this problem because of the way the cell occupancy data is organized; the total storage required for all the linked lists is fixed and known in advance.

The additional variables required to support the cell method (the three-dimensional case is shown) are

```
int *cellList, cells[NDIM + 1];
```

and memory allocation for the array `cellList` that will hold all the information associated with the linked lists is added to the function `AllocArrays`:

```
cellList = AllocVecI (nAtom + cells[1] * cells[2] *
    cells[3]);
```

Rather than use separate arrays for the two kinds of pointer, namely, those between atoms in the same cell (p) and those to the initial atom in the list belonging to each cell (f), the first `nAtom` elements in `cellList` are used for the former, and the remainder for the latter. How the list elements are accessed will be clarified by the program listing. The size of the cell array is determined in `SetParams`:

```
for (k = 1; k <= NDIM; k ++) cells[k] = region[k] / rCut;
```

We tacitly assume that it is most efficient to use the smallest cells (exceeding r_c in size) possible; only where the density is sufficiently low that the mean cell occupancy drops substantially below unity is it worth considering using larger (and hence fewer) cells.

The force computation function, including cell assignment, allowance for periodic boundaries, energy and virial calculation, is as follows:

```
   ComputeForces () {
     real dr[NDIM + 1], invWid[NDIM + 1], shift[NDIM + 1], f,
       fcVal, rr, rrCut, rri, rri3, uVal;
     int c, j1, j2, k, m1, m1X, m1Y, m1Z, m2, m2X, m2Y, m2Z,
5      n, offset,
       iofX[] = {0,0,1,1,0,-1,0,1,1,0,-1,-1,-1,0,1},
       iofY[] = {0,0,0,1,1,1,0,0,1,1,1,0,-1,-1,-1},
       iofZ[] = {0,0,0,0,0,0,1,1,1,1,1,1,1,1,1};
     rrCut = Sqr (rCut);
10   for (k = 1; k <= NDIM; k ++)
       invWid[k] = cells[k] / region[k];
     for (n = nAtom + 1; n <= nAtom + cells[1] * cells[2] *
```

```
            cells[3]; n ++) cellList[n] = 0;
        for (n = 1; n <= nAtom; n ++) {
15          c = ((int) ((r[3][n] + regionH[3]) * invWid[3]) *
                cells[2] + (int) ((r[2][n] + regionH[2]) *
                invWid[2])) * cells[1] + (int) ((r[1][n] +
                regionH[1]) * invWid[1]) + nAtom + 1;
            cellList[n] = cellList[c];
20          cellList[c] = n;
        }
        for (n = 1; n <= nAtom; n ++) {
            for (k = 1; k <= NDIM; k ++) ra[k][n] = 0.;
        }
25      uSum = 0.;     virSum = 0.;
        for (m1Z = 1; m1Z <= cells[3]; m1Z ++) {
            for (m1Y = 1; m1Y <= cells[2]; m1Y ++) {
                for (m1X = 1; m1X <= cells[1]; m1X ++) {
                    m1 = ((m1Z - 1) * cells[2] + m1Y - 1) * cells[1] +
30                      m1X + nAtom;
                    for (offset = 1; offset <= 14; offset ++) {
                        m2X = m1X + iofX[offset];     shift[1] = 0.;
                        if (m2X > cells[1]) {
                            m2X = 1;     shift[1] = region[1];
35                      } else if (m2X == 0) {
                            m2X = cells[1];     shift[1] = - region[1];
                        }
                        m2Y = m1Y + iofY[offset];     shift[2] = 0.;
                        if (m2Y > cells[2]) {
40                          m2Y = 1;     shift[2] = region[2];
                        } else if (m2Y == 0) {
                            m2Y = cells[2];     shift[2] = - region[2];
                        }
                        m2Z = m1Z + iofZ[offset];     shift[3] = 0.;
45                      if (m2Z > cells[3]) {
                            m2Z = 1;     shift[3] = region[3];
                        } else if (m2Z == 0) {
                            m2Z = cells[3];     shift[3] = - region[3];
                        }
50                      m2 = ((m2Z - 1) * cells[2] + m2Y - 1) *
                            cells[1] + m2X + nAtom;
                        j1 = cellList[m1];
                        while (j1 > 0) {
                            j2 = cellList[m2];
55                          while (j2 > 0) {
                                if (m1 != m2 || j2 < j1) {
                                    for (k = 1; k <= NDIM; k ++)
                                        dr[k] = r[k][j1] - r[k][j2] - shift[k];
                                    rr = Sqr (dr[1]) + Sqr (dr[2]) +
```

```
60          Sqr (dr[3]);
         if (rr < rrCut) {
            rri = 1. / rr;
            rri3 = rri * rri * rri;
            fcVal = 48. * rri3 * (rri3 - 0.5) * rri;
65          uVal = 4. * rri3 * (rri3 - 1.) + 1.;
            for (k = 1; k <= NDIM; k ++) {
               f = fcVal * dr[k];
               ra[k][j1] = ra[k][j1] + f;
               ra[k][j2] = ra[k][j2] - f;
70          }
            uSum = uSum + uVal;
            virSum = virSum + fcVal * rr;
         }
      }
75    j2 = cellList[j2];
   }
   j1 = cellList[j1];
} } } } }
}
```

The above function is longer and more intricate than the all-pairs
version in Chapter 2, but, as we have already indicated, it is incomparably
faster for systems of beyond minimal size; within the limits set by
numerical rounding it will of course produce the same answers. The
basic organization involves scanning cell pairs, namely each cell with
itself and with half its neighbors; for each pair of cells the atoms
contained in each are also paired to determine which of them lie within
interaction range. Much of the code is devoted to the special handling of
cells adjacent to one or more of the periodic boundaries, and there is an
implicit assumption that there are at least three cells in each direction.
The corresponding two-dimensional version is easily derived by removing
all mention of the z-components and reducing the number of cell offsets
to just five.

As indicated above, the array cellList plays a dual role: the first
part of the array consists of pointers linking different atoms belonging
to the same cell, while the remaining elements, one per cell, point to
the first atom in each cell; zero values indicate the final atom in the list
belonging to a cell and an empty cell respectively. If there are roughly as
many cells as there are atoms this array requires close to two elements
per atom. There are also three arrays, iofX, ..., that together specify the
offsets of each of the 14 neighbor cells (including the cell itself); as with
most other arrays that are declared at compile time, the zeroth element
of each of these arrays is not used.

In view of the fact that the majority of the work in this function is carried out inside a highly nested series of loops, it hardly comes as a surprise to learn that there are different ways of organizing the computation. The method used here is to scan over cells, then over offsets, and only then over cell contents; alternatives include scanning over relative cell offsets and then over cells, or scanning the atoms in the outermost loop, with inner loops that scan the neighboring cells of the cell that contains the atom together with their contents. Some computer architectures may be sensitive to the method chosen (Chapter 14), otherwise it is a matter of convenience. Since the cells are often used as part of the neighbor list method (see below), this issue is usually not critical.

3.4.3 Neighbor-list method

Only a small fraction of the atoms examined by the cell method – an average of $4\pi/81 \approx 0.16$ in three dimensions, $\pi/9 \approx 0.35$ in two – lie within interaction range. If we construct a list of such pairs from those found by the cell method, but in order to allow this list to be useful over several successive timesteps we replace r_c in the test of interatomic separation by $r_n = r_c + \Delta r$, then it should be possible to benefit from this reduced neighborhood size [ver67]. The success of the approach relies on the slowly changing microscopic environment, which implies that the list of neighbors remains valid over a number of timesteps – typically between 10 and 20 even for relatively small Δr. The fact that the list contains pairs that lie outside the interaction range ensures that over this series of timesteps no new interacting pairs can appear that are not already listed. The only disadvantage is the additional storage needed for the list of pairs; once, this might have proved an obstacle, but modern computers usually have sufficient memory for all except (possibly) the very largest of systems.

The value of Δr is inversely related to the rate at which the list must be rebuilt, and it also determines the number of extra noninteracting pairs that are included in the list; it therefore has a certain influence on both processing time and storage requirements. The decision to refresh the neighbor list is based on monitoring the maximum velocity at each step and waiting until

$$\sum_{\text{steps}} \left(\max_i |v_i| \right) > \frac{\Delta r}{2\Delta t} \tag{3.10}$$

before doing the refreshing. This criterion, which is equivalent to examining atomic displacements, errs slightly on the conservative side since it combines contributions from different atoms, but it guarantees that no interacting pairs are ever missed because atoms cannot approach from r_n to r_c during the elapsed time interval; a more precise test could be based on the accumulated motions of individual atoms, but, because the refreshing is already infrequent, the saving will be minimal. Typically, for the fastest computation at liquid density, $\Delta r \approx 0.3-0.4$.

The neighbor list itself can be represented in various ways, one of which is a simple table of atom pairs – the method used here. An alternative method used in Chapter 11 employs a separate list of neighbors for each atom; all are stored in a single array with a separate set of indices specifying the range of list entries for each atom. In either instance the cell method is used to build the neighbor list, with the cell size now being determined by the distance r_n rather than r_c (if the system is too small – relative to r_n – for the cell method to work, then the more costly all-pairs approach must be used).

The new variables required by the neighbor-list method are

```
real dispHi, rNebrShell, vvMax;
int **nebrTab, nebrNow, nebrTabFac, nebrTabLen, nebrTabMax;
```

The quantities that are set in `SetParams` are

```
nebrTabMax = nebrTabFac * nAtom;
for (k = 1; k <= NDIM; k ++)
   cells[k] = region[k] / (rCut + rNebrShell);
```

the initialization in `SetupJob` is

```
nebrNow = 1;
```

and further input data are

```
INAME (nebrTabFac),
RNAME (rNebrShell),
```

The variable `nebrTabFac` determines how much storage should be provided for the neighbor list (per atom), and `rNebrShell` is the variable corresponding to Δr. The additional memory allocated in `AllocArrays` is

```
nebrTab = AllocMatI (2, nebrTabMax);
```

The decision as to when to refresh the neighbor list is based on information about the maximum possible movement of the atoms; this is monitored in `EvalProps` using the following additional code:

```
vvMax = 0.;
for (n = 1; n <= nAtom; n ++) {
```

```
    ...
    if (vv > vvMax) vvMax = vv;
5 }
  dispHi = dispHi + sqrt (vvMax) * deltaT;
  if (dispHi > 0.5 * rNebrShell) nebrNow = 1;
```

If a refresh is due it is done during the next timestep, at the beginning of SingleStep:

```
  if (nebrNow) {
    nebrNow = 0;
    dispHi = 0.;
    BuildNebrList ();
5 }
```

Refreshing the neighbor list implies complete reconstruction. The construction function is very similar to the cell version of ComputeForces, so that the common parts can be skipped, leaving just enough to place the modifications in context. The difference is that, instead of computing the interactions, potentially interacting pairs are merely recorded in the neighbor list for subsequent processing. Note the safety check to ensure that the neighbor list does not grow beyond the storage available:

```
   BuildNebrList () {
     real dr[NDIM + 1], invWid[NDIM + 1], shift[NDIM + 1],
       rr, rrNebr;
     int c, j1, j2, k, m1, m1X, m1Y, m1Z, m2, m2X, m2Y, m2Z,
5      n, offset, iofX[] = ... ;
     rrNebr = Sqr (rCut + rNebrShell);
     for (k = 1; ...
     for (n = nAtom + 1; ...
     for (n = 1; ...
10     ... (identical to cell method)  ...
     nebrTabLen = 0;
     for (m1Z = 1; m1Z <= cells[3]; m1Z ++) {
       ... (identical to cell method)  ...
       if (rr < rrNebr) {
15       nebrTabLen = nebrTabLen + 1;
         if (nebrTabLen > nebrTabMax)
           ErrExit ("neighbor list overflow");
         nebrTab[1][nebrTabLen] = j1;
         nebrTab[2][nebrTabLen] = j2;
20     }
     ...
   }
```

The neighbor list can now be used to compute the interactions; this function is also based on the cell version of ComputeForces:

```
    ComputeForces () {
      real dr[NDIM + 1], f, fcVal, rr, rrCut, rri, rri3, uVal;
      int j1, j2, k, n;
      rrCut = Sqr (rCut);
5     for (n = 1; n <= nAtom; n ++) {
        for (k = 1; k <= NDIM; k ++) ra[k][n] = 0.;
      }
      uSum = 0.;     virSum = 0.;
      for (n = 1; n <= nebrTabLen; n ++) {
10      j1 = nebrTab[1][n];     j2 = nebrTab[2][n];
        for (k = 1; k <= NDIM; k ++) {
          dr[k] = r[k][j1] - r[k][j2];
          if (fabs (dr[k]) > regionH[k])
            dr[k] = dr[k] - SignR (region[k], dr[k]);
15      }
        rr = Sqr (dr[1]) + Sqr (dr[2]) + Sqr (dr[3]);
        if (rr < rrCut) {
          ...  (identical to cell method)  ...
      } }
20  }
```

The check for coordinate wraparound associated with periodic boundaries, using the function ApplyBoundaryCond shown in Chapter 2, is
really only necessary when the neighbor list is about to be refreshed, or
when properties that depend on the atomic coordinates are to be evaluated (while only a minor detail here, since the extra work is minimal, it
becomes a more significant issue when distributed processing is involved
– Chapter 14).

3.4.4 Further methods

For completeness we make brief mention of two additional techniques
that can prove useful, although they are not in widespread use; both aim
at reducing the amount of work required for the interaction calculations.

The replication method simplifies the calculation of interactions across
periodic boundaries by introducing copies of all atoms that are within
a distance r_n of any region boundary and placing them just outside the
simulation region adjacent to the opposite boundary. If these replica
atoms are included in the force computation the wraparound checks are
no longer required, but the cell array will have to be enlarged to include
the region that the replica atoms can occupy. The set of replica atoms
need only be rebuilt (taking care to avoid storage overflow) when the
neighbor list is refreshed, but the coordinates of these atoms are updated

at each timestep. This technique proves particularly useful for distributed and vector processing (Chapter 14), but when the computations are carried out on a single processor the gain is small and usually barely justifies the effort.

The multiple-timestep method is available for medium-range forces that extend beyond several mean atomic spacings (but excluding long-range Coulomb-like forces) [str78]. Pairs of interacting neighbors are divided into groups on the basis of their separation, and the contributions of more distant groups are evaluated at less frequent intervals. While the method has proved useful, it is essential to verify that this approximation does not adversely affect the behavior being studied.

3.4.5 Force tabulation

In most MD simulations the bulk of the computation time is spent computing interactions, and every effort is made to ensure that this is done as efficiently as possible. As an alternative to direct evaluation, interactions can be computed using a simple table lookup, possibly accompanied by interpolation for additional accuracy. Which method is faster depends on the complexity of the potential function. For the LJ case direct evaluation is likely to be more efficient, but for a potential involving, for example, exponential functions, tabulating the entire function, or at least certain parts of it, could improve performance.

The value of tabulation can depend on the computer hardware in ways that are not obvious. Just to give one example, several floating-point computations can often be carried out in the time required merely to retrieve one item at random from a large table. So, for extensive simulations some empirical investigation of this subject should prove worthwhile. If the potential function also depends on molecular orientation the lookup table becomes multi-dimensional, and storage limitations may prevent construction of a table with adequate resolution.

3.5 Integration methods

3.5.1 Selection criteria

A variety of different numerical methods is available, at least in principle, for integrating the equations of motion [pre92]. Most can be quickly dismissed for the simple reason that the heaviest component of the computation is the force evaluation, and any integration method requiring more than one such calculation per timestep is wasteful, unless it can deliver a proportionate increase in the size of the timestep Δt while maintaining the same accuracy. However, because of the strongly repulsive force at short distances in the typical LJ-based potential, there is in effect an upper bound to Δt, so that the well-known Runge–Kutta type methods are unable to enlarge the timestep beyond this limit. The same holds true for adaptive methods that change Δt dynamically to maintain a specified level of accuracy; the fact that each atom experiences a rapidly changing environment due to the local rearrangement of its neighborhood will defeat such an approach. Only two classes of method have achieved widespread use, one a low-order leapfrog technique, the other involving a predictor–corrector approach; both appear in various different but equivalent forms.

Obtaining a high degree of accuracy in the trajectories is neither a realistic nor a practical goal. As we will see below, the sharply repulsive potentials result in trajectories for which even the most minute numerical errors grow exponentially with time, rapidly overwhelming the power-law type of local error introduced by any of the numerical integrators. This is not merely a mathematical curiosity, it also corresponds to what happens in nature, and the issue of trajectory accuracy beyond several average 'collision times' is not a meaningful one. So the criteria for choosing a numerical method focus on energy conservation and on the ability to reproduce certain time- and space-dependent correlations to a sufficient degree of accuracy.

3.5.2 Leapfrog-type methods

Two very simple numerical schemes that are widely used in MD are known as the leapfrog and Verlet methods [bee76, ber86b]; they are completely equivalent algebraically. In their simplest form the methods yield coordinates that are accurate to third order in Δt, and, from the point of view of energy conservation when LJ-type potentials are involved, tend to be considerably better than the higher-order methods discussed subsequently. Their storage requirements are also minimal.

The derivation of the Verlet formula (described much earlier by Delambre [lev93]) follows immediately from the Taylor expansion of the coordinate variable – typically $x(t)$:

$$x(t + h) = x(t) + h\dot{x}(t) + (h^2/2)\ddot{x}(t) + O(h^3) \tag{3.11}$$

where t is the current time, and $h \equiv \Delta t$. Here, $\dot{x}(t)$ is the velocity component, and $\ddot{x}(t)$ the acceleration – or force $f(t)$ in reduced MD units. Note that although $\ddot{x}(t)$ has been expressed as a function of t, it is actually a known function – via the force law – of the coordinates at time t. If we subtract the corresponding expansion for $x(t-h)$ from (3.11) and rearrange, we obtain

$$x(t+h) = 2x(t) - x(t-h) + h^2\ddot{x}(t) + O(h^4) \tag{3.12}$$

The truncation error is of order $O(h^4)$ because the h^3 terms cancel. A possible disadvantage of (3.12) is that at low machine precision the h^2 term multiplying the acceleration may prove a source of inaccuracy. The velocity is not directly involved in the solution, but if required it can be obtained from

$$\dot{x}(t) = [x(t+h) - x(t-h)]/2h + O(h^2) \tag{3.13}$$

with higher-order expressions based on values from earlier timesteps available if needed, though rarely used.

The (highly intuitive [fey63]) leapfrog method is equally simple to derive. Rewrite the Taylor expansion as

$$x(t+h) = x(t) + h[\dot{x}(t) + (h/2)\ddot{x}(t)] + O(h^3) \tag{3.14}$$

The term multiplying h is just $\dot{x}(t+h/2)$, so (3.14) becomes (3.16) below. The result (3.15) is obtained by subtracting from $\dot{x}(t+h/2)$ the corresponding expression for $\dot{x}(t-h/2)$. The leapfrog integration formulae are then

$$\dot{x}(t+h/2) = \dot{x}(t-h/2) + h\ddot{x}(t) \tag{3.15}$$

$$x(t+h) = x(t) + h\dot{x}(t+h/2) \tag{3.16}$$

The fact that coordinates and velocities are evaluated at different times does not present a problem; if an estimate for $\dot{x}(t)$ is required there is a simple connection that can be expressed in either of two ways:

$$\dot{x}(t) = \dot{x}(t \mp h/2) \pm (h/2)\ddot{x}(t) \tag{3.17}$$

The initial conditions can be handled in a similar manner, although a minor inaccuracy in describing the starting state, namely, the distinction between $\dot{x}(0)$ and $\dot{x}(h/2)$, is often ignored. The implementation of this method appeared in Chapter 2.

3.5.3 *Predictor–corrector methods*

Predictor–corrector (PC) methods [gea71, bee76, ber86b] are multiple-value methods in the sense that they make use of several items of information computed at one or more earlier timesteps. In the two most familiar forms of the method there is a choice between using the acceleration values at a series of previous timesteps – the multistep (Adams) approach – or using the higher derivatives of the acceleration at the current timestep (the Nordsieck method). For methods accurate to a given power of h the two forms can be shown to be algebraically equivalent. The methods are of higher order than leapfrog, but entail a certain amount of extra computation and require storage for the additional variables associated with each atom. We will focus just on multistep methods, because derivatives of the acceleration – quantities that are not natural participants in Newtonian dynamics – are absent. The advantage of using higher derivatives is that h can easily be changed in the course of the calculation; this is never done in MD.

Since the origin of the numerical coefficients appearing in the PC formulae may seem a little mysterious we include a brief summary of the derivation. The goal is to solve the second-order differential equation

$$\ddot{x} = f(x, \dot{x}, t) \tag{3.18}$$

with $P()$ and $C()$ denoting the formulae used in the predictor and corrector steps of the calculation. The predictor step for time $t + h$ is simply an extrapolation of values computed at earlier times $t, t - h, \ldots$, namely,

$$P(x): \; x(t + h) = x(t) + h\dot{x}(t) + h^2 \sum_{i=1}^{k-1} \alpha_i f(t + [1 - i]h) \tag{3.19}$$

and for a given value of k this so-called Adams–Bashforth method (which contains the same information as a Taylor expansion) provides exact results for $x(t) = t^q$ provided $q \leq k$; in the general case the local error is $O(h^{k+1})$. In order for this to be true the coefficients $\{\alpha_i\}$ must satisfy the set of $k - 1$ equations

$$\sum_{i=1}^{k-1} (1 - i)^q \alpha_i = \frac{1}{(q + 1)(q + 2)}, \quad q = 0, \ldots, k - 2 \tag{3.20}$$

These and the subsequent sets of linear equations are readily solved; the coefficients are all rational fractions. A similar result holds for $\dot{x}$,

$$P(\dot{x}): \; h\dot{x}(t + h) = x(t + h) - x(t) + h^2 \sum_{i=1}^{k-1} \alpha_i' f(t + [1 - i]h) \tag{3.21}$$

with coefficients that satisfy equations

$$\sum_{i=1}^{k-1}(1-i)^q\alpha_i' = \frac{1}{q+2} \tag{3.22}$$

After computing the value of $f(t+h)$, using the predicted values of x and $\dot{x}$, the corrections are made with the aid of the Adams–Moulton method (which was originally formulated as a separate implicit method, but subsequently adopted for use as a means of refining the predicted estimate):

$$C(x):\ x(t+h) = x(t) + h\dot{x}(t) + h^2\sum_{i=1}^{k-1}\beta_i f(t+[2-i]h) \tag{3.23}$$

$$C(\dot{x}):\ h\dot{x}(t+h) = x(t+h) - x(t) + h^2\sum_{i=1}^{k-1}\beta_i' f(t+[2-i]h) \tag{3.24}$$

with coefficients obtained from

$$\sum_{i=1}^{k-1}(2-i)^q\beta_i = \frac{1}{(q+1)(q+2)}, \qquad \sum_{i=1}^{k-1}(2-i)^q\beta_i' = \frac{1}{q+2} \tag{3.25}$$

Note that the predicted values do not appear in the corrector formulae, except for their involvement in evaluating f. The coefficients $(\alpha_i, \ldots)$ obtained by solving these equations with $k = 4$ are embedded in the integration functions, and are also tabulated at the end of the chapter. The results are readily adapted to the multi-variable MD situation: the first part of the processing involves applying the predictor step to all the variables (atomic coordinates and velocities), followed by the force computation based on the predicted values, and finally the corrector step.

While most of the dynamical problems studied here can be expressed as second-order differential equations, there are cases where first-order equations are required. Analogous PC formulae are available for the equation

$$\dot{x} = f(x,t) \tag{3.26}$$

The predictor and corrector are

$$P(x):\ x(t+h) = x(t) + h\sum_{i=1}^{k}\alpha_i f(t+[1-i]h) \tag{3.27}$$

$$C(x):\ x(t+h) = x(t) + h\sum_{i=1}^{k}\beta_i f(t+[2-i]h) \tag{3.28}$$

with coefficients that satisfy

$$\sum_{i=1}^{k}(1-i)^q\alpha_i = \frac{1}{q+1}, \qquad \sum_{i=1}^{k}(2-i)^q\beta_i = \frac{1}{q+1} \qquad (3.29)$$

The coefficients are also tabulated at the end of the chapter and incorporated into programs used in later case studies.

The functions that use the $k = 4$ PC method for integrating the MD equations of motion follow (here the short arrays holding the coefficients are indexed from zero); the predicted velocities are not always required but are included here for use in those cases where they are:

```
   PredictorStep () {
     real cr[] = {19.,-10.,3.}, cv[] = {27.,-22.,7.},
        div = 24.;
     int k, n;
5    for (n = 1; n <= nAtom; n ++) {
       for (k = 1; k <= NDIM; k ++) {
         ro[k][n] = r[k][n];
         rvo[k][n] = rv[k][n];
         r[k][n] = r[k][n] + deltaT * rv[k][n] +
10          (deltaT * deltaT / div) * (cr[0] * ra[k][n] +
            cr[1] * ra1[k][n] + cr[2] * ra2[k][n]);
         rv[k][n] = (r[k][n] - ro[k][n]) / deltaT +
            (deltaT / div) * (cv[0] * ra[k][n] +
            cv[1] * ra1[k][n] + cv[2] * ra2[k][n]);
15          ra2[k][n] = ra1[k][n];
         ra1[k][n] = ra[k][n];
       } }
     }

   CorrectorStep () {
     real cr[] = {3.,10.,-1.}, cv[] = {7.,6.,-1.}, div = 24.;
     int k, n;
     for (n = 1; n <= nAtom; n ++) {
5      for (k = 1; k <= NDIM; k ++) {
         r[k][n] = ro[k][n] + deltaT * rvo[k][n] +
            (deltaT * deltaT / div) * (cr[0] * ra[k][n] +
            cr[1] * ra1[k][n] + cr[2] * ra2[k][n]);
         rv[k][n] = (r[k][n] - ro[k][n]) / deltaT +
10          (deltaT / div) * (cv[0] * ra[k][n] +
            cv[1] * ra1[k][n] + cv[2] * ra2[k][n]);
       } }
     }
```

Additional arrays are needed to hold the acceleration values from two earlier timesteps (times $t - h$ and $t - 2h$), and to provide temporary storage for the old coordinates and velocities from time t so that they can be overwritten by the predicted values (if the predicted velocity is not required all reference to it can be dropped). These arrays are

```
real **ra1, **ra2, **ro, **rvo;
```

and all are allocated in the function `AllocArrays` using the same calls to `AllocMatR` as for r.

In `SingleStep`, the changes necessary in order to use this method (including removal of the call to `LeapfrogStep`) are

```
  PredictorStep ();
  ApplyBoundaryCond ();
  ComputeForces ();
  CorrectorStep ();
5 ApplyBoundaryCond ();
```

The interactions are evaluated using the results of the predictor step, but are not reevaluated following the corrector; as a consequence, those properties of the system that depend on the interactions themselves, such as the pressure, are based on the predicted rather than the corrected values – the mean error should be insignificant. Variations of this method tried in the past include actually doing this second evaluation – at considerable computational cost – and applying the corrector more than once; neither were found to provide noticeable improvement in accuracy and they are not used. Two calls to the periodic boundary function `ApplyBoundaryCond` are included here: if the neighbor-list method is used the first call serves no useful purpose and can be omitted; the second call is really only necessary if the neighbor list is due for reconstruction, or if the corrected coordinates are needed for evaluating properties of the system, but because only a small amount of computation is involved it is perhaps safer to leave it in place.

3.5.4 Comparison

Because of the greater flexibility and potentially higher local accuracy, PC methods are suited to more complex problems such as rigid bodies or constrained dynamics, where greater accuracy at each timestep is desirable, or where the equations of motion include velocity-dependent forces. The leapfrog approach needs less work and reduced storage, but has the disadvantage that it must be specially adapted for different kinds of problem; leapfrog gives better energy conservation with strongly divergent LJ-type potentials at larger Δt, and because of its minimal storage needs is suitable for extremely large-scale studies where storage can become an important issue. Tests of comparative accuracy will be given in Section 3.7.

3.6 Initial state

3.6.1 Initial coordinates

If we assume that the purpose of the simulation is to study the equilibrium fluid state, then the nature of the initial configuration should have no influence whatsoever on the outcome of the simulation. In choosing the initial coordinates the usual method is to position the atoms at the sites of a lattice whose unit cell size is chosen to ensure uniform coverage of the simulation region. Typical lattices used in three dimensions are the face-centered cubic (FCC) and simple cubic, whereas in two dimensions the square and triangular lattices are used; if the goal is the study of the solid state, then this will dictate the lattice selection. There is little point in laboriously constructing a random arrangement of atoms, typically using a Monte Carlo procedure to avoid overlap, since the dynamics will produce the necessary randomization very quickly. (An obvious way of reducing equilibration time is to base the initial state on the final state of a previous run.)

The function that generates an FCC arrangement (with the option of unequal edges) follows; there are four atoms per unit cell, and the system is centered at the origin. Examples of other lattices are shown separately at the end of the chapter.

```
     InitCoords () {
       real c[NDIM + 1], gap[NDIM + 1];
       int j, k, n, nX, nY, nZ;
       for (k = 1; k <= NDIM; k ++)
5        gap[k] = region[k] / initUcell[k];
       n = 0;
       for (nZ = 1; nZ <= initUcell[3]; nZ ++) {
         c[3] = (nZ - 0.75) * gap[3] - regionH[3];
         for (nY = 1; nY <= initUcell[2]; nY ++) {
10         c[2] = (nY - 0.75) * gap[2] - regionH[2];
           for (nX = 1; nX <= initUcell[1]; nX ++) {
             c[1] = (nX - 0.75) * gap[1] - regionH[1];
             for (j = 1; j <= 4; j ++) {
               n = n + 1;
15             for (k = 1; k <= NDIM; k ++) {
                 if (j == k || j == 4) r[k][n] = c[k];
                 else r[k][n] = c[k] + 0.5 * gap[k];
       } } } } }
     }
```

For the FCC lattice, the region size evaluation in SetParams (Chapter 2) uses the expression

```
region[k] = initUcell[k] / pow (density / 4., 1./3.);
```

and the total number of atoms must be changed to

```
nAtom = 4 * initUcell[1] * initUcell[2] * initUcell[3];
```

3.6.2 *Initial velocities*

Similar considerations apply to the initial velocities, namely, that rapid equilibration renders the careful fabrication of a Maxwell distribution unnecessary. We repeat the simple scheme of Chapter 2, but now in three dimensions; the velocity magnitude is fixed, each velocity vector is assigned a random direction, and the velocities are then adjusted to ensure that the center of mass is at rest:

```
  InitVels () {
    real e[NDIM + 1], vSum[NDIM + 1];
    int k, n;
    for (k = 1; k <= NDIM; k ++) vSum[k] = 0.;
5   for (n = 1; n <= nAtom; n ++) {
      RandVec3 (e, &randSeed);
      for (k = 1; k <= NDIM; k ++) {
        rv[k][n] = vMag * e[k];
        vSum[k] = vSum[k] + rv[k][n];
10  } }
    ... (same as in two dimensions) ...
  }
```

The function makes use of RandVec3 (see Appendix) to provide unit vectors with random orientation.

3.6.3 *Initialization of integration variables*

In addition to setting the initial values for the more obvious physical quantities, the numerical integrator requires its own initializing. For the leapfrog method, if the user is prepared to forgo the minor difference between $t = 0$ and $t = \Delta t/2$ in setting the initial velocities, no further work is required, and even if this difference is not to be overlooked a single interaction computation is all that is required. For the PC method the function for initializing the additional arrays is

```
  InitAccels () {
    int k, n;
```

```
    for (n = 1; n <= nAtom; n ++) {
      for (k = 1; k <= NDIM; k ++)
5       ra[k][n] = ra1[k][n] = ra2[k][n] = 0.;
    }
  }
```

While these are of course not the correct values, there is little benefit in doing a more careful job, such as using a self-starting Runge–Kutta method for the first few timesteps. One reason for this is that the trajectories are highly sensitive to computational details such as rounding error (see below), and this has a much stronger influence than the precise details of the initial state; the other is that additional velocity adjustments are usually made early in the run to force the system to the correct temperature.

3.6.4 Temperature adjustment

Bringing the system to the required average temperature calls for velocity rescaling. If there is a gradual energy drift due to numerical integration error, further velocity adjustments will be required over the course of the run. The drift rate depends on a number of factors – the integration method, potential function, the value of Δt, and the ambient temperature.

Since T fluctuates there is no point in making adjustments based on instantaneous estimates. Instead, we can make use of the average $\langle T \rangle$ values that are already available. The temperature adjustment (or velocity rescaling) function below would therefore be called from SingleStep immediately after the call AccumProps(2) that summarizes the results:

```
  AdjustTemp () {
    real vFac;
    int k, n;
    vFac = vMag / sqrt (2. * sKinEnergy);
5   for (n = 1; n <= nAtom; n ++) {
      for (k = 1; k <= NDIM; k ++)
        rv[k][n] = rv[k][n] * vFac;
    }
  }
```

How frequently this adjustment is required, if at all, must be determined empirically; initially it should be omitted since it may interfere with energy conservation. If needed, the interval between adjustments would be specified by the variable

```
  int stepAdjustTemp;
```

the value included in the input data:

```
INAME (stepAdjustTemp),
```

and the adjustment made by a call from SingleStep:

```
if (stepCount % stepAdjustTemp == 0) AdjustTemp ();
```

(An alternative would be to automate the scheme, applying the adjustment whenever drift exceeds a given threshold.)

Forcing the system to have the correct $\langle T \rangle$ during the equilibration phase of the simulation uses separate estimates of $\langle E_k \rangle$. In SingleStep add (after the call to EvalProps):

```
    if (stepCount < stepEquil) {
      sInitKinEnergy = sInitKinEnergy + kinEnergy;
      if (stepCount % stepInitlzTemp == 0) {
        InitAdjustTemp ();
5       sInitKinEnergy = 0.;
      }
    }
```

and introduce a new function:

```
    InitAdjustTemp () {
      ...
      sInitKinEnergy = sInitKinEnergy / stepInitlzTemp;
      vFac = vMag / sqrt (2. * sInitKinEnergy);
5     ... (same as AdjustTemp) ...
    }
```

The extra variables are

```
    real sInitKinEnergy;
    int stepInitlzTemp;
```

an additional input data item:

```
    INAME (stepInitlzTemp),
```

and initialization:

```
    sInitKinEnergy = 0.;
```

3.7 Performance measurements

3.7.1 Accuracy

In order to demonstrate the accuracy of the integration methods, and the way in which accuracy depends on Δt, we will carry out a series of measurements of the energy as a function of time for the leapfrog and $k = 4$ PC integrators and several timestep values. We use the soft-sphere interaction rather than LJ in order to avoid any additional fluctuations due to the discontinuity at r_c.

Input data to the calculation are as follows:

```
runId          1
initUcell      5 5 5
density        0.8
temperature    1.0
nebrTabFac     8
rNebrShell     0.4
randSeed       17
stepEquil      0
stepInitlzTemp 9999
```

Other input data items are: deltaT varies over a $16:1$ range between 0.001 25 and 0.02, the value of stepLimit is chosen to give a total run length of 200 time units (the extreme values being 160 000 and 10 000), and stepAvg is set so that a result is output every ten time units. The initial state is a simple cubic lattice, so that $N_a = 125$, and computations are carried out in 64-bit (double) precision. This particular value of nebrTabFac is more than adequate for the soft-sphere fluid at moderate density; for an LJ fluid the value depends on r_c, with 50 or larger typically being required. A few brief test runs with the actual potential function in the density range of interest should be sufficient to determine the necessary value, including a safety margin.

The results are shown in Figure 3.2, where it is clear that the leapfrog (LF) method allows a much larger Δt for a given degree of energy conservation [ber86b]. To stress the accuracy of the method (from the energy point of view) some of these results are repeated in Table 3.1, where we also compare the effect of using 32-bit arithmetic in the leapfrog method with the other results based on 64-bit arithmetic. To a limited extent, accuracy can be sacrificed in the cause of speed, for example when the goal is a real-time demonstration, but there are limits to the size of Δt if numerical instability is to be avoided. All computations in the case studies will use 64-bit precision.

3.7.2 Reproducibility

The issue of reproducibility is tied to the rate of approach to equilibrium (or to a stationary final state). In most cases, once the system has equilibrated there will be no memory of the details of the initial state, but problems can arise in cases of very slow convergence, or where there

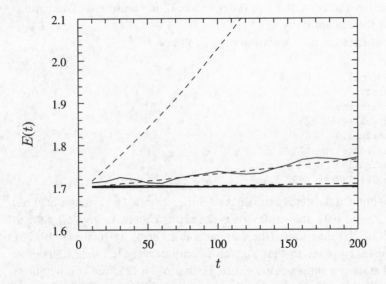

Fig. 3.2 Energy drift for different values of Δt, for both LF (solid curves, the first four of which are indistinguishable, are for $\Delta t = 0.001\,25$, 0.0025, 0.005, 0.01, 0.02) and PC (dashed, for $\Delta t = 0.001\,25$, 0.0025, 0.005) methods.

Table 3.1. *Energy conservation*

Δt:	0.00125	0.00125	0.0025	0.0025	0.005	0.005	0.01
t	LF	PC	LF	PC	LF	LF(32)	LF
10	1.7013	1.7015	1.7015	1.7032	1.7021	1.7022	1.7041
50	1.7012	1.7030	1.7017	1.7162	1.7022	1.7025	1.7048
100	1.7013	1.7049	1.7018	1.7326	1.7015	1.7026	1.7029
150	1.7012	1.7069	1.7019	1.7495	1.7026	1.7034	1.7023
200	1.7013	1.7089	1.7019	1.7671	1.7029	1.7039	1.7035

are different metastable states in which the system can become trapped. If we exclude such special circumstances, the averaged results from separate runs should agree to within the limits set by the fluctuations. As a brief demonstration we show how the kinetic energy varies with time for simulations that differ only in the choice of the initial random velocities.

The input data for this test, with $N_a = 256$, $\Delta t = 0.005$, $T = 1$, and $\rho = 0.8$, includes

```
stepAvg        200
stepEquil      1000
```

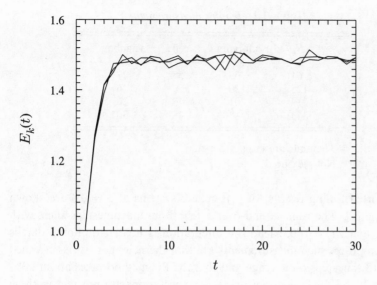

Fig. 3.3 Convergence of mean kinetic energy from different initial states.

```
stepInitlzTemp  200
stepLimit       6000
```

The runs use different values of `randSeed` (such as 17, 19, and 21), and the results are shown in Figure 3.3. Convergence of $\langle E_k \rangle$ to its final value is dominated by the temperature adjustments that are made while $t < 5$, but the differences between the runs lie within the range of the fluctuations.

3.7.3 *Efficiency*

Two methods of improving the efficiency of the force computation were described here, cells and neighbor lists. To show that these methods really do provide at least some of the promised benefits we provide a few timing comparisons for the three methods. It is the relative timings rather than the absolute values that are of most interest here, but for the record, these particular measurements were made on an Silicon Graphics Indy with a 100MHz, MIPS 4000SC processor. The major cause of any timing fluctuations is the size of the cells used in both the cell and neighbor-list methods; the fact that an integral number of cells must fit along each region edge can lead to variations in the mean cell occupancy that will affect performance, especially for small systems.

Table 3.2. *Timing measurements* [a]

N_a	all-pairs	cells	neighbors
64	63	55	27
125	120	50	26
216	201	49	26
343	$-$[b]	48	30

[a] The units are in μs/atom-step.
[b] Not measured.

Assorted timing results for soft-sphere systems at $\rho = 0.8$ are shown in Table 3.2. The runs extend over a few thousand timesteps each, with $T = 1$ and $\Delta t = 0.005$; leapfrog integration is used. Additional details emerging from such measurements are that the neighbor list is refreshed every 13 timesteps on average and that the PC method takes about 50% longer. If the theoretical performance expectations are not met in these relatively small systems, allowance should be made for contributions from other parts of the computation.

3.8 Trajectory sensitivity

One particular consequence of the numerical approach deserves special consideration. We have seen how measurements of bulk properties, such as kinetic energy, are reproducible, subject only to well-understood statistical fluctuations. Other equilibrium and steady-state properties are similarly well behaved. When it comes to the trajectories themselves it is an entirely different story: trajectories display an exponential sensitivity to even the most minute perturbation. This implies that trajectories are sensitive to the precision and rounding method used for floating-point arithmetic, and even to the exact sequence of machine instructions in the program. Short of using infinite precision (!) there is no way in which two different MD programs, or even the same MD program run on computers of different design, will yield the same trajectories (and so the Laplacian vision is laid to rest). This is hardly surprising, but it is also irrelevant, since there is no meaningful physical quantity that depends on just a single trajectory realization; all realistic measurements involve averages that conceal this sensitivity, including the transport properties based on integration along the actual trajectories (Chapter 5). This extreme sensitivity is the microscopic basis for molecular chaos that plays such an important role in statistical mechanics: though the equations of motion are time-reversible this fact turns out to be unobservable in most practical situations [orb67, lev93].

To actually measure this behavior we consider a system of $2N_a$ atoms in which odd- and even-numbered atoms form independent but identical subsystems that are assigned the same initial coordinates and velocities. One subsystem is slightly perturbed by multiplying its velocities by $1 + \epsilon s$, where ϵ is a small number and s a random value in the range $(-1, 1)$, and we then examine how the root-mean-square coordinate difference

$$\Delta r = \sqrt{\frac{1}{N_a} \sum_{i=1}^{N_a} (r_{2i} - r_{2i-1})^2} \tag{3.30}$$

varies with time. Although the study uses soft atoms and a leapfrog method subject to numerical integration error, this error is not the dominant factor, because similar results can also be obtained in hard-sphere studies that are free from integration error.

Only a few simple modifications to the MD program are required. The atoms are divided into two entirely separate subsystems, and nAtom is doubled. In BuildNebrList the criterion for selecting pairs is replaced by

```
if ((j1 - j2) % 2 == 0 && (m1 != m2 || j2 < j1))
```

Properties such as the energy can be computed for either or both subsystems by selecting which atoms contribute to the various sums. The addition to InitCoords to produce the duplicate system is just an extra copy of the code already in the function:

```
n = n + 1;
for (k = 1; k <= NDIM; k ++) ...
```

thereby ensuring that pairs of atoms $2i - 1$ and $2i$ are located at the same lattice site, and the change to InitVels is the addition of

```
  n = n + 1;
  for (k = 1; k <= NDIM; k ++) {
    rv[k][n] = vMag * e[k];
    vSum[k] = vSum[k] + rv[k][n];
5 }
```

after setting the velocity of an odd-numbered atom.

The trajectory perturbation function is

```
PerturbTrajDev () {
  int k, n;
```

```
    for (k = 1; k <= NDIM; k ++) {
      for (n = 2; n <= nAtom; n += 2) {
5       r[k][n] = r[k][n - 1];
        rv[k][n] = rv[k][n - 1] * (1. + pertTrajDev *
          (2. * RandR (&randSeed) - 1.));
    } }
    countTrajDev = 1;
10  valTrajDev[countTrajDev] = 0.;
  }
```

and trajectory analysis, allowing for periodic boundaries, is carried out
by the following function:

```
  MeasureTrajDev () {
    real d, dSum;
    int k, n;
    dSum = 0.;
5   for (k = 1; k <= NDIM; k ++) {
      for (n = 2; n <= nAtom; n += 2) {
        d = fabs (r[k][n] - r[k][n - 1]);
        if (d > regionH[k]) d = region[k] - d;
        dSum = dSum + Sqr (d);
10  } }
    countTrajDev = countTrajDev + 1;
    valTrajDev[countTrajDev] = sqrt (dSum / (0.5 * nAtom));
  }
```

The addition to SingleStep is

```
  if (stepCount == stepEquil) PerturbTrajDev ();
  if (stepCount > stepEquil && (stepCount - stepEquil)
    stepTrajDev == 0) {
    MeasureTrajDev ();
5   if (countTrajDev == limitTrajDev) {
      PrintTrajDev (stdout);
      PerturbTrajDev ();
  } }
```

and the output function is

```
  PrintTrajDev (FILE *fp) {
    real tVal;
    int n;
    for (n = 1; n <= limitTrajDev; n ++) {
5     tVal = (n - 1) * stepTrajDev * deltaT;
      fprintf (fp, "%.4e %.4e\n", tVal, valTrajDev[n]);
    }
  }
```

Additional variables used in these measurements are

```
real *valTrajDev, pertTrajDev;
int countTrajDev, limitTrajDev, stepTrajDev;
```

input data items to be added to `nameList` are

```
RNAME (pertTrajDev),
INAME (limitTrajDev),
INAME (stepTrajDev),
```

and an array allocated in `AllocArrays` is

```
valTrajDev = AllocVecR (limitTrajDev);
```

The measurements shown in Figure 3.4 are based on a soft-sphere system with $N_a = 2048$, $T = 1$, $\rho = 0.8$, and $\Delta t = 0.005$. Other input data items include

```
limitTrajDev    100
pertTrajDev     1e-6
stepEquil       3000
stepTrajDev     20
```

Three values of velocity perturbation `pertTrajDev` are used, namely, 10^{-6}, 10^{-5}, and 10^{-4}; just one set of measurements averaged over all atoms is made after allowing sufficient time for equilibration. The initial linear growth in $\log(\Delta r)$ to beyond $t = 2$ corresponds to exponential divergence ($t = 0$ on the graph is where the perturbation is applied). Once the size of the deviation reaches the atomic diameter (≈ 1) the more familiar diffusive processes (with deviations proportional to $\sqrt{t}$) take over.

3.9 Further work

3.1 Implement the two-dimensional versions of the cell and neighbor-list methods.

3.2 See whether the $k = 5$ PC method is an improvement over the $k = 4$ method used here.

3.3 Explore the use of PC methods involving derivatives of the acceleration [bee76, ber86b].

3.4 Determine the performance gain using tabulated interactions.

3.5 How is the computation speed affected by the organization and dimension of the data arrays?

3.6 What is the effect of smoothing the LJ interaction at r_c?

3.7 Investigate the use of multiple-timestep methods.

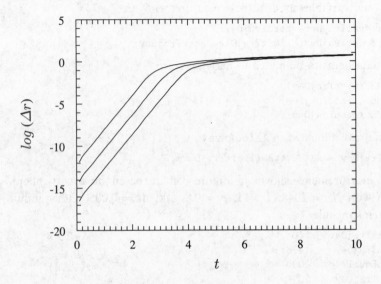

Fig. 3.4 Trajectory divergence for different initial velocity perturbations.

3.10 Additional material

3.10.1 Predictor–corrector coefficients

The coefficients for PC methods typically used in MD applications are listed here. For second-order differential equations the coefficients for $k = 4$ (incorporated in the functions listed previously) and $k = 5$ (sometimes used, although not here) are listed in Table 3.3. Table 3.4 lists the coefficients for first-order equations with $k = 3$ used in Chapters 6 and 8, as well as $k = 4$.

3.10.2 Initial coordinates

Examples of the changes to InitCoords for other types of lattice arrangement are shown here. For the simple cubic lattice where there is only a single atom in each unit cell the loop over j is removed, and the coordinate assignment is just

```
     for (nZ ...
       c[3] = (nZ - 0.5) * gap[3] - regionH[3];
       for (nY ...
         c[2] = (nY - 0.5) * gap[2] - regionH[2];
5        for (nX ...
           c[1] = (nX - 0.5) * gap[1] - regionH[1];
           n = n + 1;
           for (k = 1; k <= NDIM; k ++) r[k] [n] = c[k];
```

Table 3.3. *PC coefficients for second-order equations*

$k = 4 \ (\times 1/24)$	1	2	3	
$P(x)$:	19	-10	3	
$P(\dot{x})$:	27	-22	7	
$C(x)$:	3	10	-1	
$C(\dot{x})$:	7	6	-1	
$k = 5 \ (\times 1/360)$	1	2	3	4
$P(x)$:	323	-264	159	-38
$P(\dot{x})$:	502	-621	396	-97
$C(x)$:	38	171	-36	7
$C(\dot{x})$:	97	114	-39	8

Minor changes are also required in SetParams:

```
region[k] = initUcell[k] / pow (density, 1./3.);
...
nAtom = initUcell[1] * initUcell[2] * initUcell[3];
```

The BCC lattice has two atoms per unit cell, so the coordinates are assigned by

```
    for (nZ ...
      c[3] = (nZ - 0.75) * gap[3] - regionH[3];
      for (nY ...
        c[2] = (nY - 0.75) * gap[2] - regionH[2];
5       for (nX ...
          c[1] = (nX - 0.75) * gap[1] - regionH[1];
          for (j = 1; j <= 2; j ++) {
            n = n + 1;
            for (k = 1; k <= NDIM; k ++)
10            r[k][n] = c[k] + (j - 1) * gap[k] * 0.5;
```

and in SetParams we require

```
region[k] = initUcell[k] / pow (density / 2., 1./3.);
...
nAtom = 2 * initUcell[1] * initUcell[2] * initUcell[3];
```

The diamond lattice is a slightly more complicated form of the FCC code since the lattice is most readily defined as two staggered FCC lattices, one of which is offset along the diagonal by a quarter unit cell:

```
real ... subShift;
int ... m;
...
```

Table 3.4. *PC coefficients for first-order equations*

$k = 3 \ (\times 1/12)$	1	2	3	
$P(x)$:	23	-16	5	
$C(x)$:	5	8	-1	
$k = 4 \ (\times 1/24)$	1	2	3	4
$P(x)$:	55	-59	37	-9
$C(x)$:	9	19	-5	1

```
     for (nZ ...
5      c[3] = (nZ - 0.875) * gap[3] - regionH[3];
       for (nY ...
         c[2] = (nY - 0.875) * gap[2] - regionH[2];
         for (nX ...
           c[1] = (nX - 0.875) * gap[1] - regionH[1];
10         for (m = 1; m <= 2; m ++) {
             subShift = 0.;
             if (m == 2) subShift = 0.25;
             for (j = 1; j <= 4; j ++) {
               n = n + 1;
15             for (k = 1; k <= NDIM; k ++) {
                 if (j == k || j == 4)
                   r[k][n] = c[k] + gap[k] * subShift;
                 else r[k][n] = c[k] + gap[k] * (0.5 +
                   subShift);
20    } } }
```

The changes to SetParams are as for the BCC, but with the value 2 replaced by 8.

Returning to two-dimensional systems, the triangular lattice with two atoms per unit cell requires

```
     for (nY ...
       c[2] = (nY - 0.75) * gap[2] - regionH[2];
       for (nX ...
         c[1] = (nX - 0.75) * gap[1] - regionH[1];
5        for (j = 1; j <= 2; j ++) {
           ... (as for BCC) ...
```

Because the unit cell shape is not square the region size must be specified differently in SetParams:

```
region[1] = initUcell[1] / sqrt (density * sqrt (3.) / 2.);
region[2] = initUcell[2] / sqrt (density /
```

```
        (2. * sqrt (3.)));
   ...
5  nAtom = 2 * initUcell[1] * initUcell[2];
```

One further initial arrangement is worth including here, namely, a totally random set of initial coordinates. Though not used in the MD programs, it is useful during analysis of spatial organization, in order to contrast MD results with those of random point arrays:

```
   real randTab[101];
   int i, k, n;
   for (i = 1; i <= 100; i ++) randTab[i] = RandR (&randSeed);
   for (n = 1; n <= nAtom; n ++) {
5    for (k = 1; k <= NDIM; k ++) {
        i = (int) (100. * RandR (&randSeed)) + 1;
        r[k][n] = (randTab[i] - 0.5) * region[k];
        randTab[i] = RandR (&randSeed);
   } }
```

To reduce any unwanted correlations in the random numbers a shuffling scheme is employed: the random values are used as indices for accessing a table of random numbers, the entries of which are replaced each time they are used.

4

Equilibrium properties of simple fluids

4.1 Introduction

In this chapter we examine the behavior of systems in equilibrium; in particular, we focus on measurements of thermodynamic properties and studies of spatial structure and organization. The treatment of properties associated with the motion of atoms – the dynamical behavior – forms the subject of Chapter 5.

While basic MD simulation methods – formulating and solving the equations of motion – fall into a comparatively limited number of categories, a wide range of techniques is used to analyze the results. Rarely is the wealth of detail embodied in the atomic or molecular trajectories of particular interest in itself, and the issue is how to extract meaningful information from this vast body of data; even a small system of 10^3 structureless atoms followed over a mere 10^4 timesteps can produce up to 6×10^7 numbers – a full chronological listing of the atomic coordinates and velocities. A great deal of data averaging and filtration of various kinds is required to reduce this to a manageable and meaningful level; how this is achieved depends on the questions that the simulation is supposed to answer. Much of this processing will be carried out while the simulation is in progress, but some types of analysis are best done subsequently using data saved in the course of the simulation run; the decision is determined by the amount of work and data involved, as well as the need for active user participation in the analysis.

Averages corresponding to thermodynamic quantities in homogeneous systems at equilibrium are the easiest measurements to make. Statistical mechanics relates such MD averages to their thermodynamic counterparts, and the ergodic hypothesis can be invoked to justify equating trajectory averages to ensemble-based thermodynamic properties [mcq76].

However, the fact that statistical mechanics has no knowledge of trajectories means that it is incapable of discussing quantities that are defined in terms of atomic motion – diffusion for example. This is the strength of MD: detailed trajectory histories are available, so that not only can the quantities meaningful in a statistical mechanical framework be addressed, but so, too, can any other conceivable quantity

Some aspects of behavior, such as the structural correlations present in the fluid, ranging from the basic pair-correlation function to more subtle correlations involving both position and orientation, or the three-body correlation function, require quite heavy calculations, often rivaling the interaction computation in terms of the amount of work required. Fortunately, such calculations are not needed at each timestep since fluid structure changes only gradually; the rate of change is indeed the criterion for choosing the interval between such measurements.

If the system is spatially inhomogeneous, all quantities, from the simplest thermodynamic values onward, must be based on localized measurements. If the system is also nonstationary over time, long-term time averaging is ruled out because it would obliterate the very effects being studied. In short, the more complex the phenomenon the more demanding the measurement task. These topics will be encountered in Chapters 7 and 13.

4.2 Thermodynamic measurements

4.2.1 Relation to statistical mechanics

Measurements of equilibrium properties that are thermodynamic in nature can be regarded as exercises in numerical statistical mechanics. In such instances MD provides an alternative to Monte Carlo (MC), and if no further information is required about the system, computational efficiency alone should determine the choice of technique. While MC requires less computation per interacting atom pair because only the potential energy has to be evaluated, the number of MC cycles required to obtain uncorrelated samples (more precisely, a series of samples that are only weakly correlated) may exceed the corresponding number of MD timesteps. The reason for this is that the atomic displacements are randomly chosen in MC, and this is a less efficient way for the system to traverse configuration space than via the cooperative dynamics intrinsic to MD.

Because both the number of atoms and the total energy (assuming that numerical drift has been suppressed) are fixed in the MD simulations encountered so far, the relevant statistical-mechanical ensemble for discussing equilibrium behavior is the microcanonical (NVE) one. There is just one minor difference, in that each conserved momentum component removes one degree of freedom, but this is a negligible effect for systems of beyond minimal size.

4.2.2 Error analysis

The measurement process in MD is very similar to experiment. But the experimentalist often has the advantage of knowing that each estimate is independent, allowing well-established statistical methods to be used in the data analysis. With MD, where a series of measurements is carried out in the course of a simulation of limited duration, there is no guarantee that successive estimates are sufficiently unrelated to ensure the reliability of these simple statistical methods. Averages of directly measured quantities may not be the main problem, given an adequate run length, but statistical error estimates are particularly sensitive to correlations between samples.

We assume that the problem has been correctly formulated and implemented; errors in the results can then be categorized as follows. There are systematic errors associated with, for example, finite-size effects, interaction cutoff, and the numerical integration itself; these are an intrinsic part of the computer experiment and are reproducible. There are errors due to inadequate sampling of phase space where, especially near a thermodynamic phase boundary, or in the case of infrequently occurring events, enough of the relevant behavior fails to be sampled; this is symptomatic of poor experimental design. And finally there is statistical error due to random fluctuations in the measurements; under normal circumstances this determines the degree of confidence that can be placed in the results. Only for errors of the last kind is the usual statistical analysis applicable.

Consider a series of M measurements of some fluctuating property A in a system at equilibrium. The mean value is

$$\langle A \rangle = \frac{1}{M} \sum_{\mu=1}^{M} A_\mu \tag{4.1}$$

and if each measurement A_μ is independent, with variance

$$\sigma^2(A) = \frac{1}{M} \sum_\mu \left(A_\mu - \langle A \rangle \right)^2 = \langle A^2 \rangle - \langle A \rangle^2 \tag{4.2}$$

then the variance of the mean $\langle A \rangle$ is

$$\sigma^2(\langle A \rangle) = \sigma^2(A)/M \tag{4.3}$$

But if, as is usually the case in MD (and other) simulations, the assumed independence of the A_μ is unwarranted, $\sigma^2(\langle A \rangle)$ is liable to be underestimated because the effective number of independent measurements is considerably less than M. How the correlation between measurements affects the results can be seen by rewriting the variance correctly as

$$\sigma^2(\langle A \rangle) = \frac{1}{M}\sigma^2(A)\left[1 + 2\sum_\mu (1 - \mu/M)\phi_\mu\right] \tag{4.4}$$

where ϕ_μ is the autocorrelation function

$$\phi_\mu = \frac{\langle A_\mu A_0 \rangle - \langle A \rangle^2}{\langle A^2 \rangle - \langle A \rangle^2} \tag{4.5}$$

A detailed error analysis would involve examining ϕ_μ, but there is little need for this in practice because a much simpler method is available based on block averaging [fly89].

Assuming the A_μ to be correlated, if averages are evaluated over blocks of successive values, then as the block size increases the block averages will be decreasingly correlated; eventually, once the block length exceeds the (unknown) longest correlation time present in the data, the block averages will be independent from a statistical point of view. What is needed is a criterion for choosing the minimal necessary block length: too short a block provides little improvement over the original correlated data, too long a block reduces the number of block averages available for reliable estimation of the variance of the final result.

A very straightforward scheme is based on a series of successive block sizes $b = 1, 2, 4, \ldots$, with the upper bound being set by the total size of the data set. For each b the estimator for the variance can be shown to be

$$\sigma^2(\langle A \rangle_b) = \frac{1}{M_b - 1}\sum_{\beta=1}^{M_b}(A_\beta^2 - \langle A \rangle_b^2) \tag{4.6}$$

where M_b is the total number of blocks, A_β a typical block average, and $\langle A \rangle_b$ the overall average. Whenever the current M_b is odd, the last value is simply discarded before doubling the block size. What should happen, assuming that the total measurement period far exceeds the longest correlation time, is that the successive $\sigma^2(\langle A \rangle_b)$ increase until a plateau is eventually reached; the plateau value is the result. In less than

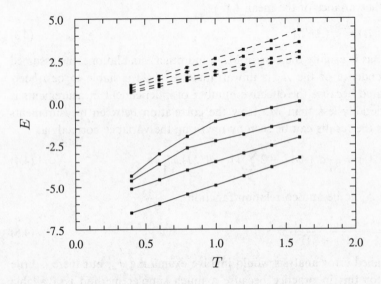

Fig. 4.1 Density and mean temperature dependence of energy for LJ (solid curves) and soft-sphere (dashed) systems; the graphs are for $\rho = 0.4, 0.6, 0.8$, and 1.0.

ideal situations where the measurement period is too short, or barely adequate in length, the plateau will either not appear at all or will be very narrow; in such cases the variance estimate is unreliable. When the method works successfully the block size at the start of the plateau is an indication of the extent to which the samples are correlated.

4.2.3 Energy

Energy measurements are the simplest, and here we briefly examine both LJ and soft-sphere systems. Leapfrog integration is used with $\Delta t = 0.005$; for the LJ system we use a cutoff $r_c = 2.2$. The temperature fluctuates, and in three dimensions we have $\langle E_k \rangle = 3\langle T \rangle/2$. Figure 4.1 shows the dependence of E on ρ and $\langle T \rangle$ as measured in a series of runs that include the following input data:

```
initUcell        5 5 5
stepAvg          1000
stepEquil        1000
stepInitlzTemp   200
stepLimit        3000
```

An FCC lattice is used, so that $N_a = 500$. For the LJ case, this equilibration period is too short for the smaller ρ and T values, so the data is changed to

```
stepEquil        4000
stepLimit        6000
```

For careful quantitative studies the results should be examined closely when deciding on the run length and equilibration period.

In the microcanonical ensemble, thermodynamic quantities based on fluctuations adopt a different form from the canonical ensemble. The most familiar such quantity is the constant volume specific heat $C_V = (\partial E/\partial T)_V$. It is usually defined in terms of energy fluctuations, namely,

$$C_V = \frac{N_a}{k_B T^2} \langle \delta E^2 \rangle \tag{4.7}$$

where $\langle \delta E^2 \rangle = \langle E^2 \rangle - \langle E \rangle^2$, but while this is appropriate in the canonical ensemble, for MD we have $\langle \delta E^2 \rangle = 0$. Instead, it can be shown [leb67] that the relevant fluctuations to consider are those of E_k or E_u individually (they are identical), and that the specific heat is

$$C_V = \frac{3k_B}{2} \left(1 - \frac{2N_a \langle \delta E_k^2 \rangle}{3(k_B T)^2} \right)^{-1} \tag{4.8}$$

Either this directly measurable result or numerical differentiation of the $E(T)$ graph – strictly speaking, $E(\langle T \rangle)$ – could be used for estimating C_V.

4.2.4 Equation of state

Pressure is obtained from the virial expression in Chapter 2; while it can also be expressed in terms of momentum transferred across an arbitrary plane, there is little reason to resort to such a definition that only uses information from a fraction of the system and is therefore subject to larger fluctuations. The virial definition assumes the presence of hard walls responsible for imposing the external pressure, but the result is equally applicable in the case of periodic boundaries [erp77].

Pressure measurements for the runs described above are shown in Figure 4.2. Negative pressure is an indication that the system is being held at too low a density, and in a sufficiently large system separation into distinct liquid (or solid) and vapor phases occurs. A more extensive analysis of this kind would lead to the complete equation of state [nic79]. In the LJ case, when a 'real' substance is being modeled, the values of

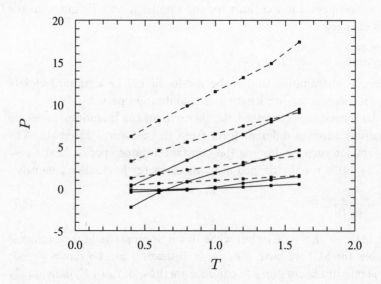

Fig. 4.2 Density- and temperature-dependence of pressure for LJ and soft-sphere systems.

both E and P can be corrected [vog85] to compensate for the truncation at r_c – as in (4.15).

Finite-size effects are already relatively small at $N_a = 500$, at least for positive pressure (and away from the critical point). For example, consider the LJ fluid at $\rho = 0.8$ and $T = 1$. The result obtained in this case is $P = 1.92$, $\sigma(P) = 0.13$, based on a single average over 2000 steps. For the case where $N_a = 2048$ an average over 4000 steps leads to $P = 2.03$, $\sigma(P) = 0.07$. Thus even this very rough comparison suggests that size-dependence will normally only be an issue if high-quality estimates are required.

The pressure measurements provide an opportunity to demonstrate the block averaging method for estimating the variance of the mean described earlier. Here we consider the soft-sphere fluid with $N_a = 500$, $\rho = 0.8$, and $T = 1$. The pressure measurements are governed by the following input data:

```
stepAvg        1
stepEquil      1000
stepLimit      17384
```

and the results are sufficient for 14 doublings of the block size starting from $b = 1$. Table 4.1 shows the outcome of this analysis, and reveals

Table 4.1. *Block-averaged estimates of*
$\sigma(\langle P \rangle)$

b	M_b	$\sigma(\langle P \rangle)$
1	16384	0.0012
2	8192	0.0017
4	4096	0.0023
8	2048	0.0031
16	1024	0.0039
32	512	0.0044
64	256	0.0046
128	128	0.0047
256	64	0.0040
512	32	0.0044
1024	16	0.0045
2048	8	0.0050
4096	4	0.0033

that convergence occurs at a block size of $b = 32$ (M_b is the number of blocks). The fact that $\sigma(\langle P \rangle) \approx 4\sigma(P)$ (the value at $b = 1$) should serve as a reminder that closely spaced measurements are strongly correlated.

4.3 Structure

4.3.1 Radial distribution function

The fluid state is characterized by the absence of any permanent structure. There are, nevertheless, well-defined structural correlations which can be measured experimentally to provide important details about the average molecular organization [mcq76, han86b]. The treatment of structural correlation (in the canonical ensemble) begins with the completely general pair-distribution function:

$$g(r_1, r_2) = \frac{N_a(N_a - 1) \int dr_3 \ldots r_{N_a} \exp(-U(r_1, \ldots, r_{N_a})/k_B T)}{\rho^2 \int dr_1 \ldots r_{N_a} \exp(-U(r_1, \ldots, r_{N_a})/k_B T)} \qquad (4.9)$$

where the integral in the denominator is just the partition function, and the integral in the numerator differs only in that r_1 and r_2 are excluded from the integration. In the case of spatially homogeneous systems, only relative separation is meaningful, leading to a sum over atom pairs:

$$g(r) = \frac{2V}{N^2} \left\langle \sum_{i<j} \delta(r - r_{ij}) \right\rangle \qquad (4.10)$$

and if the system is also isotropic the function can be averaged over angles without loss of information. The result is the radial distribution function $g(r)$ – RDF for short – a function that describes the spherically averaged local organization around any given atom; $g(r)$ plays a central role in liquid state physics and all functions that depend on the pair separation (such as potential energy and pressure) can be expressed in terms of integrals involving $g(r)$.

The definition of $g(r)$ implies that $\rho g(r)dr$ is proportional to the probability of finding an atom in the volume element dr at a distance r from a given atom, and (in three dimensions) $4\pi\rho g(r)r^2 \Delta r$ is the mean number of atoms in a shell of radius r and thickness Δr surrounding the atom. The RDF is related to the experimentally measurable structure factor $S(k)$ by Fourier transformation – $S(k)$ is a key quantity in interpreting X-ray scattering measurements. The general result, not assuming isotropy, is

$$S(k) = 1 + \rho \int g(r)\exp(-i k \cdot r)dr \tag{4.11}$$

and for isotropic liquids this simplifies to

$$S(k) = 1 + 4\pi\rho \int \frac{\sin kr}{kr} g(r)r^2 dr \tag{4.12}$$

Equation (4.12) provides an important link between MD simulation and the real world. The MD approach can of course provide the answer to any question about structure, such as the nature of spatial correlations between atoms taken three at a time; while information of this type can prove useful in trying to understand behavior, since the corresponding experimental data is unobtainable comparison is impossible.

From the definition of $g(r)$ in (4.10) it is apparent that the RDF can be measured [rah64, ver68] using a histogram of discretized pair separations. If h_n is the number of atom pairs (i, j) for which $(n-1)\Delta r \leq r_{ij} < n\Delta r$ (h_n is $N_a/2$ times the mean number of neighbors in the shell), then, assuming that Δr is sufficiently small, we have the result

$$g(r_n) = \frac{V h_n}{2\pi N_a^2 r_n^2 \Delta r} \tag{4.13}$$

where $r_n = (n - 1/2)\Delta r$. If the RDF measurements extend out to a maximum range r_d the required number of histogram bins is $r_d/\Delta r$. The two-dimensional version is

$$g(r_n) = \frac{A h_n}{\pi N_a^2 r_n \Delta r} \tag{4.14}$$

The normalization factors ensure that $g(r \to \infty) = 1$, although periodic boundaries limit the range r_d to no more than half the smallest edge of the simulation region, with wraparound used in evaluating interatomic distances.

The RDF computation has much in common with the interaction calculation, and the cell method should be used whenever r_d is less than a quarter of the region size. Otherwise all pairs must be considered, and this is the version shown here; quite accurate results can in fact be obtained from a relatively small number of measurements so that the overall computational cost is not excessive:

```
     EvalRdf () {
         real dr[NDIM + 1], deltaR, normFac, rr, rrRange;
         int j1, j2, k, n;
         countRdf = countRdf + 1;
5        if (countRdf == 1) {
             for (n = 1; n <= sizeHistRdf; n ++) histRdf[n] = 0.;
         }
         rrRange = Sqr (rangeRdf);
         deltaR = rangeRdf / sizeHistRdf;
10       for (j1 = 1; j1 <= nAtom - 1; j1 ++) {
             for (j2 = j1 + 1; j2 <= nAtom; j2 ++) {
                 for (k = 1; k <= NDIM; k ++) {
                     dr[k] = r[k][j1] - r[k][j2];
                     if (fabs (dr[k]) > regionH[k]) dr[k] = dr[k] -
15                       SignR (region[k], dr[k]);
                 }
                 rr = Sqr (dr[1]) + Sqr (dr[2]) + Sqr (dr[3]);
                 if (rr < rrRange) {
                     n = (int) (sqrt (rr) / deltaR) + 1;
20                   histRdf[n]= histRdf[n] + 1.;
       } } }
         if (countRdf == limitRdf) {
             normFac = region[1] * region[2] * region[3] / (2. *
                 pi * pow (deltaR, 3.) * nAtom * nAtom * countRdf);
25           for (n = 1; n <= sizeHistRdf; n ++)
                 histRdf[n] = histRdf[n] * normFac / Sqr (n - 0.5);
             PrintRdf (stdout);
             countRdf = 0;
         }
30   }
```

In addition to computing the discretized version of the RDF for the current state of the system, the above function also accumulates the average over a series of such 'snapshots', as well as initializing the calculation and producing the final output when sufficient data have been collected; the decision whether to initialize or prepare the final summary is based on the value of countRdf. This is the three-dimensional version of the computation; the changes for two dimensions are minor.

New quantities introduced here are

```
real *histRdf, rangeRdf;
int countRdf, limitRdf, sizeHistRdf, stepRdf;
```

and the additional input data items are

```
RNAME (rangeRdf),
INAME (limitRdf),
INAME (sizeHistRdf),
INAME (stepRdf),
```

Memory allocation is carried out in AllocArrays:

```
histRdf = AllocVecR (sizeHistRdf);
```

and the measurement counter is initialized in SetupJob:

```
countRdf = 0;
```

The addition to SingleStep to request RDF processing is

```
if (stepCount >= stepEquil && (stepCount - stepEquil) %
    stepRdf == 0) EvalRdf ();
```

and the output function is simply

```
  PrintRdf (FILE *fp) {
    real rBin;
    int n;
    fprintf (fp, "rdf\n");
5   for (n = 1; n <= sizeHistRdf; n ++) {
      rBin = (n - 0.5) * rangeRdf / sizeHistRdf;
      fprintf (fp, "%8.4f %8.4f\n", rBin, histRdf[n]);
    }
  }
```

The RDF results shown here are obtained from soft-sphere runs using the following input data:

```
initUcell     8 8 8
density       1.0
temperature   1.0
deltaT        0.005
stepAvg       1000
```

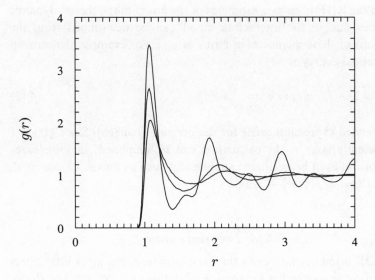

Fig. 4.3 Radial distribution function for soft spheres at $\rho = 0.6$, 0.8, and 1.0.

```
stepEquil         2000
stepInitlzTemp    200
stepLimit         17000
limitRdf          100
rangeRdf          4.0
sizeHistRdf       200
stepRdf           50
```

An FCC initial state is used, so that $N_a = 2048$. Three values of ρ are used, namely, 0.6, 0.8, and 1.0. The results appearing in Figure 4.3 are those obtained during the last 5000 steps of each run; the way in which structure emerges as ρ increases is clearly visible.

For a simple monatomic fluid $g(r)$ shows how, on average, the neighborhood seen by an atom consists of concentric shells of atoms with well-defined radii. As the density increases, these shells become distorted, an effect reflected in the RDF by additional peaks that appear once the lattice structure of the imminent solid phase begins to make its presence felt. The fact that in the liquid all correlation is lost beyond a few atomic diameters confirms the absence of any long-range positional order and suggests a picture in which atoms can regard their more distant neighbors as a smeared-out continuum, a useful idealization when trying to construct simple liquid models.

Once the RDF is known, estimates of the errors in the thermodynamic properties due to the interaction cutoff can be determined from the definitions of these quantities in terms of $g(r)$. For example, the error in the potential energy is

$$\Delta E_u = 2\pi\rho \int_{r_c}^{\infty} g(r)u(r)r^2 dr \tag{4.15}$$

and a related expression exists for the pressure [han86b]. Since $g(r) \approx 1$ at sufficiently large r, the calculation can be simplified; in some cases the error can even be evaluated analytically, such as for the LJ potential, where $\Delta E_u = 8\pi\rho(1/9r_c^9 - 1/3r_c^3)$.

4.3.2 Long-range order

The RDF primarily addresses the local structure, but gives little direct information as to whether long-range crystalline order exists. The sharpness of the RDF peaks and the presence of additional peaks at positions indicative of specific lattices provide indirect evidence that is better appreciated once the existence of crystalline order has been established by other means.

Long-range order corresponds to the presence of lattice structure and is the quantity underlying X-ray scattering measurements from crystalline materials. The local density at a point r can be expressed as a sum over atoms:

$$\rho(r) = \sum_{j=1}^{N_a} \delta(r - r_j) \tag{4.16}$$

and its Fourier transform is simply

$$\rho(k) = \frac{1}{N_a} \sum_{j=1}^{N_a} \exp(-ik \cdot r_j) \tag{4.17}$$

In a calculation of $|\rho(k)|$ designed to test for the presence of long-range order k should be chosen to be a reciprocal lattice vector of the ordered state; this can be any linear combination of the vectors appropriate for the expected FCC lattice, so we choose $k = (2\pi/l)(1, -1, 1)$, where l is the unit cell edge. If the system is almost fully ordered $|\rho(k)| \approx 1$, but in the disordered liquid state $|\rho(k)| = O(N_a^{-1/2})$.

The function for evaluating long-range order, assuming (for convenience) all region edges to be the same length, is

```
EvalLatticeCorr () {
   real kVec[NDIM + 1], si, sr, t;
   int k, n;
   kVec[1] = 2. * pi * initUcell[1] / region[1];
5  kVec[2] = - kVec[1];
   kVec[3] = kVec[1];
   sr = si = 0.;
   for (n = 1; n <= nAtom; n ++) {
     t = 0.;
10   for (k = 1; k <= NDIM; k ++) t = t + kVec[k] * r[k][n];
     sr = sr + cos (t);     si = si + sin (t);
   }
   latticeCorr = sqrt (Sqr (sr) + Sqr (si)) / nAtom;
}
```

No averaging over separate measurements is included, but this could easily be added. One additional variable is introduced here:

```
real latticeCorr;
```

The function is called prior to the call to `PrintSummary` and the value of `latticeCorr` is added to the output.

One point must be kept in mind when studying solidification in a finite system, namely, that the best results will be obtained if the region size and shape allow the formation of an integral number of unit cells along each lattice direction. Any mismatch will introduce imperfections of one kind or another into the ordered state, leading to a reduction in the apparent long-range order.

In Figure 4.4 we show how long-range order varies with time during the early stages of runs begun in the ordered state; we use the same system as for the RDF studies but without any initial temperature adjustment. The four density values shown are between 0.8 and 1.1. At the larger densities a moderate to high degree of order persists throughout the observation period (although this is not a guarantee of what might happen over much longer times), whereas at the lowest density the long-range order rapidly vanishes.

4.3.3 Packing studies

There are many reasons for seeking information about local atomic organization that is more detailed than the RDF can provide. In simple fluids the motivation is to understand better how atoms are arranged, and what distinguishes the average packing from the fully ordered crystalline

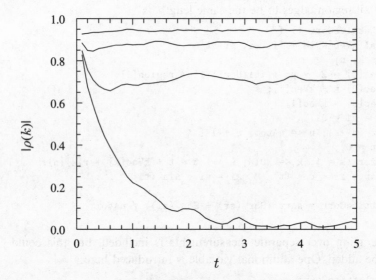

Fig. 4.4 Time dependence of long-range order in soft-sphere systems that start in an ordered state; the results are for densities 0.8, 0.9, 1.0, and 1.1.

state. In more complex systems the same packing questions can be asked in order to gain more specific information about molecular organization; for example, an estimate of the exposed surface of part of a large molecule can be important for studies of chemical reactivity.

The means of describing the spatial organization of what sometimes amounts to little more than a random array of atoms is far from obvious. The most widely used method is based on a Voronoi subdivision [hsu79, cap81, rap83, med90], in which each atom is surrounded by a convex polyhedron constructed using certain prescribed rules. The result of this construction process is the partitioning of space into a set of polyhedra, with all points that are closer to a particular atom than to any other belonging to its polyhedron. Atoms can then be regarded as adjacent if their polyhedra share a common face – in this way it is possible to define the neighborhood of an atom uniquely. The polyhedra themselves are also of considerable interest since the interactions can influence their geometrical properties.

The Voronoi analysis will be carried out separately from the MD run to demonstrate how this kind of post-processing is done in general; in view of the complexity of the Voronoi analysis it is clearly desirable to keep it distinct from the simulation. Every so often a snapshot containing sufficient information to reproduce the atomic configuration is written to a disk file; this will provide the raw data for subsequent analysis. The following line is added to main following the call to SingleStep:

```
if (stepCount % stepConfig == 0) PutConfig ();
```

where stepConfig specifies the number of timesteps between snapshots. The definition

```
int stepConfig;
```

and input data item

```
INAME (stepConfig),
```

are also required. The listing of PutConfig appears at the end of the chapter.

Construction of Voronoi polyhedra is an exercise in computational geometry, and is by far the longest and most complex of the analysis programs used in these case studies. There are various ways of dealing with this problem [bro78, fin79, tan83]: the version described here computes each polyhedron separately, but does the job with a constant computational effort that is independent of the total number of atoms. Periodic boundaries are assumed.

A concise summary of the method follows. The first step for each atom is to generate a list of its neighbors ordered by distance. A large tetrahedron is then constructed as a generous overestimate of the eventual polyhedron; portions of this polyhedron will be removed in the course of the computation until what remains at the end is the Voronoi polyhedron for that atom. The computation begins with the initial tetrahedron and carries out the following sequence of operations for each neighbor in the list until none remain that could possibly alter the polyhedron shape:

(a) compute the bisecting plane between the atom of interest and the neighbor;
(b) determine which polyhedron vertices lie on the far side of the plane;
(c) determine which edges and faces are cut by the plane;
(d) compute the locations of the intercepts of the plane with each cut edge;
(e) update the description of each cut face and determine which faces are deleted from the polyhedron entirely;
(f) add the new vertices, edges, and face to the polyhedron;

(g) remove deleted vertices, edges, and faces from the polyhedron description;

(h) locate the most distant vertices in the new and cut faces.

When the process terminates, a test is made to ensure that nothing remains of the initial faces; any remnants are symptomatic of a poor choice of initial polyhedron. Measurements made on the resulting polyhedron include vertex, edge, and face counts, as well as its volume and surface area.

Assuming that a list of atoms to be tested during the construction of the polyhedron for a particular atom has already been prepared, the following function shows how the computation is organized:

```
AnalVorPoly () {
    int nF;
    Sort (distSq, siteSeq, nTestSites);
    InitVorPoly ();
5   for (curSite = 1; curSite < nTestSites; curSite ++) {
        if (distSq[siteSeq[curSite]] >= 4. * vDistSqMax) break;
        siteB = testSites[siteSeq[curSite]];
        nVDel = 0;    nENew = 0;    nEDel = 0;
        nECut = 0;    nFDel = 0;    nFCut = 0;
10      BisectPlane ();
        if (nVDel > 0) ProcDelVerts ();
        if (nECut > 0) ProcCutEdges ();
        if (nFCut > 0) ProcCutFaces ();
        if (nENew > 0) ProcNewVerts ();
15      if (nFCut > 0) ProcNewFace ();
        RemoveOld ();
        if (nFCut > 0) FindDistVerts ();
    }
    for (nF = 1; nF <= 4; nF ++)
20      if (fStat[nF] != 0) ErrExit ("incomplete");
    PolyGeometry ();
    PolySize ();
}
```

The Voronoi construction task involves a great many small details, and for this reason the complete program is actually longer than a typical MD program. Because of its length, the remainder of the listing is relegated to the end of the chapter.

The sample results shown Table 4.2 are obtained from $N_a = 864$ soft-sphere systems started at $T = 1$ but without any temperature adjustment, and at three different ρ. Three sets of configurations are recorded at 1000-step intervals for use in the analysis. To allow comparison with

Table 4.2. *Properties of Voronoi polyhedra* [a] *for soft-sphere systems and for a random point set*

ρ	n_v	n_e	n_f	n_{ef}
1.0	24.167	36.250	14.083	5.144
	24.176	36.264	14.088	5.145
	24.204	36.306	14.102	5.145
0.8	25.296	37.944	14.648	5.172
	25.213	37.819	14.606	5.170
	25.231	37.847	14.616	5.171
0.6	25.801	38.701	14.900	5.182
	25.898	38.847	14.949	5.185
	25.810	38.715	14.905	5.183
random	27.024	40.536	15.512	5.188
	26.836	40.254	15.418	5.181
	27.176	40.764	15.588	5.194

[a] Results shown are mean numbers of vertices (n_v), edges (n_e), and faces (n_f) per polyhedron, and mean number of edges per face (n_{ef}); three sets of results are shown for each case.

the behavior of random systems, results from totally random 1000-point arrays (generated by a special version of InitCoords in Chapter 3) are included. The trends in the results are clearly visible.

4.3.4 Clusters

Cluster formation in fluids is frequently a subject of interest, both because clustering is a real physical process and because some models attribute special properties to clusters. In either case it is important to be able to identify atoms belonging to common clusters and to measure various cluster properties. Here we focus on clusters appearing in instantaneous snapshots of a soft-sphere system, but a deeper analysis might also need to consider time-dependent behavior such as cluster growth rates or lifetimes in systems where attractive interactions actually bind atoms together.

Different criteria are available for determining whether an atom belongs to a cluster. One option is to consider the energy that binds (assuming interactions with an attractive component) an atom to other atoms already in the cluster. An alternative method requiring less computation (that for attractive pair potentials is essentially the same) is to base the criterion on the interatomic distance, so that if atom *i* is already in the cluster, atom j will also be included if $r_{ij} < r_d$, where r_d is the chosen threshold separation; we will adopt this definition here. The value of r_d would typically be based on some energy condition, but this does not affect the technique. If there is no physical reason for preferring a particular value of r_d, the sensitivity of the results to a change in r_d should of course be examined.

This analysis will also be carried out separately from the MD run; the user will want to try different r_d values, so it is more sensible to have the MD configurations available for immediate analysis. The configuration data is input to the analysis program using the function GetConfig listed at the end of the chapter. Cluster construction begins by determining those atom pairs that are separated by less than r_d; the function used for this is derived from BuildNebrList (Chapter 3), with cell size (use of cells is optional) based on the variable rClust that corresponds to r_d:

```
BuildClusters () {
   real ... rrClust;
   rrClust = Sqr (rClust);
   ...
5  if (rr < rrClust) AddBondedPair (j1, j2);
   ...
}
```

The task of adding atoms to clusters and merging existing clusters that are found to share a common member is carried out by the function AddBondedPair below. The atoms belonging to each cluster are stored in a linked list. Several arrays are used in the process: the elements of clustHead point to the first atom in each cluster, clustNext contains pointers from one atom to the next, clustSize eventually contains the number of atoms in each cluster, and inClust records the cluster to which each atom belongs. Tracing the detailed logic of this function is left as an exercise for the reader:

```
AddBondedPair (int j1, int j2) {
   int cBig, cSmall, m, mp, nc, nc1, nc2;
   if (inClust[j1] == 0) {
     if (inClust[j2] == 0) {
5      nClust = nClust + 1;
       inClust[j1] = nClust;    inClust[j2] = nClust;
       clustSize[nClust] = 2;   clustHead[nClust] = j1;
       clustNext[j1] = j2;      clustNext[j2] = 0;
     } else {
10     nc = inClust[j2];
       clustSize[nc] = clustSize[nc] + 1;
       inClust[j1] = nc;
```

```
              clustNext[j1] = clustHead[nc];      clustHead[nc] = j1;
           }
15      } else {
           if (inClust[j2] == 0) {
              nc = inClust[j1];
              clustSize[nc] = clustSize[nc] + 1;
              inClust[j2] = nc;
20            clustNext[j2] = clustHead[nc];      clustHead[nc] = j2;
           } else {
              nc1 = inClust[j1];    nc2 = inClust[j2];
              if (nc1 != nc2) {
                 if (clustSize[nc1] <= clustSize[nc2]) cBig = nc2;
25               else cBig = nc1;
                 cSmall = nc1 + nc2 - cBig;
                 m = clustHead[cSmall];
                 while (m > 0) {
                    inClust[m] = cBig;      mp = m;
30                  m = clustNext[m];
                 }
                 clustNext[mp] = clustHead[cBig];
                 clustHead[cBig] = clustHead[cSmall];
                 clustSize[cBig] = clustSize[cBig] +
35                  clustSize[cSmall];
                 clustSize[cSmall] = 0;
        } } }
   }
```

Prior to starting cluster construction a little preparation is required:

```
   InitClusters () {
      int k, n;
      for (n = 1; n <= nAtom; n ++) inClust[n] = 0;
      nClust = 0;
5     for (k = 1; k <= NDIM; k ++) regionH[k] = ...
   }
```

After the work is complete the clusters can be reindexed to remove any reference to those clusters that were absorbed by others during construction:

```
   CompressClusters () {
      int j, m, nc;
      nc = 0;
      for (j = 1; j <= nClust; j ++) {
5        if (clustSize[j] > 0) {
            nc = nc + 1;
            clustSize[nc] = clustSize[j];
            clustHead[nc] = clustHead[j];
```

```
         m = clustHead[nc];
10       while (m > 0) {
             inClust[m] = nc;      m = clustNext[m];
   } } }
   nClust = nc;
 }
```

The variables used in this analysis program include

```
real **r, region[NDIM + 1], regionH[NDIM + 1], rClust,
   sSum, ssSum;
int *cellList, *clustHead, *clustNext, *clustSize, *inCell,
   *inClust, cells[NDIM + 1], bigSize, blockNum, blockSize,
5  nAtom, nCellEdge, nClust, nSingle, runId;
char *progId = "ss";
```

All the arrays used to hold cluster data are of size nAtom, to allow for
the extreme situation where all atoms form their own clusters. The size
of the cell array is based on the number of cells per edge:

```
nCellEdge = region[1] / rClust;
```

Array allocation uses the standard method (AllocArrays), with nAtom
and region obtained from the configuration data file; the variables
blockNum, blockSize, runId and progId are needed for accessing this
file (see the listing of GetConfig in Section 4.5).

Once generation is complete the analysis of both geometric and spatial
properties of the clusters can be carried out. Spatial properties of
the clusters include the radius of gyration and moments of the mass
distribution; such studies involve calculations similar to those used for
polymer chains in Chapter 9 and will not be considered here. Other
measurements are of a more geometrical flavor; examples are included in
the following function used to count the number of isolated atoms, find
the cluster with the most atoms, and evaluate the mean and standard
deviation of the cluster-size distribution:

```
AnalClusterSize () {
   int cBig, nc, ncUse;
   ncUse = 0;      cBig = 1;
   sSum = ssSum = 0.;
5  for (nc = 1; nc <= nClust; nc ++) {
      if (clustSize[nc] > clustSize[cBig]) cBig = nc;
      if (clustSize[nc] > 1) {
         ncUse = ncUse + 1;
         sSum = sSum + clustSize[nc];
10       ssSum = ssSum + Sqr (clustSize[nc]);
   } }
```

```
     bigSize = clustSize[cBig];
     nSingle = nAtom - sSum;
     if (ncUse > 0) {
15     sSum = sSum / ncUse;
       ssSum = sqrt (ssSum / ncUse - Sqr (sSum));
     }
   }
```

More complex aspects, such as the number of ways in which atoms are linked into a cluster, or the topology of the link network, can also be explored using the data available.

The main program used in the cluster analysis is

```
   main (int argc, char **argv) {
     runId = atoi (argv[1]);
     rClust = atof (argv[2]);
     SetupFiles ();
5    blockNum = -1;
     while (GetConfig ()) {
       InitClusters ();
       BuildClusters ();
       CompressClusters ();
10     AnalClusterSize ();
       printf ("%d %d %d %.1f %.1f\n",
           nSingle, nClust, bigSize, sSum, ssSum);
     }
   }
```

The values of `runId` and `rClust` must be supplied when the program is run.

Examples of cluster properties are shown in Table 4.3. The configuration data produced by the $\rho = 0.8$ Voronoi run is used here as well. The results of analyzing three different realizations are shown for various values of the cluster cutoff distance r_d. Percolation theory can be used to explain the changing behavior as r_d is varied, and also to inspire other kinds of cluster analysis [sta92].

4.4 Further work

4.1 Compare the specific heats obtained from the kinetic energy fluctuations and from dE/dT.

4.2 Examine the errors in energy and pressure due to truncating the LJ interaction.

4.3 Study the soft-sphere equation of state near the melting transition. What kind of transition occurs?

Table 4.3. *Cluster properties*[a] *for a soft-sphere fluid*

r_d	n_i	n_c	s_m	$\langle s \rangle$
1.02	454	155	9	2.6
	492	137	11	2.7
	492	142	12	2.6
1.04	245	158	35	3.9
	284	145	37	4.0
	266	148	29	4.0
1.06	102	78	405	9.8
	118	59	533	12.6
	131	56	409	13.1
1.08	41	14	784	58.8
	52	12	779	67.7
	60	16	765	50.3
1.10	14	6	835	141.7
	17	3	842	282.3
	15	3	845	283.0

[a] Results shown are numbers of isolated atoms (n_i) and multi-site clusters (n_c), size of the largest cluster (s_m), and mean cluster size ($\langle s \rangle$); three sets of values are shown for each r_d.

4.4 The possible existence of a hexatic phase in two-dimensional liquids – one in which there is a certain long-range orientational order although no translational order – has been explored using MD [abr86]; look into the subject.

4.5 Examine the difference between the LJ and soft-sphere RDFs.

4.6 Extend the structural analysis to consider three-body correlations [vog84, bar88]; for example, study the distribution of angles subtended by pairs of neighbors (suitably defined) of each atom.

4.7 The Voronoi analysis is greatly simplified when applied to two-dimensional systems (the name of Dirichlet is associated with the problem); generate and analyze some typical soft-disk configurations.

4.8 Examine the cluster distributions for the two-dimensional case from the point of view of percolation theory [hey89].

4.9 Apply cluster analysis to the LJ fluid; here, unlike soft spheres, the binding energy can be computed for each cluster and the study of cluster formation takes on physical meaning.

4.5 Additional material

4.5.1 Configuration snapshot files

Snapshots of the system configuration are stored on disk by the following function:

```
    PutConfig () {
      int blockSize, fOk, k, n;
      short *rI;
      fOk = 1;
5     blockSize = (NDIM + 1) * sizeof (real) +
          3 * sizeof (int) + nAtom * NDIM * sizeof (short);
      if ((fp = fopen (fileName[FL_CONFIG], "a"))) {
        fwrite (&blockSize, sizeof (int), 1, fp);
        fwrite (&timeNow, sizeof (real), 1, fp);
10      fwrite (&stepCount, sizeof (int), 1, fp);
        fwrite (&region[1], sizeof (real), NDIM, fp);
        fwrite (&nAtom, sizeof (int), 1, fp);
        rI = AllocVecS (nAtom);
        for (k = 1; k <= NDIM; k ++) {
15        for (n = 1; n <= nAtom; n ++)
            rI[n] = 32767. * (r[k][n] / region[k] + 0.5);
          fwrite (&rI[1], sizeof (short), nAtom, fp);
        }
        FreeVecS (rI);
20      if (ferror (fp)) fOk = 0;
        fclose (fp);
      } else fOk = 0;
      if (! fOk) ErrExit ("write config data");
    }
```

Note the use of binary data rather than human-readable text; this reduces the storage requirements substantially, although when in this form data files are not readily transportable between different kinds of computers. To further reduce storage the coordinate data is scaled to fit into a two-byte word, rather than the four or eight bytes used for floating-point variables in the MD code itself; the consequent loss of precision is acceptable for the kinds of analysis that we carry out on the data. Standard Unix file functions are used. The variable blockSize is set to the total number of bytes written per snapshot, allowing more flexibility in the applications that process the data later. The name of the output file is stored as one of the entries in the character-string array fileName, and the function SetupFiles must be called at the start of the job to handle file-related matters – details appear in the Appendix.

Reading the data into an analysis program uses the following function;

the variable blockNum indicates which set of configuration data from the file is to be read, in effect allowing random access:

```
    int GetConfig () {
      int fOk, k, n;
      short *rI;
      fOk = 1;
5     if (blockNum == -1) {
        if (! (fp = fopen (fileName[FL_CONFIG], "r"))) fOk = 0;
      } else {
        fseek (fp, blockNum * blockSize, 0);
        blockNum = blockNum + 1;
10    }
      if (fOk) {
        fread (&blockSize, sizeof (int), 1, fp);
        if (feof (fp)) return (0);
        fread (&timeNow, sizeof (real), 1, fp);
15      ... (more data written by PutConfig) ...
        fread (&nAtom, sizeof (int), 1, fp);
        if (blockNum == -1) {
          SetCellSize ();
          AllocArrays ();
20        blockNum = 1;
        }
        rI = AllocVecS (nAtom);
        for (k = 1; k <= NDIM; k ++) {
          fread (&rI[1], sizeof (short), nAtom, fp);
25        for (n = 1; n <= nAtom; n ++)
            r[k][n] = (rI[n] / 32767. - 0.5) * region[k];
        }
        FreeVecS (rI);
        if (ferror (fp)) fOk = 0;
30    }
      if (! fOk) ErrExit ("read config data");
      return (1);
    }
```

The first time this function is called blockNum has the value -1; as part of the initial call the function AllocArrays is used to allocate the necessary storage, and if cells are needed by the analysis SetCellSize must be called just prior to this.

4.5.2 *Voronoi analysis*

The functions used in constructing the Voronoi polyhedra and evaluating some of their properties are described here. Because of the rather complex nature of the algorithm the details can be handled in a variety of ways; this is one such possibility. For brevity we omit checks on array overflow and other potential problems, although such safety measures should be included to help detect programming or runtime errors.

The method for determining which atoms can contribute to a particular polyhedron assumes that the region has been subdivided into cells. The atoms required are obtained by first scanning a range of cells around the one containing the atom under examination, then sorting the atoms found into ascending distance order and placing the ordered list of atom indices in the array siteSeq (the call to Sort in AnalVorPoly can use any standard sorting function – see Appendix). Here we only scan neighbor cells, but the range could be extended:

```
   FindTestSites (int na) {
     real d, dd;
     int c, cnX, cnY, cnZ, cX, cY, cZ, i, k, ofX, ofY, ofZ;
     cX = inCell[na] % cells[1];
5    cY = (inCell[na] / cells[1]) % cells[2];
     cZ = inCell[na] / (cells[1] * cells[2]);
     nTestSites = 0;
     for (ofZ = -1; ofZ <= 1; ofZ ++) {
       cnZ = (cZ + ofZ + cells[3]) % cells[3];
10     for (ofY = -1; ofY <= 1; ofY ++) {
         cnY = (cY + ofY + cells[2]) % cells[2];
         for (ofX = -1; ofX <= 1; ofX ++) {
           cnX = (cX + ofX + cells[1]) % cells[1];
           c = (cnZ * cells[2] + cnY) * cells[1] + cnX +
15           nAtom + 1;
           i = cellList[c];
           while (i > 0) {
             nTestSites = nTestSites + 1;      dd = 0.;
             for (k = 1; k <= NDIM; k ++) {
20             d = r[k][na] - r[k][i];
               if (fabs (d) > regionH[k])
                   d = d - SignR (region[k], d);
               dd = dd + Sqr (d);
             }
25           testSites[nTestSites] = i;
             distSq[nTestSites] = dd;
             i = cellList[i];
   } } } }
   }
```

The polyhedron used to start the calculation is a tetrahedron. The following function specifies the vertex coordinates and initializes all the data needed to describe the structure of the polyhedron during its subsequent modification. Linked lists are used in this work:

```
      InitVorPoly () {
      real r2, r6, vPosI[4][3] =
         {-1.,-1.,-1., 1.,-1.,-1., 0.,2.,-1., 0.,0.,3.};
      int m, n, nE, nF, nV, s,
5        vValI[] = {1,3,6,1,2,5,2,3,4,4,5,6},
         eFacesI[] = {1,4,1,2,1,3,2,3,2,4,3,4},
         eVertsI[] = {1,2,2,3,1,3,3,4,2,4,1,4},
         fListEI[] = {1,2,3,5,4,2,3,4,6,6,5,1},
         fListVI[] = {1,2,3,2,4,3,1,3,4,1,4,2};
10    r2 = sqrt (2.) * rangeLim;    r6 = sqrt (6.) * rangeLim;
      siteA = testSites[siteSeq[1]];
      eLast = 6;    fLast = 4;    vLast = 4;    m = 0;
      for (nV = 1; nV <= vLast; nV ++) {
         vPos[1][nV] = r[1][siteA] + vPosI[nV - 1][0] * r6;
15       vPos[2][nV] = r[2][siteA] + vPosI[nV - 1][1] * r2;
         vPos[3][nV] = r[3][siteA] + vPosI[nV - 1][2] *
            rangeLim;
         vDistSq[nV] = 0.375 * Sqr (rangeLim);    vStat[nV] = 2;
         for (n = 1; n <= 3; n ++) {
20          m = m + 1;    vEdges[n][nV] = vValI[m - 1];
      } }
      vDistSqMax = vDistSq[1];
      for (nE = 1; nE <= eLast; nE ++) {
         eVerts[1][nE] = eVertsI[2 * nE - 2];
25       eFaces[1][nE] = eFacesI[2 * nE - 2];
         eVerts[2][nE] = eVertsI[2 * nE - 1];
         eFaces[2][nE] = eFacesI[2 * nE - 1];
         eStat[nE] = 3;
      }
30    for (s = 1; s <= MAX_FLIST - 1; s ++)
         fListLink[s] = s + 1;
      s = 1;    m = 0;
      for (nF = 1; nF <= fLast; nF ++) {
         fVfar[nF] = fListVI[m];    fStat[nF] = 3;
35       fPtr[nF] = s;
         for (n = 1; n <= 3; n ++) {
            fListV[s] = fListVI[m];    fListE[s] = fListEI[m];
            m = m + 1;    s = s + 1;
         }
40       fListLink[s - 1] = fPtr[nF];
      }
      fListLast = s - 1;
      }
```

Several arrays appear in the course of the program that are used to describe the geometrical details of the polyhedron as it is being constructed. The vertex coordinates are stored in vPos, and the squared distance of each vertex from the atom in vDistSq. To simplify the program the data representation assumes that there are exactly three edges attached to each vertex; this excludes certain regular lattice arrangements, as well as the extremely rare case of numerical degeneracy (one candidate for the missing safety checks); the identities of the three edges terminating at a vertex appear in vEdges. The identities of the two vertices asociated with each edge are stored in eVerts, and the two faces that are joined along each polyhedron edge are in eFaces. With each face is associated a circular list (a linked list whose final element points back to the start [knu68]) that itemizes, in order of appearance, the edges and vertices defining its boundary; these are stored in fListE and fListV, with fListLink providing the pointers that connect the items (fListLast is the last list-storage element in use), and fPtr pointing to the first item. fVFar identifies the vertex in the face that is furthest from the atom. The reader will detect a certain amount of redundancy in this information, but having it all readily accessible simplifies the computation. The values used to initialize all these arrays (in InitVorPoly) define the starting tetrahedron.

A number of other arrays will also be introduced at this point. Three of them, vStat, eStat and fStat, are used as status indicators for the vertices, edges, and faces; as the construction progresses they show whether the elements still belong to the polyhedron, have been deleted, or are about to change status. Elements that have been identified as deleted by the plane currently under consideration are listed in vDel, eDel and fDel, while eCut and fCut identify edges and faces that are only cut (intersected) by the current plane, and eNew identifies new edges that are in the process of being added to the polyhedron.

Bisection of the line between the atom and one of its neighbors to produce a possible new face for the polyhedron is carried out by the following function; the coefficients of the plane equation are placed in the array fParam. Allowance is made for periodic boundaries. The vector operations appearing in the function implement the results:

(a) the equation of the plane bisecting r_{ij} is $(r_i - r_j) \cdot p = (r_i^2 - r_j^2)/2$, or, more concisely, $a \cdot p = b$;

(b) the equation of the edge joining vertices v_1 and v_2 is $p = p_{v_1} + \alpha(p_{v_2} - p_{v_1})$, where $0 \le \alpha \le 1$;

(c) the intercept between the plane and the edge, if there is one, occurs when $\alpha = (b - a \cdot p_{v_1})/a \cdot (p_{v_2} - p_{v_1})$.

```
    BisectPlane () {
      real dShift, d1, d2, d3;
      int k, nV;
      d1 = 0.;    fParam[4] = 0.;
 5    for (k = 1; k <= NDIM; k ++) {
        fParam[k] = r[k][siteB] - r[k][siteA];
        dShift = 0.;
        if (fabs (fParam[k]) > regionH[k])
          dShift = SignR (region[k], fParam[k]);
10      fParam[k] = fParam[k] - dShift;
        fParam[4] = fParam[4] +
          Sqr (r[k][siteB] - dShift) - Sqr (r[k][siteA]);
        d1 = d1 + fParam[k] * r[k][siteA];
      }
15    fParam[4] = fParam[4] * 0.5;
      for (nV = 1; nV <= vLast; nV ++) {
        if (vStat[nV] != 0) {
          d2 = 0.;
          for (k = 1; k <= NDIM; k ++)
20          d2 = d2 + fParam[k] * vPos[k][nV];
          if (d1 != d2) {
            d3 = (fParam[4] - d1) / (d2 - d1);
            if (d3 > 0. && d3 < 1.) {
              nVDel = nVDel + 1;    vDel[nVDel] = nV;
25            vStat[nV] = 1;
    } } } }
    }
```

Several functions (called in succession from AnalVorPoly) then deal with those vertices, edges, and faces of the polyhedron that are added, deleted, or modified, as a result of including this new face. The first of these functions determines the edges and faces affected by the deleted vertices:

```
    ProcDelVerts () {
      int e, m, n, nV;
      for (nV = 1; nV <= nVDel; nV ++) {
        for (m = 1; m <= 3; m ++) {
 5        e = vEdges[m][vDel[nV]];    eStat[e] = eStat[e] - 1;
          if (eStat[e] == 2) {
            nECut = nECut + 1;    eCut[nECut] = e;
          } else {
            nEDel = nEDel + 1;    eDel[nEDel] = e;
10        }
          for (n = 1; n <= 2; n ++) {
            if (fStat[eFaces[n][e]] == 3) {
              nFCut = nFCut + 1;    fCut[nFCut] = eFaces[n][e];
```

```
              fStat[eFaces[n][e]] = 2;
15    } } } }
    }
```

The next function deals with the edges that have been cut by the plane;
the intersection points will become vertices of the polyhedron:

```
    ProcCutEdges () {
      real d, dt1, dt2;
      int k, nd, nE, vt1, vt2;
      for (nE = 1; nE <= nECut; nE ++) {
5       if (eStat[eCut[nE]] == 2) {
          eStat[eCut[nE]] = 3;
          vt1 = eVerts[1][eCut[nE]];
          vt2 = eVerts[2][eCut[nE]];
          dt1 = 0.;    dt2 = 0.;
10        for (k = 1; k <= NDIM; k ++) {
            dt1 = dt1 + fParam[k] * vPos[k][vt1];
            dt2 = dt2 + fParam[k] * vPos[k][vt2];
          }
          if (vStat[vt1] == 1) nd = 1;
15        else if (vStat[vt2] == 1) nd = 2;
          vLast = vLast + 1;
          vStat[vLast] = 2;    vDistSq[vLast] = 0.;
          d = (fParam[4] - dt1) / (dt2 - dt1);
          for (k = 1; k <= NDIM; k ++) {
20          vPos[k][vLast] = d * vPos[k][vt2] +
              (1. - d) * vPos[k][vt1];
            vDistSq[vLast] = vDistSq[vLast] +
              Sqr (vPos[k][vLast] - r[k][siteA]);
          }
25        eVerts[nd][eCut[nE]] = vLast;
          vEdges[1][vLast] = eCut[nE];
          vEdges[2][vLast] = 0;    vEdges[3][vLast] = 0;
      } }
    }
```

The faces cut by the plane are now examined; if a face is not completely
eliminated, its lists of boundary edges and vertices are updated to account
for the changes:

```
    ProcCutFaces () {
      int faceGone, nF, s, s1, s2, s3, s4, vDelCount, v1, v2;
      fLast = fLast + 1;    eLastP = eLast;
      for (nF = 1; nF <= nFCut; nF ++) {
5       s = fPtr[fCut[nF]];    faceGone = 0;
        while (vStat[fListV[s]] != 2 && faceGone == 0) {
          s = fListLink[s];
```

```
            if (s == fPtr[fCut[nF]]) faceGone = 1;
         }
10       if (faceGone == 1) {
            nFDel = nFDel + 1;
            fDel[nFDel] = fCut[nF];      fStat[fCut[nF]] = 1;
         } else {
            fStat[fCut[nF]] = 3;      fPtr[fCut[nF]] = s;
15          s1 = s;      s2 = fListLink[s1];
            while (vStat[fListV[s2]] == 2) {
               s1 = s2;      s2 = fListLink[s1];
            }
            vDelCount = 1;
20          s3 = s2;      s4 = fListLink[s3];
            while (vStat[fListV[s4]] != 2) {
               vDelCount = vDelCount + 1;
               s3 = s4;      s4 = fListLink[s3];
            }
25          v1 = eVerts[1][fListE[s1]] +
                 eVerts[2][fListE[s1]] - fListV[s1];
            v2 = eVerts[1][fListE[s3]] +
                 eVerts[2][fListE[s3]] - fListV[s4];
            eLast = eLast + 1;
30          fListV[s3] = v2;
            if (vDelCount == 1) {
               fListLast = fListLast + 1;
               s = fListLast;      fListLink[s1] = s;
               fListLink[s] = s2;
35             fListV[s] = v1;      fListE[s] = eLast;
            } else {
               fListV[s2] = v1;      fListE[s2] = eLast;
               if (vDelCount > 2) fListLink[s2] = s3;
            }
40          eVerts[1][eLast] = v1;      eVerts[2][eLast] = v2;
            eFaces[1][eLast] = fCut[nF];
            eFaces[2][eLast] = fLast;
            nENew = nENew + 1;
            eNew[nENew] = eLast;      eStat[eLast] = 2;
45    } }
      }
```

A little extra processing is required for the newly added vertices:

```
   ProcNewVerts () {
     int nE, v;
     for (nE = 1; nE <= nENew; nE ++) {
       if (eNew[nE] > eLastP) {
5        v = eVerts[1][eNew[nE]];
         if (vEdges[2][v] == 0) vEdges[2][v] = eNew[nE];
```

```
        else vEdges[3][v] = eNew[nE];
        v = eVerts[2][eNew[nE]];
        if (vEdges[2][v] == 0) vEdges[2][v] = eNew[nE];
10      else vEdges[3][v] = eNew[nE];
    } }
  }
```

Likewise for new faces:

```
  ProcNewFace () {
    int e, n, nE, v;
    for (n = 1; n <= nENew; n ++) {
      fListLast = fListLast + 1;
5     if (n == 1) {
        e = eNew[1];      fPtr[fLast] = fListLast;
        v = eVerts[1][e];
      } else {
        nE = 2;      e = eNew[nE];
10      while (eVerts[1][e] != v && eVerts[2][e] != v ||
            eStat[e] == 3) {
          nE = nE + 1;      e = eNew[nE];
        } }
      fListV[fListLast] = v;
15    v = eVerts[1][e] + eVerts[2][e] - v;
      fListE[fListLast] = e;      eStat[e] = 3;
      }
    fStat[fLast] = 3;      fListLink[fListLast] = fPtr[fLast];
    fDist[fLast] = 0.5 * sqrt (distSq[siteSeq[curSite]]);
20 }
```

Deleted vertices, edges and faces are then flagged appropriately:

```
  RemoveOld () {
    int n;
    for (n = 1; n <= nVDel; n ++) vStat[vDel[n]] = 0;
    for (n = 1; n <= nEDel; n ++)
5     if (eStat[eDel[n]] == 1) eStat[eDel[n]] = 0;
    for (n = 1; n <= nFDel; n ++) fStat[fDel[n]] = 0;
  }
```

Keeping track of the most distant vertex in each face, as well as the furthest vertex of all, simplifies the task of determining whether a given plane could become a face of the polyhedron:

```
  FindDistVerts () {
    real dd;
    int nF, s;
    fCut[nFCut + 1] = fLast;
5   for (nF = 1; nF <= nFCut + 1; nF ++) {
      if (fStat[fCut[nF]] != 0) {
```

```
        s = fPtr[fCut[nF]];     dd = vDistSq[fListV[s]];
        fVfar[fCut[nF]] = fListV[s];     s = fListLink[s];
        while (s != fPtr[fCut[nF]]) {
10        if (vDistSq[fListV[s]] > dd) {
            dd = vDistSq[fListV[s]];
            fVfar[fCut[nF]] = fListV[s];
          }
          s = fListLink[s];
15  } } }
    vDistSqMax = 0.;
    for (nF = 1; nF <= fLast; nF ++) {
      if (fStat[nF] != 0 && vDistSqMax < vDistSq[fVfar[nF]])
        vDistSqMax = vDistSq[fVfar[nF]];
20  }
  }
```

Evaluation of the geometrical properties of the current polyhedron is as follows. Here the four quantities computed (indexed 1–4) are the numbers of vertices, edges, and faces, and the average number of edges per face. These results will later be combined with those from other polyhedra to produce averages for the entire system:

```
PolyGeometry () {
  int n, nE, nF, nV, s;
  for (n = 1; n <= 4; n ++) polyGeom[n] = 0.;
  for (nV = 1; nV <= vLast; nV ++) {
5   if (vStat[nV] != 0) polyGeom[1] = polyGeom[1] + 1.;
  }
  for (nE = 1; nE <= eLast; nE ++) {
    if (eStat[nE] != 0) polyGeom[2] = polyGeom[2] + 1.;
  }
10  for (nF = 1; nF <= fLast; nF ++) {
    if (fStat[nF] != 0) {
      polyGeom[3] = polyGeom[3] + 1.;
      polyGeom[4] = polyGeom[4] + 1.;
      s = fListLink[fPtr[nF]];
15      while (s != fPtr[nF]) {
        polyGeom[4] = polyGeom[4] + 1.;
        s = fListLink[s];
      } } }
    polyGeom[4] = polyGeom[4] / polyGeom[3];
20 }
```

The surface area and volume of the polyhedron are computed by the function below; these results will also be used in producing averages. The area of a single (convex) face f of the polyhedron is just the sum of the areas of the triangles into which it can be decomposed,

$A_f = \sum_i |(r_{i+1} - r_1) \times (r_i - r_1)|/2$, and the volume is $V = \sum_f d_f A_f/3$, where d_f is the distance of the face from the atom position.

```
   PolySize () {
      real a, aa, d1, d2, d3, d4;
      int  k, ka, kb, nF, s, v1, v2;
      polyArea = 0.;    polyVol = 0.;
5     for (nF = 1; nF <= fLast; nF ++) {
         if (fStat [nF] != 0) {
            aa = 0.;
            for (k = 1; k <= 3; k ++) {
               s = fPtr [nF];    v1 = fListV [s];
10             s = fListLink [s];     v2 = fListV [s];
               ka = k % 3 + 1;    kb = ka % 3 + 1;
               d1 = vPos [ka] [v2] - vPos [ka] [v1];
               d2 = vPos [kb] [v2] - vPos [kb] [v1];
               a = 0.;    s = fListLink [s];
15             while (s != fPtr [nF]) {
                  v2 = fListV [s];
                  d3 = vPos [ka] [v2] - vPos [ka] [v1];
                  d4 = vPos [kb] [v2] - vPos [kb] [v1];
                  a = a + d1 * d4 - d2 * d3;
20                s = fListLink [s];    d1 = d3;    d2 = d4;
               }
               aa = aa + Sqr (a);
            }
            a = sqrt (aa);
25          polyArea = polyArea + a / 2.;
            polyVol = polyVol + fDist [nF] * a / 6.;
      } }
   }
```

A list of the (global) variables used in the program follows:

```
   real **r, **vPos, *distSq, *fDist, *vDistSq, fParam [5],
      polyGeom [5], polyGeomSum [3] [5], polyAreaSum [3],
      polyVolSum [3], region [4], regionH [4], cellRatio,
      eulerSum, fracPolyVol, polyArea, polyVol, rangeLim,
5     regionVol, vDistSqMax;
   int **eFaces, **eVerts, **vEdges, *cellList, *eCut, *eDel,
      *eNew, *eStat, *fCut, *fDel, *fListE, *fListLink,
      *fListV, *fPtr, *fStat, *fVfar, *inCell, *siteSeq,
      *testSites, *vDel, *vStat, cells [4], blockNum,
10    blockSize, curSite, eLast, eLastP, fLast, fListLast,
      nAtom, nCells, nECut, nEDel, nENew, nFCut, nFDel,
      nTestSites, nVDel, runId, siteA, siteB, vLast;
   char *progId = "ss";
```

Several parameters are used to set the sizes of the arrays, namely,

```
  #define MAX_EDGE    200
  #define MAX_FACE     50
  #define MAX_FLIST   500
  #define MAX_ITEM     50
5 #define MAX_VERT    200
```

The values are larger than necessary (for simplicity, the storage used by deleted items is not reused), but if safety checks are added to the program all risk of array overflow can be avoided.

Array allocation is carried out by a new version of AllocArrays. The length of the array eStat is MAX_EDGE, the lengths of fDist, fPtr, fStat and fVfar is MAX_FACE, for vDistSq and vStat it is MAX_VERT, for fListE, fListV and fListLink it is MAX_FLIST, for eCut, eDel, eNew, fCut, fDel and vDel the length is MAX_ITEM, and for distSq, inCell, siteSeq and testSites it is nAtom. The arrays cellList and r are the same as in the MD programs. Finally, there are the two-dimensional arrays:

```
  vPos = AllocMatR (NDIM, MAX_VERT);
  vEdges = AllocMatI (3, MAX_VERT);
  eVerts = AllocMatI (2, MAX_EDGE);
  eFaces = AllocMatI (2, MAX_EDGE);
```

The first call to GetConfig calls AllocArrays, and also the following function to set the cell size; the prescription is somewhat arbitrary, with cellRatio used to adjust the number of cells per edge to ensure that the full complement of neighbors is found:

```
  SetCellSize () {
    int k;
    for (k = 1; k <= NDIM; k ++)
      cells[k] = region[k] * cellRatio;
5   nCells = cells[1] * cells[2] * cells[3];
    for (k = 1; k <= NDIM; k ++) regionH[k] = ...
  }
```

The main program for the Voronoi calculation follows. The input file is selected as in the MD programs. The function SubdivCells contains code borrowed from ComputeForces that assigns atoms to cells and also saves the cell numbers in inCell. (Any additional data input by GetConfig that is not required in this program should be suitably disposed of.)

```
  main (int argc, char **argv) {
    int k, n, na;
    cellRatio = 0.6;
    runId = atoi (argv[1]);
```

```
5    SetupFiles ();
     blockNum = -1;
     while (GetConfig ()) {
       regionVol = region[1] * region[2] * region[3];
       SubdivCells ();
10     rangeLim = region[1];
       for (k = 1; k <= 2; k ++) {
         polyAreaSum[k] = polyVolSum[k] = 0.;
         for (n = 1; n <= 4; n ++) polyGeomSum[k][n] = 0.;
       }
15     for (na = 1; na <= nAtom; na ++) {
         FindTestSites (na);
         AnalVorPoly ();
         polyAreaSum[1] = polyAreaSum[1] + polyArea;
         polyAreaSum[2] = polyAreaSum[2] + Sqr (polyArea);
20       ... (ditto for polyVolSum and polyGeomSum) ...
       }
       fracPolyVol = polyVolSum[1] / regionVol;
       polyAreaSum[1] = polyAreaSum[1] / nAtom;
       polyAreaSum[2] = sqrt (polyAreaSum[2] / nAtom -
25         Sqr (polyAreaSum[1]));
       ... (ditto for polyVolSum and polyGeomSum) ...
       for (k = 1; k <= 2; k ++) {
         polyAreaSum[k] = polyAreaSum[k] /
           pow (regionVol, 2./3.);
30       polyVolSum[k] = polyVolSum[k] / regionVol;
       }
       eulerSum = polyGeomSum[1][1] + polyGeomSum[1][3] -
         polyGeomSum[1][2];
       ... (print the results) ...
35   }
   }
```

Two quantities evaluated here serve as checks on the computation: the sum of the volumes of the polyhedra must of course be identical to the region volume (otherwise it is likely that insufficient neighbors are being examined), and the value of eulerSum should be exactly 2, a familiar result from graph theory.

5

Dynamical properties of simple fluids

5.1 Introduction

In this chapter we encounter measurements of a type displaying some of the unique capabilities of MD. Because complete trajectories are available it is no more difficult to measure time-dependent properties, both in and out of equilibrium, than it is to measure thermodynamic and structural properties at equilibrium. Here we concentrate on properties defined in terms of time-dependent correlation functions at the atomic level – the dynamic structure factor and transport coefficients such as the shear viscosity are examples. All the analysis is incorporated into the simulation program, but it would of course be possible (though extremely storage intensive) to store the required trajectory data for subsequent processing.

5.2 Transport coefficients

5.2.1 Background

Transport coefficients describe the material properties of a fluid within the framework of continuum fluid dynamics. Discrete atoms play no role whatsoever in the continuum picture, but this does not seriously limit the enormous range of practical engineering applications of the continuum approach. The most familiar of the transport coefficients are those applicable to simple fluids; these are the diffusion coefficient, the shear and bulk viscosities, and the thermal conductivity. Others transport coefficients appear when dealing with more complex fluids, such as those containing more than one species, or those with novel rheological behavior. In many problems the transport coefficients are assumed to be experimentally determined constants, depending only on the temperature and density of the fluid, which themselves are often

114

assumed constant for a given problem, but in more complex situations transport coefficients can depend on local behavior, an example being the dependence of shear viscosity on the velocity gradient.

While statistical mechanics focuses its attention on equilibrium systems, and there is no corresponding general theory for systems away from equilibrium, linear response theory [mcq76, han86b] describes the reaction of an equilibrium system to a small external perturbation and defines generalized 'susceptibilities' that are expressed in terms of various equilibrium correlation functions. The transport coefficients we will be discussing here can be expressed in a similar fashion [hel60, mcq76], despite the fact that there are no obvious mechanical perturbations corresponding to the concentration, velocity, and thermal gradients associated with the underlying transport processes (we will return to this subject in Chapter 7).

Each transport coefficient can be derived directly from one of the continuum equations of fluid dynamics (such as the Navier–Stokes equation) after taking the long wavelength (small-k) limit of the Fourier-transformed form of the equation. The eventual result of the derivation is a direct relation between a macroscopic transport coefficient and the time integral of a particular microscopic autocorrelation function measured in an equilibrium system; such correlations are not directly accessible to experiment.

The alternative, and from the historical point of view original, definition of a transport coefficient, namely, the constant factor relating the response of a system to an imposed driving force – such as the Newtonian definition of shear viscosity, or Fourier's law of heat transport – implies a nonequilibrium system. Such an approach is also feasible within the MD framework; there are, however, certain technical details that must be addressed in order to do such simulations, and we will deal with this approach in Chapter 7.

5.2.2 Diffusion

In a continuous system the diffusion coefficient D is defined by Fick's law relating mass flow to density gradient [mcq76]:

$$\rho u = -D\nabla\rho \tag{5.1}$$

where $u(r,t)$ is the local velocity and $\rho(r,t)$ the local density or concentration, so that the time-evolution of ρ is described by the equation

$$\frac{\partial\rho}{\partial t} = D\nabla^2\rho \tag{5.2}$$

This result applies both to the diffusion of one species through another, and to self-diffusion within a single species. At the discrete particle level ρ is just

$$\rho(r,t) = \sum_{j=1}^{N_a} \delta(r - r_j(t)) \tag{5.3}$$

Then, for large t (compared to the 'collision interval', a rather vague but intuitively obvious period of time where soft potentials are involved) we have the Einstein expression [mcq76]:

$$D = \lim_{t\to\infty} \frac{1}{6N_a t} \left\langle \sum_{j=1}^{N_a} [r_j(t) - r_j(0)]^2 \right\rangle \tag{5.4}$$

Note that for a finite system t cannot become too large because the allowed displacements are bounded; eventually this asymptotic result will break down, so that after reaching a plateau D will begin to drop to zero.

For periodic boundaries we need the 'true' atomic displacements $r'_j(t)$, from which the effects of wraparound have been removed. If we assume that the displacement per timestep is small relative to the system size (as it always is), then the two sets of coordinate components are related by

$$r'_{xj}(t) = r_{xj}(t) + \text{nint}\big([r'_{xj}(t - \Delta t) - r_{xj}(t)]/L_x\big)L_x \tag{5.5}$$

where $\text{nint}(x)$ is the nearest integer to x, and $r'_j(0) = r_j(0)$. In (5.4) $\langle\ldots\rangle$ implies an average over a sufficiently large number of (in principle) independent samples.

The alternative Green–Kubo expression [mcq76] is based on the integrated velocity autocorrelation function:

$$D = \frac{1}{3N_a} \int_0^\infty \left\langle \sum_{j=1}^{N_a} v_j(t) \cdot v_j(0) \right\rangle dt \tag{5.6}$$

The two definitions are completely equivalent.

A reliable estimate of D, as well as the other transport coefficients discussed subsequently, requires that the trajectories be computed relatively accurately for as long as the velocities remain correlated. As pointed out in Chapter 3, the main source of uncertainty in the trajectories is the strongly repulsive potential, and not the truncation error of the numerical method used for the differential equations. The former is a real physical effect that influences the velocity correlations in a way that mimics nature, so that the velocities remain correlated until overwhelmed by the noise inherent in the trajectories.

5.2.3 Shear viscosity

The shear viscosity η is defined by the Navier–Stokes equation [mcq76]:

$$\rho \left(\frac{\partial}{\partial t} + \boldsymbol{u} \cdot \nabla \right) \boldsymbol{u} = \eta \nabla^2 \boldsymbol{u} + \left(\frac{\eta}{3} + \eta_v \right) \nabla (\nabla \cdot \boldsymbol{u}) - \nabla P \qquad (5.7)$$

Another transport coefficient also appears in this equation, the bulk viscosity η_v, but it will not be studied here. Theory then leads to an expression analogous to the Einstein diffusion formula (5.4) [hel60, mcq76] (see [all93a] however), namely,

$$\eta = \lim_{t \to \infty} \frac{1}{6 k_B T V t} \left\langle \sum_{x<y} \left[\sum_j m_j r_{xj}(t) v_{yj}(t) - \sum_j m_j r_{xj}(0) v_{yj}(0) \right]^2 \right\rangle \qquad (5.8)$$

where $\sum_{x<y}$ denotes a sum over the three pairs of distinct vector components (xy, yz, and zx) used to improve the statistics. The formula shows how η characterizes the rate at which some component (typically y) of momentum diffuses in a perpendicular (x) direction. While this result bears a certain formal similarity to the diffusion expression, the conspicuous difference is that, here, a single sum unites the contributions from all atoms, whereas with diffusion each atom contributes individually (in short, the square of a sum as opposed to a sum of squares). This expression turns out to be unusable with periodic boundaries because they violate the translational invariance assumed in the derivation [all93a].

The alternative Green–Kubo form, based on the integrated autocorrelation function of the pressure tensor, does not experience this problem. The definition is [mcq76]

$$\eta = \frac{V}{3 k_B T} \int_0^\infty \left\langle \sum_{x<y} P_{xy}(t) P_{xy}(0) \right\rangle dt \qquad (5.9)$$

where

$$P_{xy} = \frac{1}{V} \left[\sum_j m_j v_{xj} v_{yj} + \frac{1}{2} \sum_{i \neq j} r_{xij} f_{yij} \right] \qquad (5.10)$$

is a component of the pressure tensor. Evaluation of the second term in P_{xy} can be carried out along with the force computation, treating periodic boundaries in the normal way. For pair potentials (such as LJ) in which $\boldsymbol{f}_{ij} = f(r_{ij})\boldsymbol{r}_{ij}/r_{ij}$, it is clear that $P_{xy} = P_{yx}$. Averaging over vector components is again used to improve the statistics.

5.2.4 Thermal conductivity

The equation for heat transfer derived from Fourier's law [mcq76], assuming that the process involves thermal conduction alone and that there is no convection (implying mass flow), is

$$\rho C_V \frac{\partial E}{\partial t} = \lambda \nabla^2 E \tag{5.11}$$

and the resulting diffusion-type formula for the thermal conductivity is

$$\lambda = \lim_{t \to \infty} \frac{1}{6k_B T^2 V t} \left\langle \sum_x \left[\sum_j r_{xj}(t)e_j(t) - \sum_j r_{xj}(0)e_j(0) \right]^2 \right\rangle \tag{5.12}$$

where

$$e_j = \tfrac{1}{2}mv_j^2 + \tfrac{1}{2}\sum_{i(\neq j)} u(r_{ij}) - \langle e \rangle \tag{5.13}$$

is the instantaneous excess energy of atom j, $\langle e \rangle$ is the mean energy, and $\sum_x$ is a sum over vector components. The periodic boundary limitation also applies here, but there is an alternative form based on the integrated heat flux autocorrelation function:

$$\lambda = \frac{V}{3k_B T^2} \int_0^\infty \langle \boldsymbol{S}(t) \cdot \boldsymbol{S}(0) \rangle \, dt \tag{5.14}$$

where

$$\boldsymbol{S} = \frac{1}{V} \left[\sum_j e_j \boldsymbol{v}_j + \tfrac{1}{2} \sum_{i \neq j} \boldsymbol{r}_{ij}(\boldsymbol{f}_{ij} \cdot \boldsymbol{v}_j) \right] \tag{5.15}$$

(The expression for $\boldsymbol{S}$ should also include a term proportional to $\sum_j \boldsymbol{v}_j \langle h \rangle$, where $\langle h \rangle$ is the mean enthalpy per atom, but, provided the total momentum is zero, the term can be dropped.)

5.3 Measuring transport coefficients

5.3.1 Direct evaluation of diffusion

We now turn to the practical side of studying the transport coefficients, beginning with the simplest example, the diffusion coefficient D based on the Einstein definition (5.4). The computations will introduce a standardized framework that can be used for all measurements extending over a series of timesteps, with each such calculation including initialization, the actual process of making and accumulating measurements at evenly spaced time intervals, and a final summary. An important feature of these computations is that the samples are overlapped to provide extra results, and this calls for additional storage and bookkeeping. While the overlap increases the correlation between successive samples, with similar consequences for error estimates as described in Chapter 4, it can improve the quality of the results without extending the duration of the run; ideally overlap should be confined to time intervals over which the correlation between measurements has decayed to a comparatively small value.

The measurements entail following the atomic trajectories over a sufficient number of timesteps to obtain convergence of (5.4) to asymptotic behavior. New sets of diffusion measurements are begun at fixed time intervals, so that several sets of measurements based on different time origins will be in progress simultaneously because of the overlapped measurements – see Figure 5.1. To be more specific, a total of nValDiffuse measurements contribute to the set used to produce a single (unaveraged) estimate of D, there are nBuffDiffuse sets of data being collected at any time (except for very early in the run), each occupying a separate storage buffer, and measurements are made every stepDiffuse timesteps. The last of the parameters governing the data collection is limitDiffuseAv, which specifies the total number of individual estimates used to produce the averaged value of D. Each set of measurements is carried out completely independently, and they are combined only when complete, with an element in the array indexDiffuse used to keep track of the number of measurements currently in that set; this array is initialized in a particular way (by the function InitDiffusion shown below) to ensure that the measurements are evenly spaced. Given the values of these parameters the total run length needed for a given number of measurements can easily be determined; only complete sets of results are used, and partially filled buffers are discarded at the end of the run.

The functions appearing below measure the squared displacement of each atom from each of several reference points (or origins), one per set, making allowance for periodic boundaries, and produce a series of estimates that for sufficiently large time intervals should converge to D. The array rDiffuseOrg stores the origins for each set of measurements, and rDiffuseTrue accumulates the components of the 'true' displacement of each atom after removing any wraparound effects. To avoid an excess of array indices in the program we have limited the arrays to just

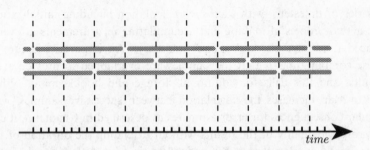

Fig. 5.1 Use of overlapped data collection for time-dependent properties; the measurements at any instant contribute to several sets of results (shown as shaded lines), with the dashed lines marking the different time origins.

two dimensions only; where more indices are required they have been combined.

The measurement functions are as follows:

```
   EvalDiffusion () {
     int k, n, nb, ni, nk;
     for (nb = 1; nb <= nBuffDiffuse; nb ++) {
       indexDiffuse[nb] = indexDiffuse[nb] + 1;
5      if (indexDiffuse[nb] <= 0) continue;
       if (indexDiffuse[nb] == 1) {
         for (n = 1; n <= nAtom; n ++) {
           for (k = 1; k <= NDIM; k ++) {
             nk = NDIM * (n - 1) + k;
10           rDiffuseTrue[nb][nk] = r[k][n];
             rDiffuseOrg[nb][nk] = r[k][n];
       } } }
       ni = indexDiffuse[nb];
       rDiffuse[nb][ni] = 0.;
15     for (n = 1; n <= nAtom; n ++) {
         for (k = 1; k <= NDIM; k ++) {
           nk = NDIM * (n - 1) + k;
           rDiffuseTrue[nb][nk] = r[k][n] +
             Nint ((rDiffuseTrue[nb][nk] - r[k][n]) /
20           region[k]) * region[k];
           rDiffuse[nb][ni] = rDiffuse[nb][ni] +
             Sqr (rDiffuseTrue[nb][nk] -
             rDiffuseOrg[nb][nk]);
       } } }
25   AccumDiffusion ();
   }
```

```
   AccumDiffusion () {
     real fac;
     int j, nb;
     for (nb = 1; nb <= nBuffDiffuse; nb ++) {
5      if (indexDiffuse[nb] == nValDiffuse) {
         for (j = 1; j <= nValDiffuse; j ++)
           rDiffuseAv[j] = rDiffuseAv[j] + rDiffuse[nb][j];
         indexDiffuse[nb] = 0;
         countDiffuseAv = countDiffuseAv + 1;
10       if (countDiffuseAv == limitDiffuseAv) {
           fac = 1. / (NDIM * 2 * nAtom * stepDiffuse *
             deltaT * limitDiffuseAv);
           for (j = 2; j <= nValDiffuse; j ++)
             rDiffuseAv[j] = rDiffuseAv[j] * fac / (j - 1);
15         PrintDiffusion (stdout);
           ZeroDiffusion ();
     } } }
   }
```

The functions that initialize and reset the calculation and output the results follow. Note the values initially assigned to indexDiffuse that determine the spacing between measurements: using negative initial values delays the start of data collection for each set of measurements until the appropriate moment:

```
   InitDiffusion () {
     int nb;
     for (nb = 1; nb <= nBuffDiffuse; nb ++)
       indexDiffuse[nb] = - (nb - 1) * nValDiffuse /
5         nBuffDiffuse;
     ZeroDiffusion ();
   }

   ZeroDiffusion () {
     int j;
     countDiffuseAv = 0;
     for (j = 1; j <= nValDiffuse; j ++) rDiffuseAv[j] = 0.;
5  }

   PrintDiffusion (FILE *fp) {
     real tVal;
     int j;
     fprintf (fp, "diffusion\n");
5    for (j = 1; j <= nValDiffuse; j ++) {
       tVal = (j - 1) * stepDiffuse * deltaT;
       fprintf (fp, "%8.4f %8.4f\n", tVal, rDiffuseAv[j]);
     }
   }
```

Incorporating the above functions into the MD program requires several additions. The measurement function is called from `SingleStep` at regular intervals after equilibration by

```
if (stepCount >= stepEquil && (stepCount - stepEquil)
    stepDiffuse == 0) EvalDiffusion ();
```

and program initialization (`SetupJob`) includes

```
InitDiffusion ();
```

The new quantities associated with these calculations are

```
real **rDiffuse, **rDiffuseOrg, **rDiffuseTrue,
    *rDiffuseAv;
int *indexDiffuse, countDiffuseAv, limitDiffuseAv,
    nBuffDiffuse, nValDiffuse, stepDiffuse;
```

the measurement parameters input to the program are

```
INAME (limitDiffuseAv),
INAME (nBuffDiffuse),
INAME (nValDiffuse),
INAME (stepDiffuse),
```

and the necessary arrays are allocated by `AllocArrays` (note the various array sizes):

```
  rDiffuse = AllocMatR (nBuffDiffuse, nValDiffuse);
  rDiffuseOrg = AllocMatR (nBuffDiffuse, NDIM * nAtom);
  rDiffuseTrue = AllocMatR (nBuffDiffuse, NDIM * nAtom);
  rDiffuseAv = AllocVecR (nValDiffuse);
5 indexDiffuse = AllocVecI (nBuffDiffuse);
```

5.3.2 *Diffusion from the velocity autocorrelation function*

The alternative approach to measuring the diffusion coefficient is based on the integrated velocity autocorrelation function (5.6). Considerations governing the use of overlapped samples discussed previously also apply; the work itself is organized in a similar way, and even the new variables have corresponding names. Each of the data collection buffers requires its own individual copy of the atom velocities at the start of the measurement period; these reference values are stored in `rvAcfOrg`. The autocorrelation function itself is constructed in the array `rvAcf`. The general technique described here forms the basis for studying the remaining transport coefficients later in this section.

The calculation is carried out by the following functions, and in this description we assume that a leapfrog integrator is used, so that velocities must be adjusted by a half-timestep (this can be removed if inappropriate):

```
EvalVacf () {
   int k, n, nb, ni;
   for (nb = 1; nb <= nBuffAcf; nb ++) {
      indexAcf[nb] = indexAcf[nb] + 1;
 5    if (indexAcf[nb] <= 0) continue;
      if (indexAcf[nb] == 1) {
        for (n = 1; n <= nAtom; n ++) {
          for (k = 1; k <= NDIM; k ++)
             rvAcfOrg[nb][NDIM * (n - 1) + k] = rv[k][n] -
10               0.5 * ra[k][n] * deltaT;
      } }
      ni = indexAcf[nb];    rvAcf[nb][ni] = 0.;
      for (n = 1; n <= nAtom; n ++) {
        for (k = 1; k <= NDIM; k ++)
15          rvAcf[nb][ni] = rvAcf[nb][ni] +
               rvAcfOrg[nb][NDIM * (n - 1) + k] * (rv[k][n] -
               0.5 * ra[k][n] * deltaT);
   } }
   AccumVacf ();
20 }

   AccumVacf () {
      real fac;
      int j, nb;
      for (nb = 1; nb <= nBuffAcf; nb ++) {
 5     if (indexAcf[nb] == nValAcf) {
         for (j = 1; j <= nValAcf; j ++)
            rvAcfAv[j] = rvAcfAv[j] + rvAcf[nb][j];
         indexAcf[nb] = 0;    countAcfAv = countAcfAv + 1;
         if (countAcfAv == limitAcfAv) {
10          fac = 1. / (NDIM * nAtom * limitAcfAv);
            diffuseAcfInt = fac * stepAcf * deltaT *
               Integrate (rvAcfAv, nValAcf);
            for (j = 2; j <= nValAcf; j ++)
               rvAcfAv[j] = rvAcfAv[j] / rvAcfAv[1];
15          rvAcfAv[1] = 1.;
            PrintVacf (stdout);
            ZeroVacf ();
      } } }
   }
```

The function `Integrate` (see Appendix) computes the integral of its first argument using a simple method such as the trapezoidal rule.

Other required functions are

```
  InitVacf () {
    int nb;
    for (nb = 1; nb <= nBuffAcf; nb ++)
        indexAcf[nb] = - (nb - 1) * nValAcf / nBuffAcf;
5   ZeroVacf ();
  }

  ZeroVacf () {
    int j;
    countAcfAv = 0;
    for (j = 1; j <= nValAcf; j ++) rvAcfAv[j] = 0.;
5 }

  PrintVacf (FILE *fp) {
    real tVal;
    int j;
    fprintf (fp, "velocity acf\n");
5   for (j = 1; j <= nValAcf; j ++) {
      tVal = (j - 1) * stepAcf * deltaT;
      fprintf (fp, "%8.4f %8.4f\n", tVal, rvAcfAv[j]);
    }
    fprintf (fp, "diffusion acf integral: %8.3f\n",
10      diffuseAcfInt);
  }
```

To incorporate the measurements into the MD program add the following statement to `SingleStep`:

```
  if (stepCount >= stepEquil && (stepCount - stepEquil)
      stepAcf == 0) EvalVacf ();
```

and a call to the initialization function from `SetupJob`:

```
  InitVacf ();
```

The additional variables used are

```
  real **rvAcf, **rvAcfOrg, *rvAcfAv, diffuseAcfInt;
  int *indexAcf, countAcfAv, limitAcfAv, nBuffAcf, nValAcf,
    stepAcf;
```

the measurement parameters input to the program are

```
  INAME (limitAcfAv),
  INAME (nBuffAcf),
  INAME (nValAcf),
  INAME (stepAcf),
```

and the array allocations (AllocArrays) are

```
rvAcf = AllocMatR (nBuffAcf, nValAcf);
rvAcfOrg = AllocMatR (nBuffAcf, NDIM * nAtom);
indexAcf = AllocVecI (nBuffAcf);
rvAcfAv = AllocVecR (nValAcf);
```

5.3.3 Shear viscosity and thermal conductivity

These computations are also based on the appropriate autocorrelation functions and closely follow the treatment used to compute D from the velocity autocorrelation. Indeed, to simplify matters we will assume that all three transport coefficients are computed together and identical parameters govern the measurements. In order to compute the quantities involved in the autocorrelation functions certain additions must be made to the interaction calculations. Reduced MD units are used.

In the formulae (5.10) and (5.15) for the pressure tensor and heat current used in the transport coefficient definitions, sums over products such as $r_{xij}f_{yij}$, $r_{xij}f_{yij}v_{yj}$, and $e_j v_{xj}$ appear; these terms should be evaluated at the same time as the forces. Additional arrays are needed to save sums of the form $\sum_i r_{xij}f_{yij}$ and the values of e_j separately for each atom; these two arrays are

```
real **rfAtom, *enAtom;
```

and are allocated in AllocArrays by

```
rfAtom = AllocMatR (9, nAtom);
enAtom = AllocVecR (nAtom);
```

(As mentioned earlier, for pair potentials depending only on r_{ij}, the relation $r_{xij}f_{yij} = r_{yij}f_{xij}$ applies; thus the matrix represented by rfAtom is symmetric and only half the off-diagonal terms are actually required in this case.) In ComputeForces the following computations appear after the virial evaluation:

```
    int ... kk, k1, k2;
    ...
    enAtom[j1] = enAtom[j1] + uVal;
    enAtom[j2] = enAtom[j2] + uVal;
5   kk = 0;
    for (k1 = 1; k1 <= 3; k1 ++) {
      for (k2 = 1; k2 <= 3; k2 ++) {
        kk = kk + 1;
        f = fcVal * dr[k1] * dr[k2];
10      rfAtom[kk][j1] = rfAtom[kk][j1] + f;
        rfAtom[kk][j2] = rfAtom[kk][j2] + f;
```

```
    } }
```

Initialization of the arrays takes place at the start of `ComputeForces`:

```
for (n = 1; n <= nAtom; n ++) {
  enAtom[n] = 0.;
  for (k = 1; k <= 9; k ++) rfAtom[k][n] = 0.;
}
```

New quantities needed for the autocorrelation function computations are

```
real **thermAcf, **thermAcfOrg, *thermAcfAv, thermAcfInt,
   **viscAcf, **viscAcfOrg, *viscAcfAv, viscAcfInt;
```

and the arrays are allocated (in `AllocArrays`) by

```
  thermAcf = AllocMatR (nBuffAcf, nValAcf);
  thermAcfOrg = AllocMatR (nBuffAcf, NDIM);
  thermAcfAv = AllocVecR (nValAcf);
  viscAcf = ... (same as thermAcf) ...
5 viscAcfOrg = ...
  viscAcfAv = ...
```

The function `EvalVacf` is modified to incorporate the additional data collection:

```
  real thermVec[NDIM + 1], v[NDIM + 1], viscVec[NDIM + 1];
  int ... j;
  ...
  for (k = 1; k <= NDIM; k ++) viscVec[k] = thermVec[k] = 0.;
5 for (n = 1; n <= nAtom; n ++) {
    for (k = 1; k <= NDIM; k ++)
      v[k] = rv[k][n] - 0.5 * ra[k][n] * deltaT;
    viscVec[1] = viscVec[1] + v[2] * v[3] +
      0.5 * rfAtom[6][n];
10  viscVec[2] = viscVec[2] + v[3] * v[1] +
      0.5 * rfAtom[3][n];
    viscVec[3] = viscVec[3] + v[1] * v[2] +
      0.5 * rfAtom[2][n];
    for (k = 1; k <= NDIM; k ++)
15    enAtom[n] = enAtom[n] + Sqr (v[k]);
    for (k = 1; k <= NDIM; k ++) {
      thermVec[k] = thermVec[k] + 0.5 * v[k] * enAtom[n];
      for (j = 1; j <= NDIM; j ++)
        thermVec[k] = thermVec[k] + 0.5 * v[j] *
20        rfAtom[NDIM * (k - 1) + j][n];
  } }
  for (nb = 1; nb <= nBuffAcf; nb ++) {
    ...
    if (indexAcf[nb] == 1) {
```

```
25    for (k = 1; k <= NDIM; k ++) {
        viscAcfOrg[nb][k] = viscVec[k];
        thermAcfOrg[nb][k] = thermVec[k];
    } }
    viscAcf[nb][ni] = thermAcf[nb][ni] = 0.;
30  for (k = 1; k <= NDIM; k ++) {
      viscAcf[nb][ni] = viscAcf[nb][ni] +
        viscAcfOrg[nb][k] * viscVec[k];
      thermAcf[nb][ni] = thermAcf[nb][ni] +
        thermAcfOrg[nb][k] * thermVec[k];
35  }
  }
```

For the actual evaluation of the autocorrelation integrals and the transport coefficients the following additions are made to AccumVacf:

```
    if (indexAcf[nb] == nValAcf) {
      ...
      for (j = 1; j <= nValAcf; j ++) {
        viscAcfAv[j] = viscAcfAv[j] + viscAcf[nb][j];
5       thermAcfAv[j] = thermAcfAv[j] + thermAcf[nb][j];
      }
      ...
      if (countAcfAv == limitAcfAv) {
        ...
10      fac = density / (3. * temperature * nAtom *
          limitAcfAv);
        viscAcfInt = fac * stepAcf * deltaT *
          Integrate (viscAcfAv, nValAcf);
        for (j = 2; j <= nValAcf; j ++)
15        viscAcfAv[j] = viscAcfAv[j] / viscAcfAv[1];
        viscAcfAv[1] = 1.;
        fac = density / (3. * Sqr (temperature) * nAtom *
          limitAcfAv);
        thermAcfInt = fac * stepAcf * deltaT *
20        Integrate (thermAcfAv, nValAcf);
        ... (normalize thermAcfAv as above) ...
```

Finally, the function ZeroVacf requires

```
    for (j = 1; j <= nValAcf; j ++)
      viscAcfAv[j] = thermAcfAv[j] = 0.;
```

and in PrintVacf the quantities viscAcfAv[j] and thermAcfAv[j] are added to the output loop, and the integrated values viscAcfInt and thermAcfInt are also output.

5.4 Space–time correlation functions

5.4.1 Definitions

The experimental significance of time-dependent correlation functions is that spectroscopic techniques, of which neutron scattering is one example, actually measure the spectra of microscopic dynamical quantities. The MD approach provides equivalent information directly from the trajectories, and comparison with experiment can be made by carrying out a Fourier analysis of the simulation results – in a sense this amounts to performing the experiment on the model system. Such correlation functions span the entire range of length and time scales, from slow long-wavelength modes at the hydrodynamic limit, right down to the atomic level [boo91].

To link the discrete atomistic picture with the continuum view of a system described in terms of smoothly varying scalar and vector fields, a function such as the number density (for a single species this is just the mass density in dimensionless MD units) at a point r at time t is expressed as a sum over atoms, as in (5.3):

$$\rho(r,t) = \sum_j \delta(r - r_j(t)) \tag{5.16}$$

In a practical sense this defines the local density in terms of the average occupancy of a small volume of space situated at r and measured over a short time interval. There are of course fluctuations as atoms enter and leave the volume, but these can be reduced by using larger volumes and/or longer time intervals. The definition satisfies the obvious requirement that matter is conserved:

$$\int \rho(r,t)dr = N_a \tag{5.17}$$

Space- and time-dependent density correlations are described by means of the van Hove correlation function [han86b], defined as

$$G(r,t) = \frac{1}{N_a}\left\langle \int \rho(r'+r,t)\rho(r',0)dr' \right\rangle \tag{5.18}$$

$$= \frac{1}{N_a}\left\langle \sum_{ij} \delta(r + r_i(0) - r_j(t)) \right\rangle \tag{5.19}$$

where homogeneity is assumed in order to carry out the r' integration in (5.18). The double summation can be divided into two parts, $G(r,t) =$

$G_s(r,t) + G_d(r,t)$, where

$$G_s(r,t) = \frac{1}{N_a} \left\langle \sum_j \delta(r + r_j(0) - r_j(t)) \right\rangle \tag{5.20}$$

is the probability of an atom being displaced by a distance r during time t, and $G_d(r,t)$ contains the remaining terms of the double sum. The $t = 0$ limits of G_s and G_d are $G_s(r,0) = \delta(r)$ and $G_d(r,0) = \rho g(r)$; in the limits $r \to \infty$ or $t \to \infty$, $G_s \to 1/V$, $G_d \to \rho$. There have been attempts in the past to establish a functional relationship between G_s and G_d, but these have been unsuccessful because of the wealth of dynamical detail that must be discarded.

The Fourier transform of the density is

$$\rho(k,t) = \int \rho(r,t) \exp(-ik \cdot r) dr = \sum_j \exp(-ik \cdot r_j(t)) \tag{5.21}$$

and the so-called intermediate scattering function is defined by

$$F(k,t) = \int G(r,t) \exp(-ik \cdot r) dr = \frac{1}{N_a} \langle \rho(k,t)\rho(-k,0) \rangle \tag{5.22}$$

Note the connection to the static structure factor, $F(k,0) = S(k)$. The dynamic structure factor is defined as

$$S(k,\omega) = \frac{1}{2\pi} \int_{-\infty}^{\infty} F(k,t) \exp(i\omega t) dt \tag{5.23}$$

This quantity satisfies the obvious sum rule $\int_{-\infty}^{\infty} S(k,\omega)d\omega = S(k)$.

If l and τ are the mean free path and collision time (suitably interpreted when dealing with continuous potentials), then the regime $kl \ll 1$ and $\omega\tau \ll 1$, or, equivalently, wavelength $\gg l$ and timescale $\gg \tau$, is where the behavior can be described by continuum fluid mechanics and the underlying atomic nature of the fluid is totally hidden. The ability to use MD to study the behavior across a range of scales provides the bridge between atomistic and macroscopic worlds.

While the local density conveys information about the distribution of atoms, it is equally possible to examine local variations in the motion of the atoms. The definition of the particle current (or momentum current for a single atomic species using MD units) is [han86b]

$$\pi(r,t) = \sum_j v_j \delta(r - r_j(t)) \tag{5.24}$$

and its Fourier transform is

$$\pi(\mathbf{k}, t) = \sum_j \mathbf{v}_j \exp(-i\mathbf{k} \cdot \mathbf{r}_j(t)) \tag{5.25}$$

The spatial correlation functions of the components of the current vector are defined as

$$C_{\alpha\beta}(\mathbf{k}, t) = \frac{k^2}{N_a} \langle \pi_\alpha(\mathbf{k}, t) \pi_\beta(-\mathbf{k}, 0) \rangle \tag{5.26}$$

For isotropic fluids, symmetry considerations lead to an expression in terms of longitudinal and transverse currents (the directions are relative to $\mathbf{k}$):

$$C_{\alpha\beta}(\mathbf{k}, t) = \frac{k_\alpha k_\beta}{k^2} C_L(\mathbf{k}, t) + \left(\delta_{\alpha\beta} - \frac{k_\alpha k_\beta}{k^2}\right) C_T(\mathbf{k}, t) \tag{5.27}$$

and by setting $\mathbf{k} = k\mathbf{e}_z$ we obtain

$$C_L(\mathbf{k}, t) = \frac{k^2}{N_a} \langle \pi_z(\mathbf{k}, t) \pi_z(-\mathbf{k}, 0) \rangle \tag{5.28}$$

$$C_T(\mathbf{k}, t) = \frac{k^2}{2N_a} \langle \pi_x(\mathbf{k}, t) \pi_x(-\mathbf{k}, 0) + \pi_y(\mathbf{k}, t) \pi_y(-\mathbf{k}, 0) \rangle \tag{5.29}$$

The dynamic structure factor is related to the Fourier transform of the transverse current, $S(\mathbf{k}, \omega) = C_T(\mathbf{k}, \omega)/\omega^2$.

In the large-k (continuum) limit, the form of $S(\mathbf{k}, \omega)$ is known [han86b]. The function is of course symmetric in ω and there are Lorentzian-shaped Brillouin peaks at $\omega = \pm v_s k$, where v_s is the adiabatic speed of sound, and a Rayleigh peak at $\omega = 0$. The width of each peak is proportional to k^2; the Rayleigh peak width is also proportional to the thermal diffusivity $(\lambda/\rho C_P)$, and the Brillouin peak width is proportional to the sound attenuation coefficient (a quantity expressible in terms of transport coefficients and specific heats). Note that wraparound effects can occur for times $t > L/v_s$; if significant, this sets an upper bound to the timescales that can be examined, and hence a lower limit to ω. The values of $\mathbf{k}$ that can be examined are restricted to vectors with components that are multiples of $2\pi/L$, and the larger the region size L the smaller the k-values that can be achieved.

5.4.2 Computational methods

The MD evaluation of $S(\mathbf{k}, \omega)$ is based on the Fourier transform of $F(\mathbf{k}, t)$. This in turn can be expressed (5.22) either as the Fourier transform of a discretized form of the van Hove correlation function – an extension of the method used in Chapter 4 for $g(r)$ – or as the time-correlation of the Fourier-transformed density. The latter is clearly preferable since it requires a great deal less work. Evaluation of $\rho(k,t)$ is based directly on a sum over atoms, as in the study of long-range order in Chapter 4 (but only for a single k-value there). An alternative for very large systems that is not explored here is to first evaluate a coarse-grained density function $\rho(r,t)$ based on a grid with suitable spatial resolution, and then use a discrete (preferably fast) Fourier transform to obtain $\rho(k,t)$. Some of the detail is lost when grid averages are used, but this affects results at short distances – typically of the order of the grid spacing itself – while details of the more interesting long-range behavior are preserved.

Further simplification is possible when studying isotropic systems, since the function of interest is the spherically averaged quantity $S(k,\omega)$, and it is therefore sufficient to consider k-vectors in a very limited number of directions. Averaging over several spatially equivalent directions will improve the statistics, so that if we confine our attention to k-vectors along the coordinate axes, the computation requires evaluation of $\rho(k,t)$ for the three vectors $k = (k,0,0)$, $(0,k,0)$, and $(0,0,k)$. If we assume a cubic region, periodic boundaries restrict the allowed values of k to integer multiples of $2\pi/L$.

5.4.3 Program details

We now turn to the details of computing the density and current correlation functions. The technique of overlapped measurements introduced earlier in the chapter is also used here, and the variables involved in data collection follow a similar naming convention.

Because of the considerable amount of data that must be collected and our policy (introduced previously) of avoiding a surfeit of array indices, we begin with some remarks on how the computations and data are organized. The calculation starts by evaluating the Fourier sums for the density and the three current components (one longitudinal and two transverse) along each of the three k-directions; these complex-valued results are stored in the array cfVal, whose index is determined by a combination of the direction of k, the value of k, and the type of quantity (there are four – the density and the three current components – and storage must be provided for a total of 24 real numbers for each value of k considered). The contributions to the correlation functions computed by a single call to EvalSpacetimeCorr are placed in the array spacetimeCorr. The results for each time offset are stored sequentially;

for each offset there are three values (each a complex number) for each k, namely, $C_L(k,t)$, $C_T(k,t)$, and $F(k,t)$. The number of different k-values used is denoted by nFunCorr.

The contributions to all the (overlapped) correlation function measurements in progress at a given instant are evaluated by the following function. A cubic region shape and leapfrog integration are assumed. The recurrence relations

$$\sin n\theta = 2\cos\theta \sin(n-1)\theta - \sin(n-2)\theta$$
$$\cos n\theta = 2\cos\theta \cos(n-1)\theta - \cos(n-2)\theta$$

(5.30)

are used for evaluating sines and cosines of multiple angles.

```
   EvalSpacetimeCorr () {
     real cosV, cosV0, cosV1, cosV2, kVal, sinV,
       sinV1, sinV2, w;
     int j, k, m, n, nb, nc, ni, nv;
5    for (j = 1; j <= 24 * nFunCorr; j ++) cfVal[j] = 0.;
     kVal = 2. * pi / region[1];
     for (n = 1; n <= nAtom; n ++) {
       j = 1;
       for (k = 1; k <= NDIM; k ++) {
10       for (m = 1; m <= nFunCorr; m ++) {
           if (m == 1) {
             cosV = cos (kVal * r[k][n]);
             sinV = sin (kVal * r[k][n]);
             cosV0 = cosV;
15         } else if (m == 2) {
             cosV1 = cosV;    sinV1 = sinV;
             cosV = 2. * cosV0 * cosV1 - 1.;
             sinV = 2. * cosV0 * sinV1;
           } else {
20           cosV2 = cosV1;    sinV2 = sinV1;
             cosV1 = cosV;    sinV1 = sinV;
             cosV = 2. * cosV0 * cosV1 - cosV2;
             sinV = 2. * cosV0 * sinV1 - sinV2;
           }
25         for (nc = 1; nc <= 4; nc ++) {
             if (nc <= 3) w = rv[nc][n] -
               0.5 * ra[nc][n] * deltaT;
             else w = 1.;
             cfVal[j] = cfVal[j] + w * cosV;
30           cfVal[j + 1] = cfVal[j + 1] + w * sinV;
             j = j + 2;
     } } } }
     for (nb = 1; nb <= nBuffCorr; nb ++) {
       indexCorr[nb] = indexCorr[nb] + 1;
```

```
35     if (indexCorr[nb] <= 0) continue;
       ni = 3 * nFunCorr * (indexCorr[nb] - 1);
       if (indexCorr[nb] == 1) {
         for (j = 1; j <= 24 * nFunCorr; j ++)
           cfOrg[nb][j] = cfVal[j];
40     }
       for (j = 1; j <= 3 * nFunCorr; j ++)
         spacetimeCorr[nb][ni + j] = 0.;
       j = 1;
       for (k = 1; k <= NDIM; k ++) {
45       for (m = 1; m <= nFunCorr; m ++) {
           for (nc = 1; nc <= 4; nc ++) {
             nv = 3 * m + ni;
             if (nc <= 3) {
               w = Sqr (kVal * m);
50             if (nc == k) nv = nv - 2;
               else {
                 nv = nv - 1;     w = w * 0.5;
               }
             } else w = 1.;
55           spacetimeCorr[nb][nv] = spacetimeCorr[nb][nv] +
               w * cfVal[j] * cfOrg[nb][j] + cfVal[j + 1] *
               cfOrg[nb][j + 1];
             j = j + 2;
       } } }  }
60   AccumSpacetimeCorr ();
   }
```

Accumulating the averages is the task of the following function; nVal-
Corr is the number of time offsets:

```
   AccumSpacetimeCorr () {
     int j, nb;
     for (nb = 1; nb <= nBuffCorr; nb ++) {
       if (indexCorr[nb] == nValCorr) {
5        for (j = 1; j <= 3 * nFunCorr * nValCorr; j ++)
           spacetimeCorrAv[j] = spacetimeCorrAv[j] +
             spacetimeCorr[nb][j];
         indexCorr[nb] = 0;     countCorrAv = countCorrAv + 1;
         if (countCorrAv == limitCorrAv) {
10         for (j = 1; j <= 3 * nFunCorr * nValCorr; j ++)
             spacetimeCorrAv[j] = spacetimeCorrAv[j] /
               (3. * nAtom * limitCorrAv);
           PrintSpacetimeCorr (stdout);
           ZeroSpacetimeCorr ();
15   } } }
   }
```

The additional variables are

```
real **cfOrg, **spacetimeCorr, *cfVal, *spacetimeCorrAv;
int *indexCorr, countCorrAv, limitCorrAv, nBuffCorr,
   nFunCorr, nValCorr, stepCorr;
```

new input values are

```
  INAME (limitCorrAv),
  INAME (nBuffCorr),
  INAME (nFunCorr),
  INAME (nValCorr),
5 INAME (stepCorr),
```

and additional array allocations:

```
  cfOrg = AllocMatR (nBuffCorr, 24 * nFunCorr);
  cfVal = AllocVecR (24 * nFunCorr);
  indexCorr = AllocVecI (nBuffCorr);
  spacetimeCorr = AllocMatR (nBuffCorr, 3 * nFunCorr *
5   nValCorr);
  spacetimeCorrAv = AllocVecR (3 * nFunCorr * nValCorr);
```

The call to the analysis function in `SingleStep` is

```
if (stepCount > stepEquil && (stepCount - stepEquil)
   stepCorr == 0) EvalSpacetimeCorr ();
```

and for initialization (`SetupJob`):

```
InitSpacetimeCorr ();
```

Finally, the initialization and output functions are

```
  InitSpacetimeCorr () {
    int j;
    for (j = 1; j <= nBuffCorr; j ++)
      indexCorr[j] = - (j - 1) * nValCorr / nBuffCorr;
5   ZeroSpacetimeCorr ();
  }

  ZeroSpacetimeCorr () {
    int j;
    countCorrAv = 0;
    for (j = 1; j <= 3 * nFunCorr * nValCorr; j ++)
5     spacetimeCorrAv[j] = 0.;
  }

  PrintSpacetimeCorr (FILE *fp) {
    real tVal;
    int j, k, n, nn;
    char *header[] = {"cur-long", "cur-trans", "density"};
5   fprintf (fp, "space-time corr\n");
```

```
     for (k = 1; k <= 3; k ++) {
       fprintf (fp, "%s\n", header[k - 1]);
       for (n = 1; n <= nValCorr; n ++) {
         tVal = (n - 1) * stepCorr * deltaT;
10       fprintf (fp, "%7.3f", tVal);
         nn = 3 * nFunCorr * (n - 1);
         for (j = k; j <= 3 * nFunCorr; j += 3)
           fprintf (fp, " %8.4f", spacetimeCorrAv[nn + j]);
         fprintf (fp, "\n");
15   } }
     }
```

In order to evaluate $S(k, \omega)$ (5.23) for a given k at a total of n_ω frequency values (including $\omega = 0$) it is necessary to collect $n_\omega + 1$ equally spaced measurements of $F(k, t)$. Since $F(k, t)$ is an even function of t, these results are reflected about $t = 0$, providing a total of $2n_\omega$ values – the first and last measurements are used only once each. Then the Fourier transform is carried out using all $2n_\omega$ values (this would normally be some power of two, so that an FFT approach can be used), and the discrete form of the function $S(k, \omega)$ appears as the real part of the first n_ω terms [pre92]. This analysis is carried out separately from the run itself; a program for doing this, as well as for tabulating the normalized functions $F(k, t)/F(k, 0)$, appears at the end of the chapter. The current correlation functions are treated in the same way.

5.5 Measurements

5.5.1 *Velocity autocorrelation function*

The velocity autocorrelation functions shown in Figure 5.2 are computed during soft-sphere runs that include the following input data:

```
initUcell        5 5 5
temperature      0.5
deltaT           0.005
stepEquil        2000
stepInitlzTemp   100
limitAcfAv       200
nBuffAcf         10
nValAcf          200
stepAcf          3
```

and values of density between 0.6 and 1.0. Leapfrog integration is used. The initial state is an FCC lattice, so that $N_a = 500$. A single set of results based on 200 sets of partially overlapped measurements

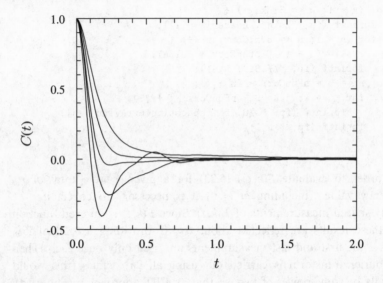

Fig. 5.2 Soft-sphere velocity autocorrelation functions for $\rho = 0.6, 0.7, 0.8, 0.9$, and 1.0.

is produced during a run of 15 000 timesteps. The negative correlations observed at higher densities (both for LJ and hard-sphere systems) were one of the important revelations of MD [rah64, ald67].

Given the exponential sensitivity of the trajectories to any numerical error (Chapter 3) it is important to establish that results such as the velocity autocorrelation function are fully reproducible. Here we show one example to confirm that this is indeed the case. The system used is the same as above, with $\rho = 0.9$, but with different values of Δt; measurement intervals and run lengths are adjusted accordingly. Figure 5.3 shows the results; in order to resolve the extremely small differences beyond $t = 1$ the vertical scale is enlarged by a factor of 100.

5.5.2 Transport coefficients

Diffusion coefficient measurements using (5.4) for the same system are shown in Figure 5.4. The input data include

```
limitDiffuseAv  200
nBuffDiffuse     10
nValDiffuse     200
stepDiffuse       3
```

At $\rho = 0.9$ the values drop to zero (there is no diffusion), while at

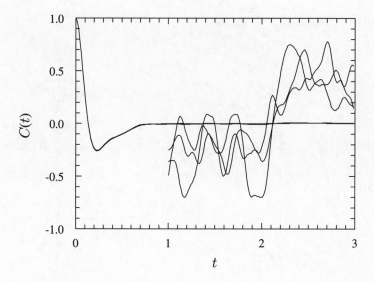

Fig. 5.3 Velocity autocorrelation functions for $\rho = 0.9$ computed using $\Delta t = 0.005$, 0.0025, and 0.00125; for $t > 1$ the results are also shown with a vertical 100-fold magnification.

smaller ρ they appear to asymptote to increasingly larger values, although convergence also slows and longer measurements are seen to be necessary at $\rho = 0.6$ and 0.7.

Estimates for D based on (5.6) can be obtained by integrating the velocity autocorrelation function shown in Figure 5.2. The results, without any attempt at error estimation, for $\rho = 0.9$, 0.8, 0.7, and 0.6 are 0, 0.039, 0.068, and 0.106 respectively; these can be compared with the (partly converged) D values of Figure 5.4.

In Figure 5.5 we show the three autocorrelation functions whose integrals yield D, η, and λ, namely, velocity, pressure tensor, and heat current. The system used here has $N_a = 864$, $\rho = 0.8$, and $T = 1$. To improve the quality of the results, the computation is run for almost a half million timesteps, with the following measurement parameters in the input data:

```
stepEquil      4000
limitAcfAv     500
nBuffAcf       30
nValAcf        600
stepAcf        3
```

This yields 15 sets of autocorrelation results; the leapfrog integrator

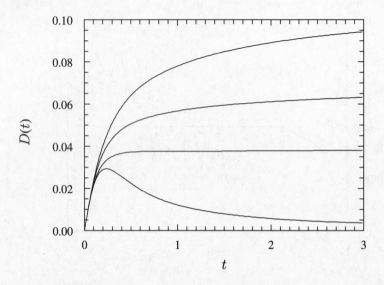

Fig. 5.4 Diffusion coefficient measurements for $\rho = 0.9$, 0.8, 0.7, and 0.6.

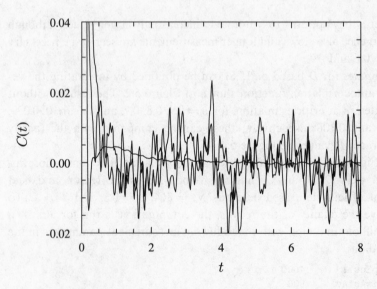

Fig. 5.5 Velocity, pressure tensor, and heat current autocorrelation functions; the vertical scale has been expanded to show the noise present in the results.

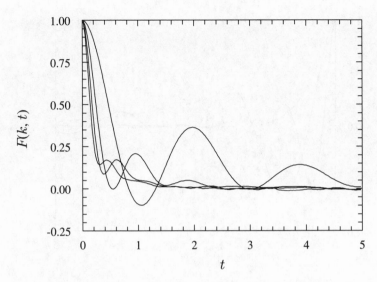

Fig. 5.6 Normalized intermediate scattering function $F(k,t)$ for the four smallest k-values; the decay becomes slower as $k \to 0$.

Table 5.1. *Transport coefficients (c) from the integrated autocorrelation functions*

c	$\langle c \rangle$	$\sigma(c)$	$\sigma(c)/\langle c \rangle$
D	0.0814	0.0012	0.0147
η	1.6318	0.6528	0.4001
λ	6.0323	2.1997	0.3647

Table 5.2. *Effect of truncating the integration after n values*

n	c	$\langle c \rangle$	$\sigma(c)$	$\sigma(c)/\langle c \rangle$
70	D	0.0777	0.0005	0.0064
	η	1.7091	0.3156	0.1847
	λ	6.1934	0.5510	0.0890
140	D	0.0804	0.0005	0.0056
	η	1.6044	0.3704	0.2309
	λ	6.0153	0.8837	0.1469
210	D	0.0812	0.0006	0.0078
	η	1.6087	0.5072	0.3153
	λ	6.3181	0.9379	0.1484

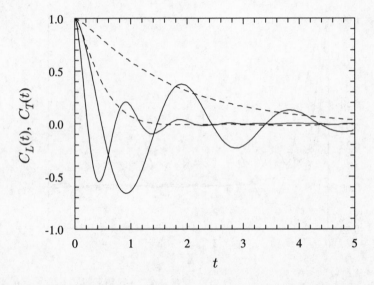

Fig. 5.7 Normalized longitudinal (solid curve) and transverse (dashed) current corrrelation functions $C_L(k,t)$ and $C_T(k,t)$ for the two smallest k-values.

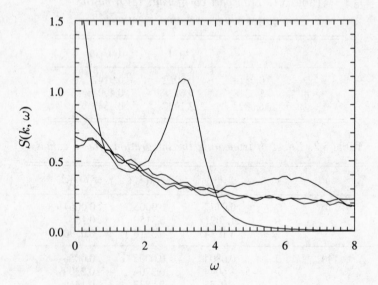

Fig. 5.8 Dynamic structure factor $S(k,\omega)$ for the four smallest k-values.

produces practically no energy drift (a mere one part in 2000) over the entire run. There is a clear difference between the smooth velocity autocorrelation function which involves separate contributions from each atom, and the other two functions that are comparatively noisy because the entire system must be considered to obtain a single measurement.

Evaluation of the transport coefficients by integrating these autocorrelation functions over the entire range to $t = 9$ leads to the results shown in Table 5.1. The uncertainty in the estimates of η and λ is considerable, but can be reduced if it is realized that the noise in the integrands makes a substantial contribution to the error without improving the estimate of the mean [lev87]. To show the potential for improvement, Table 5.2 lists the results obtained by terminating the integration after the first 70 values, corresponding to $t \approx 1$, as well as after two and three times this number of values. Further information on transport coefficient calculations using these methods can be found in [lev73, sch85, lev87, vog87].

5.5.3 Space–time correlations

Sample results for space–time correlation functions are obtained from a single run of a soft-sphere system with $N_a = 2048$, $T = 0.7$, and $\rho = 0.84$. The state point is chosen to be fairly close to published results for the LJ fluid [sch86]. The input data include

```
limitCorrAv    500
nBuffCorr      80
nFunCorr       4
nValCorr       1025
stepCorr       5
```

A run of a little over 4×10^5 timesteps produces 13 sets of results.

Figure 5.6 shows the normalized intermediate scattering function $F(k, t)$ for the four smallest values of k. The normalized current corrrelation functions $C_L(k, t)$ and $C_T(k, t)$ are shown in Figure 5.7, with results for the two smallest k-values included in both cases. The absence of any structure in C_T is expected for normal liquids that do not support shear waves; the peaks in C_L correspond to sound propagation. Finally, the dynamic structure factor $S(k, \omega)$, with its expected peaks, is shown in Figure 5.8. In general, the soft-sphere results resemble those for the LJ case [sch86] (allowing for the different time units); they can also be compared with hard-sphere results [all83]. Additional correlation functions are treated in [des88].

5.6 Further work

5.1 All the transport coefficient values have been given in reduced MD units; convert them to physical units and compare with the experimental argon values.

5.2 Extend the diffusion measurements until all converge.

5.3 Compare the transport coefficients for soft-sphere and LJ fluids under similar conditions.

5.4 The large-t behavior of the autocorrelation functions is a subject of particular interest [erp85, erp88]; examine the kind of decay that occurs and whether it depends on some power of t (rather than the exponential decay that might have been naïvely expected). What kind of cooperative motion is responsible for this behavior in the case of the velocity autocorrelation function [ald70b]?

5.5 The bulk viscosity is another transport coefficient; measure it using the appropriate autocorrelation function [lev73, lev87].

5.6 New transport coefficients appear when binary fluids are involved; consider the possibilities [vog88].

5.7 The space–time correlations show propagating longitudinal modes (sound waves): is it possible to observe similar transverse modes (shear waves) at sufficiently high density [lev87]?

5.7 Additional material

The program listed here is used for analyzing the space–time correlation results; after averaging the data produced by the run, it either generates the normalized time-dependent correlations (and error estimates, here unused) or the Fourier transforms. The program also demonstrates a general approach to organizing analysis in which only selected data are extracted from the job output file based on the headings accompanying the data.

Whether the real-space or Fourier version of the program is run depends on doFourier, and for the latter an optional windowing function [pre92] is available; such options could have been included as input data to the program instead. The same applies to cutoffs for limiting the output and the specification of how much of the early data to skip. Here the only input is the run identifier used to select the data file for processing; this file is assumed to contain a full copy of the program output (the Appendix includes discussion of file management). The macro NAMEVAL is used to locate specific data items at the start of the data file. The function FtFastComplex (see Appendix) does a fast Fourier transform (FFT) and overwrites the original data with the result (the length of the processed array must be a power of 2):

```
   #define NAMEVAL(x) \
      if (! strncmp (bp, #x, strlen (#x))) {          \
        bp += strlen (#x);    x = strtod (bp, &bp); }
   #define BUFF_LEN    1024
 5 char *header[] = {"cur-long", "cur-trans", "density"},
      *txtCorr = "space-time corr";

   main (int argc, char **argv) {
     char *bp, buff[BUFF_LEN];
10   real **corrSum, **corrSumSq, *work, damp, deltaT,
         deltaTCorr, omegaMax, pi, tMax, w, x;
     int doFourier, doWindow, id, j, k, kk, n, nData,
         nFunCorr, nSet, nSetSkip, nV, nValCorr, stepCorr;
     FILE *fpIn;
15   id = atoi (argv[1]);
     doFourier = 1;    doWindow = 0;
     nSetSkip = 1;
     omegaMax = 10.;    tMax = 5.;
     sprintf (buff, "ss%02dlog.data", id);
20   if (! (fpIn = fopen (buff, "r"))) {
       printf ("no file\n");
       exit (0);
     }
     while (1) {
25     bp = fgets (buff, BUFF_LEN, fpIn);
       if (*bp == '-') break;
       NAMEVAL (deltaT);     NAMEVAL (nValCorr);
       NAMEVAL (nFunCorr);     NAMEVAL (stepCorr);
     }
30   deltaTCorr = stepCorr * deltaT;
     corrSum = AllocMatR (3, nFunCorr * nValCorr);
     corrSumSq = AllocMatR (3, nFunCorr * nValCorr);
     for (j = 1; j <= 3; j ++) {
       for (n = 1; n <= nFunCorr * nValCorr; n ++)
35       corrSum[j][n] = corrSumSq[j][n] = 0.;
     }
     work = AllocVecR (4 * (nValCorr - 1));
     nData = 0;    nSet = 0;
     while (1) {
40     if (! (bp = fgets (buff, BUFF_LEN, fpIn))) break;
       if (! strncmp (bp, txtCorr, strlen (txtCorr))) {
         nSet = nSet + 1;
         if (nSet < nSetSkip) continue;
```

```
            nData = nData + 1;
45          for (j = 1; j <= 3; j ++) {
              bp = fgets (buff, BUFF_LEN, fpIn);
              for (n = 1; n <= nValCorr; n ++) {
                bp = fgets (buff, BUFF_LEN, fpIn);
                w = strtod (bp, &bp);
50              for (k = 1; k <= nFunCorr; k ++) {
                  w = strtod (bp, &bp);
                  kk = (k - 1) * nValCorr;
                  corrSum[j][kk + n] = corrSum[j][kk + n] + w;
                  corrSumSq[j][kk + n] = corrSumSq[j][kk + n] +
55                    Sqr (w);
          } } } } }
            fclose (fpIn);
            printf ("%d\n", nData);
            for (j = 1; j <= 3; j ++) {
60            for (n = 1; n <= nFunCorr * nValCorr; n ++) {
                corrSum[j][n] = corrSum[j][n] / nData;
                corrSumSq[j][n] = sqrt (corrSum[j][n] / nData -
                  Sqr (corrSum[j][n]));
            } }
65          if (doFourier) {
              for (j = 1; j <= 3; j ++) {
                for (k = 1; k <= nFunCorr; k ++) {
                  kk = (k - 1) * nValCorr;
                  for (n = 1; n <= nValCorr; n ++) {
70                  if (doWindow) damp = (nValCorr - n + 1) /
                      (nValCorr + 0.5);
                    else damp = 1.;
                    work[2 * n - 1] = corrSum[j][kk + n] * damp;
                  }
75                for (n = nValCorr + 1; n <= 2 * nValCorr - 2; n ++)
                    work[2 * n - 1] = work[2 * (2 *
                      nValCorr - n) - 1];
                  for (n = 1; n <= 2 * nValCorr - 2; n ++)
                    work[2 * n] = 0.;
80                FtFastComplex (work, 2 * nValCorr - 2, 1);
                  for (n = 1; n <= nValCorr; n ++)
                    corrSum[j][kk + n] = work[2 * n - 1];
              } }
              pi = 4. * atan (1.);
85            if (omegaMax > pi / deltaTCorr)
                omegaMax = pi / deltaTCorr;
              nV = nValCorr * omegaMax / (pi / deltaTCorr);
            } else {
              for (j = 1; j <= 3; j ++) {
90              for (k = 1; k <= nFunCorr; k ++) {
```

```
          kk = (k - 1) * nValCorr;
          for (n = 2; n <= nValCorr; n ++)
            corrSum[j][kk + n] = corrSum[j][kk + n] /
              corrSum[j][kk + 1];
95        corrSum[j][kk + 1] = 1.;
      } }
      if (tMax > (nValCorr - 1) * deltaTCorr)
        tMax = (nValCorr - 1) * deltaTCorr;
      nV = nValCorr * tMax / ((nValCorr - 1) * deltaTCorr);
100  }
    for (j = 1; j <= 3; j ++) {
      printf ("%s\n", header[j - 1]);
      for (n = 1; n <= nV; n ++) {
        if (doFourier) x = (n - 1) * omegaMax / nV;
105      else x = (n - 1) * deltaTCorr;
        printf ("%9.4f", x);
        for (k = 1; k <= nFunCorr; k ++)
          printf (" %9.4f", corrSum[j][(k - 1) *
            nValCorr + n]);
110      printf ("\n");
    } }
  }
```

6

Alternative ensembles

6.1 Introduction

The equations of motion used in MD are based on Newtonian mechanics; in this way MD mimics nature. If one adopts the purely mechanical point of view there is little more to be said, but if a broader perspective is permitted and MD is regarded as a tool for generating equilibrium states satisfying certain specified requirements, then it is possible to modify the dynamics and address a broader range of problems. But at the outset it must be emphasized that no physical meaning is attributed to the actual dynamics, and the approach is merely one of computational convenience for generating particular equilibrium thermodynamic states, although – and this is not an attempt to extract any such meaning – the deviations of the motion from the truly Newtonian may in fact be extremely small.

Conventional MD differs from most experimental studies in that it is the energy and volume that are fixed, rather than temperature and pressure. In statistical mechanical terms, MD produces microcanonical (NVE) ensemble averages, whereas constant-temperature experiments correspond to the canonical (NVT) ensemble; if constant pressure is imposed as well, as is generally the case in the laboratory, it is the isothermal–isobaric (NPT) ensemble that is the relevant one. While the choice of ensemble is usually one of convenience at the macroscopic level since (away from the critical point) thermal fluctuations are small, for the microscopic systems studied by MD the fluctuations of nonregulated quantities can be sufficiently large to make precise measurement difficult. Modifying the dynamics allows MD to model the equilibrium behavior of such ensembles directly.

We will describe two different approaches to the problem. One employs a feedback mechanism for correcting deviations in the controlled parameter (for example, temperature) from the preset mean value; the value itself fluctuates, but the size of the fluctuations can be regulated. The other method ensures that the controlled parameter is strictly constant, apart for numerical drift, by augmenting the equations of motion with suitable mechanical constraints; thus temperature can be held constant by introducing a constraint that fixes the kinetic energy.

There are other ways to change the ensemble, such as by coupling the system to a constant-temperature heat bath, and even by simply resetting the kinetic energy at each timestep. The former requires a stochastic mechanism for adjusting velocities in order to reproduce the effect of a heat bath [and80], but this violates the deterministic nature of the dynamics. The latter method is sufficiently crude not to merit consideration, although when it comes to introducing hard walls into the simulation (Chapter 7) the same idea can be adopted, but the justification is of course entirely different.

6.2 Feedback methods

6.2.1 Controlled temperature

The mechanism for feedback regulation of temperature rests on the idea that because the temperature is proportional to the mean-square velocity it ought to be possible to vary the temperature by adjusting the rate at which time progresses [nos84a]. A new dynamical variable s is introduced into the Lagrangian in a manner that is equivalent to rescaling the unit of time, and extra terms are added in just the way needed to obtain the desired behavior. There are now two distinct time variables: the real, or physical, time t', and a scaled, or virtual, time t; the relation between them is through their differentials:

$$dt = s(t') \, dt' \tag{6.1}$$

The Lagrangian for this rather unusual 'extended' system is written as

$$\mathcal{L} = \tfrac{1}{2} m s^2 \sum_i \dot{r}_i^2 - \sum_{i<j} u(r_{ij}) + \tfrac{1}{2} M_s \dot{s}^2 - n_f T \ln s \tag{6.2}$$

where T is the required temperature, $n_f = 3N_a + 1$ is the number of degrees of freedom (which could be reduced by three to account for momentum conservation), M_s plays the role of a mass that is needed in order to construct an equation of motion for the new 'coordinate' s, and the dot stands for the derivative d/dt – note that (6.2) is defined in terms of the virtual time. The Lagrange equations of motion that are obtained by the standard procedure are

$$\ddot{r}_i = \frac{1}{ms^2}F_i - \frac{2\dot{s}}{s}\dot{r}_i \tag{6.3}$$

$$M_s\ddot{s} = ms\sum_i \dot{r}_i^2 - \frac{n_f T}{s} \tag{6.4}$$

Because the relationship between t and t' depends on the entire history of the system, namely,

$$t = \int s(t')dt' \tag{6.5}$$

it is more convenient if the equations are transformed to use physical time units [nos84b, hoo85]; henceforth the dot will be used to denote d/dt', and the equations can be rewritten as

$$\ddot{r}_i = \frac{1}{m}F_i - \frac{\dot{s}}{s}\dot{r}_i \tag{6.6}$$

$$\ddot{s} = \frac{\dot{s}^2}{s} + \frac{G_1 s}{M_s} \tag{6.7}$$

where

$$G_1 = m\sum_i \dot{r}_i^2 - n_f T \tag{6.8}$$

The first of these equations of motion resembles the conventional Newtonian equation with an additional friction-like term, though not true friction because the coefficient can be of either sign; the second equation defines the feedback mechanism by which s is varied to regulate temperature.

The motivation for the $\ln s$ term in the Lagrangian can now be appreciated. Assume that it is replaced by a general function $w(s)$; since s is finite, the time average of $\ddot{s}$ must vanish, implying that

$$m\left\langle \sum_i \dot{r}_i^2/s \right\rangle = \langle dw/ds \rangle \tag{6.9}$$

The left-hand side is just $n_f\langle T/s \rangle$, so that if we equate the actual values rather than just the averages we find that $w(s) = n_f T \ln s$.

The equilibrium averages of the physical system can be shown to be those of the canonical ensemble at temperature T [nos84a]. In order to establish this result the microcanonical partition function of the extended system is simply integrated over the s variable and what remains is the canonical partition function. The temperature itself is not constant, however, but the negative feedback acting through s ensures that the fluctuations are limited and the mean value is equal to T.

The Hamiltonian of the extended system:

$$\mathscr{H} = \tfrac{1}{2}m\sum_i \dot{r}_i^2 + \sum_{i<j} u(r_{ij}) + \tfrac{1}{2}M_s(\dot{s}/s)^2 + n_f T \ln s \qquad (6.10)$$

is conserved since there are no time-dependent external forces, and this provides a useful check on the accuracy of the numerical solution. The Hamiltonian itself has no physical significance; its first two terms represent the energy of the physical system, but their sum is free to fluctuate. The quantity M_s is a parameter whose value must be determined empirically; M_s itself has no particular physical meaning and is simply a part of the computational technique. In principle, the value of M_s does not affect the final equilibrium results, but it does influence their accuracy and reliability, because if the kinetic energy fluctuations are allowed to become too large it is rather difficult to think of the system existing at a particular temperature. For small variations in s the characteristic period of the fluctuations is [nos84a]

$$\tau_s = 2\pi\sqrt{M_s\langle s\rangle^2/2n_f T} \qquad (6.11)$$

and the simulation must extend over many such periods to prevent these fluctuations from influencing the results adversely.

An MD program demonstrating temperature feedback will not be shown separately but will be combined with the version that incorporates pressure feedback as well. This method is described below.

6.2.2 Controlled pressure and temperature

While the connection between time and temperature just introduced is not reminiscent of any physical mechanism, pressure can be adjusted by altering the container volume. In the MD context this is achieved by a uniform isotropic volume change brought about by rescaling the atomic coordinates [and80]. A more useful method emerges if this is combined with the temperature feedback; the appropriate Lagrangian treatment leads to a system whose behavior corresponds to the isothermal–isobaric (NPT) ensemble.

We consider a cubic simulation region with a volume that is allowed to vary. Scaled coordinates r are introduced that span the unit cube and are related to the physical coordinates r' by

$$r = r'/V^{1/3} \qquad (6.12)$$

The same scaled time variable introduced previously is also used here, so that now, both V and s are treated as supplementary dynamical variables. The Lagrangian for this system is a generalization of (6.2), with additional terms designed to ensure the correct pressure feedback mechanism [nos84b] (and once again the dot denotes d/dt):

$$\mathscr{L} = \tfrac{1}{2}mV^{2/3}s^2\sum_i \dot{r}_i^2 - \sum_{i<j}u(V^{1/3}r_{ij}) + \tfrac{1}{2}M_s\dot{s}^2 + \tfrac{1}{2}M_v\dot{V}^2$$
$$- n_f T \ln s - PV \tag{6.13}$$

where P and T are the required (externally imposed) values, and M_v is another generalized mass parameter. Roughly speaking, M_v can be regarded as the mass of a piston that could have been used to regulate pressure by altering the volume, but because of the need to avoid explicit walls the effect of a sliding piston is achieved by means of a uniform volume change; a real piston would also introduce undesirable effects such as pressure waves. The first term in the Lagrangian is the kinetic energy, though not that of the physical system obtained by the substitution $\dot{r}' = V^{1/3}\dot{r} + (\dot{V}/3V^{2/3})r$, but, rather, it is a value based on velocities measured relative to the rate at which the region size changes (the $\dot{V}$ term is dropped) [and80]; removal of the flow component of the atomic velocities is essential to ensure the correct definition of temperature.

The Lagrange equations of motion, in scaled variables, are

$$\ddot{r}_i = \frac{1}{mV^{1/3}s^2}F_i - \left(\frac{2\dot{s}}{s} + \frac{2\dot{V}}{3V}\right)\dot{r}_i \tag{6.14}$$

$$M_s\ddot{s} = mV^{2/3}s\sum_i \dot{r}_i^2 - \frac{n_f T}{s} \tag{6.15}$$

$$M_v\ddot{V} = \frac{ms^2}{3V^{1/3}}\sum_i \dot{r}_i^2 + \frac{1}{3V^{2/3}}\sum_{i<j}r_{ij}\cdot f_{ij} - P \tag{6.16}$$

Returning to physical time units, with the dot now denoting d/dt', we obtain

$$\ddot{r}_i = \frac{1}{mV^{1/3}}F_i - \left(\frac{\dot{s}}{s} + \frac{2\dot{V}}{3V}\right)\dot{r}_i \tag{6.17}$$

$$\ddot{s} = \frac{\dot{s}^2}{s} + \frac{G_1 s}{M_s} \tag{6.18}$$

$$\ddot{V} = \frac{\dot{s}\dot{V}}{s} + \frac{G_2 s^2}{3M_v V} \tag{6.19}$$

where

$$G_1 = mV^{2/3} \sum_i \dot{r}_i^2 - n_f T \tag{6.20}$$

$$G_2 = mV^{2/3} \sum_i \dot{r}_i^2 + V^{1/3} \sum_{i<j} \mathbf{r}_{ij} \cdot \mathbf{f}_{ij} - 3PV \tag{6.21}$$

Scaled coordinates have been retained since they are more convenient from a computational point of view (see further).

When the dynamics are supplemented by these two extra degrees of freedom the equilibrium averages of the physical system can be shown to be those of the NPT ensemble [nos84b]. The Hamiltonian is

$$\mathcal{H} = \tfrac{1}{2}mV^{2/3} \sum_i \dot{r}_i^2 + \sum_{i<j} u(V^{1/3} r_{ij}) + \tfrac{1}{2}M_s(\dot{s}/s)^2 + \tfrac{1}{2}M_v(\dot{V}/s)^2 \tag{6.22}$$
$$+ n_f T \ln s + PV$$

and, though conserved, it is once again not a physically meaningful quantity. The method of establishing a reasonable value for M_v is again empirical; for small variations in V the characteristic period is [nos83]

$$\tau_v = 2\pi \sqrt{M_v \langle \delta V \rangle^2 / T} \tag{6.23}$$

Periodic boundaries are most readily handled when the problem is expressed in terms of scaled coordinates because the simulation region is then a fixed unit cube; use of physical variables introduces unnecessary complications when handling boundary crossings because velocities and accelerations must be adjusted in addition to the coordinates [eva84]. When working with the PC method, the conversion to physical coordinates needed for the interaction calculations can overwrite the scaled coordinates because the predicted values are not needed for the corrector computation. Since the volume varies, provision must be made for an adjustable number of cells for use in the interaction calculations; for simplicity we use the cell method without a neighbor list.

In SingleStep the following additional code is required – the insertion positions should be obvious:

```
PredictorStepPT ();
ApplyBoundaryCond ();
UpdateCellSize ();
UnscaleCoords ();
5  ...
ComputeDerivsPT ();
   ...
CorrectorStepPT ();
```

The changes to EvalProps (assuming for simplicity $n_f = 3N_a$) are

```
pressure = (vvSum + virSum) / (3. * varV);
totEnergy = totEnergy + (0.5 * (massS * Sqr (varSv) +
   massV * Sqr (varVv)) / Sqr (varS) + extPressure *
   varV) / nAtom + 3. * temperature * log (varS);
```

where totEnergy is the Hamiltonian (6.22). Note that UpdateCellSize is called immediately after the predictor calculation because no neighbor list is used and the cells must be reconstructed every timestep.

The new variables needed are

```
real extPressure, g1Sum, g2Sum, massS, massV, varS, varSa,
   varSa1, varSa2, varSo, varSv, varSvo, varV, varVa,
   varVa1, varVa2, varVo, varVv, varVvo;
int maxEdgeCells;
```

where varS and varV correspond to s and V, and the various suffixes denote derivatives and earlier values (for the PC method) in the same way as for r. The extra input data consists of

```
RNAME (extPressure),
RNAME (massS),
RNAME (massV),
```

and additional initialization functions are called from SetupJob:

```
InitFeedbackVars ();
ScaleCoords ();
ScaleVels ();
```

The maximum size of the cell array is set in SetParams; the calculation assumes a cubic region, and we allow for a reasonable (but arbitrarily chosen) amount of expansion beyond the initial region size:

```
maxEdgeCells = cells[1] * 1.3;
```

Storage allocation in AllocArrays to provide a cell array based on this maximum size requires the change

```
cellList = AllocVecI (nAtom + maxEdgeCells * maxEdgeCells *
   maxEdgeCells);
```

Computing the right-hand sides of the feedback equations is as follows:

```
  ComputeDerivsPT () {
    real aFac, vFac;
    int k, n;
    vvSum = 0.;
5   for (n = 1; n <= nAtom; n ++) {
      for (k = 1; k <= NDIM; k ++)
```

```
          vvSum = vvSum + Sqr (rv[k][n]);
      }
      vvSum = vvSum * pow (varV, 2./3.);
10    g1Sum = vvSum - 3. * nAtom * temperature;
      g2Sum = vvSum + virSum - 3. * extPressure * varV;
      aFac = pow (varV, -1./3.);
      vFac = - varSv / varS - 2. * varVv / (3. * varV);
      for (n = 1; n <= nAtom; n ++) {
15      for (k = 1; k <= NDIM; k ++)
            ra[k][n] = aFac * ra[k][n] + vFac * rv[k][n];
      }
      varSa = Sqr (varSv) / varS + g1Sum * varS / massS;
      varVa = varSv * varVv / varS + g2Sum * Sqr (varS) /
20        (3. * massV * varV);
  }
```

The second-order feedback equations for *s* and *V* are solved using the same PC method as the equations of motion:

```
  PredictorStepPT () {
    real cr[] = ...
    int k;
    varSo = varS;     varSvo = varSv;
5   varVo = varV;     varVvo = varVv;
    varS = varS + deltaT * varSv +
        (deltaT * deltaT / div) * (cr[0] * varSa +
        cr[1] * varSa1 + cr[2] * varSa2);
    varSv = (varS - varSo) / deltaT +
10      (deltaT / div) * (cv[0] * varSa +
        cv[1] * varSa1 + cv[2] * varSa2);
    varV = varV + deltaT * varVv +
        (deltaT * deltaT / div) * (cr[0] * varVa +
        cr[1] * varVa1 + cr[2] * varVa2);
15  varVv = (varV - varVo) / deltaT +
        (deltaT / div) * (cv[0] * varVa +
        cv[1] * varVa1 + cv[2] * varVa2);
    varSa2 = varSa1;     varVa2 = varVa1;
    varSa1 = varSa;     varVa1 = varVa;
20  for (k = 1; k <= NDIM; k ++) {
      region[k] = pow (varV, 1. / NDIM);
      regionH[k] = 0.5 * region[k];
    }
  }

  CorrectorStepPT () {
    real cr[] = ...
    int k;
    varS = varSo + deltaT * varSvo +
```

```
5        ... (as above) ...
    varSv = ...
    varV = varVo + deltaT * varVvo +
        ... (as above) ...
    varVv = ...
10   for (k = 1; k <= NDIM; k ++) {
        region[k] = ... ;    regionH[k] = ... ;
    }
}
```

Initialization of the extra variables is as follows:

```
InitFeedbackVars () {
    varS = 1.;      varSv = 0.;
    varSa = varSa1 = varSa2 = 0.;
    varV = pow (region[1], 3.);
5   varVv = 0.;
    varVa = varVa1 = varVa2 = 0.;
}
```

Coordinate and velocity rescaling (just one example is shown, the others are similar) and adjusting the cell size are simply

```
ScaleCoords () {
    real fac;
    int k, n;
    fac = pow (varV, -1. / NDIM);
5   for (n = 1; n <= nAtom; n ++) {
        for (k = 1; k <= NDIM; k ++) r[k][n] = r[k][n] * fac;
    }
}

UpdateCellSize () {
    int k;
    for (k = 1; k <= NDIM; k ++) {
        cells[k] = region[k] / rCut;
5       if (cells[k] > maxEdgeCells) cells[k] = maxEdgeCells;
    }
}
```

The function EvalProps needs a minor addition to allow for the scaled velocities (the form of the kinetic energy term in (6.22) determines the way vvSum must be calculated):

```
vvSum = vvSum * pow (varV, 2./3.);
```

and ApplyBoundaryCond must be altered to use scaled coordinates:

```
if (r[k][n] >= 0.5) r[k][n] = r[k][n] - 1.;
else if (r[k][n] < -0.5) r[k][n] = r[k][n] + 1.;
```

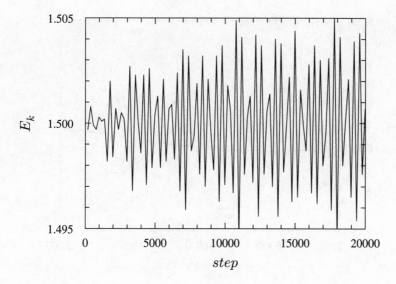

Fig. 6.1 Kinetic energy fluctuations for PT-feedback simulation with $M_s = 1$.

The output of the instantaneous region edge length `region[1]` should be added to `PrintSummary`.

The results shown here are based on a soft-sphere system with the following input data:

```
initUcell      5 5 5
density        0.8
temperature    1.0
deltaT         0.001
extPressure    6.5
massS          0.1
massV          0.01
stepAvg        200
```

The values used for M_v are 0.01, 0.1, and 1.0, with just the single value used for M_s; the value of Δt depends to some extent on both these mass parameters, but here the drift in the total Hamiltonian over 20 000 timesteps is less than one part in 5000. The initial state is an FCC lattice, so that $N_a = 500$.

The fluctuating kinetic energy for the case $M_v = 1$ is shown in Figure 6.1. The average value over 18 000 timesteps, ignoring the first 2000 steps, is the expected $\langle E_k \rangle = 1.5000$, with $\sigma(E_k) = 0.0028$.

Pressure could be displayed similarly. A more revealing result, however, is the way the region edge length varies with time; this is shown for

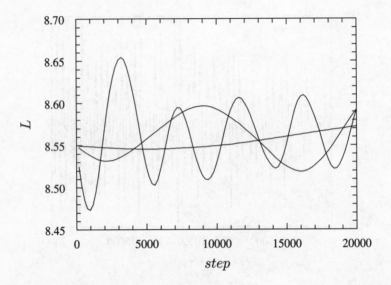

Fig. 6.2 Region edge fluctuations; the curves are for $M_v = 0.01$, 0.1, and 1.0 –
the smaller M_v the higher the frequency.

Table 6.1. *Pressure estimates*

M_v	$\langle P \rangle$	$\sigma(P)$
1.00	6.572	0.144
0.10	6.552	0.255
0.01	6.478	0.313

different values of M_v in Figure 6.2. The corresponding pressure results
for all three values of M_v are listed in Table 6.1. It is clear that in the
case of $M_v = 1$ the run is not long enough, so that even though the
apparent $\sigma(P)$ is the smallest the estimate of $\langle P \rangle$ could be incorrect due
to inadequate sampling. The additional degrees of freedom used in the
feedback methods introduce their own timescales that must be taken into
account in determining the run length.

6.2.3 *Controlled pressure with variable region shape*

In the above treatment of pressure feedback the simulation volume retained its cubic form, so that changes consist of uniform contractions and
expansions. The method is readily extended to the case of a simulation

region in which the lengths and directions of the edges are allowed to vary independently, subject to uniform external pressure [par80, nos83] (an even more general case where the applied stress components are specified separately can also be handled [par81]). The more flexible approach allows for the size and shape changes needed to accommodate lattice formation on freezing and for the study of structural phase transitions between different crystalline states. We will outline the mathematical formulation of the problem (omitting temperature feedback) but not use it in a case study.

Once again, scaled coordinates are introduced, but are now defined using a more general linear transformation:

$$r' = Hr \tag{6.24}$$

where the transformation matrix $H = (h_{\mu\nu}) = (c_1, c_2, c_3)$ is defined in terms of the vectors $\{c_\mu\}$ specifying the edges of the MD region, and the volume is $V = c_1 \cdot c_2 \times c_3 = \det H$. A metric tensor $G = H^T H$ can be introduced, so that $r'^2_{ij} = r^T_{ij} G r_{ij}$. The scaled coordinates span the unit cube and periodic images have coordinates $H(r + (n_x, n_y, n_z))$. Note the standard relation between spatial derivatives in the two coordinate systems, $\partial/\partial r' = (H^T)^{-1} \partial/\partial r$. The distortion of the simulation region is limited by the requirement that the interaction range does not exceed half the smallest region dimension.

The Lagrangian for this system is

$$\mathcal{L} = \tfrac{1}{2} m \sum_i \dot{r}_i^T G \dot{r}_i - \sum_{i<j} u(H r_{ij}) + \tfrac{1}{2} M_v \sum_{\mu\nu} \dot{h}_{\mu\nu}^2 - PV \tag{6.25}$$

so that the Lagrange equations for the coordinates are

$$\ddot{r}_i = H^{-1} F_i / m - G^{-1} \dot{G} \dot{r}_i \tag{6.26}$$

In the isotropic case this reduces to the earlier result, because $H^{-1} = V^{1/3} I$ and $G^{-1} \dot{G} = 2\dot{V}/3V$. The Lagrange equations for the components of H are

$$M_v \ddot{h}_{\mu\nu} = m \sum_i [H \dot{r}_i]_\mu \dot{r}_{vi} + \sum_{i<j} f_{\mu ij} r_{vij} - P \frac{\partial V}{\partial h_{\mu\nu}} \tag{6.27}$$

If we introduce an additional matrix

$$U = (u_{\mu\nu}) = \left(\frac{\partial V}{\partial h_{\mu\nu}} \right) = V(H^{-1})^T = (c_2 \times c_3, c_3 \times c_1, c_1 \times c_2) \tag{6.28}$$

then because

$$r_{vi} = V^{-1} \sum_{\mu} u_{\mu v} [Hr_i]_{\mu} \qquad (6.29)$$

the equation of motion (6.27) can be expressed concisely as

$$M_v \ddot{H} = (P - PI)U \qquad (6.30)$$

where P is the pressure tensor.

6.3 Constraint methods

6.3.1 Constant temperature

The alternative to feedback control is the use of mechanical constraints. Enforcing constant temperature amounts to introducing a nonholonomic constraint into the equations of motion in order to fix the kinetic energy – in effect this serves as a mathematical thermostat [hoo82, eva83a]. The justification for this arises not from Hamilton's variational principle, but from another formulation of mechanics known as Gauss's principle of least constraint [eva83b], which states that $\sum_i m_i(\ddot{r}_i - F_i/m_i)^2$ is minimized by the constrained motion. If the nonholonomic constraints are nonlinear but homogeneous functions of velocity (as is the case here), the results are formally the same as those which the variational principle would have produced [ray72]. The equilibrium properties of this isothermal system can be shown to be those of the canonical ensemble [eva84], but the dynamics must once again be interpreted with care since the motion is no longer Newtonian.

Since there are $3N_a$ degrees of freedom (we ignore the three lost to momentum conservation), the constraint equation designed to ensure constant temperature is

$$\tfrac{1}{2}m \sum_{i=1}^{N_a} \dot{r}_i^2 = N_a E_k \qquad (6.31)$$

The constrained equation of motion is

$$\ddot{r}_i = F_i/m + \alpha \dot{r}_i \qquad (6.32)$$

and since $\dot{E}_k = 0$, or equivalently $\sum_i \dot{r}_i \cdot \ddot{r}_i = 0$, it follows that the value of the Lagrange multiplier α is

$$\alpha = -\frac{\sum_i \dot{r}_i \cdot F_i}{m \sum_i \dot{r}_i^2} \tag{6.33}$$

If the thermostat is used together with the PC method the following function should be called from `SingleStep` immediately after the force evaluation:

```
ApplyThermostat () {
   real s1, s2, vFac;
   int k, n;
   s1 = s2 = 0.;
5  for (n = 1; n <= nAtom; n ++) {
      for (k = 1; k <= NDIM; k ++) {
         s1 = s1 + rv[k][n] * ra[k][n];
         s2 = s2 + Sqr (rv[k][n]);
      } }
10 vFac = - s1 / s2;
   for (n = 1; n <= nAtom; n ++) {
      for (k = 1; k <= NDIM; k ++)
         ra[k][n] = ra[k][n] + vFac * rv[k][n];
   }
15 }
```

The constant-temperature approach can also be combined with the leapfrog integrator, but because of the nature of the leapfrog method the isothermal condition is actually embedded into the integration procedure [bro84]. The isothermal version of the leapfrog velocity equation is readily seen to be (for $m = 1$)

$$\dot{r}_i(t + h/2) = (1 + \alpha h)\dot{r}_i(t - h/2) + h(1 + \alpha h/2)F_i(t) \tag{6.34}$$

where $\alpha = -\sum_i \dot{r}_i(t) \cdot F_i(t)/\sum_i \dot{r}_i(t)^2$, and $\dot{r}_i(t) = \dot{r}_i(t - h/2) + (h/2)F_i(t)$. The other leapfrog equation is unaltered. The isothermal version of the leapfrog function is then

```
LeapfrogStep () {
   real a, s1, s2, t;
   int k, n;
   s1 = 0.;    s2 = 0.;
5  for (n = 1; n <= nAtom; n ++) {
      for (k = 1; k <= NDIM; k ++) {
         t = rv[k][n] + 0.5 * deltaT * ra[k][n];
         s1 = s1 + t * ra[k][n];    s2 = s2 + Sqr (t);
      } }
10 a = - s1 / s2;
   for (n = 1; n <= nAtom; n ++) {
```

```
        for (k = 1; k <= NDIM; k ++) {
          rv[k][n] = (1. + a * deltaT) * rv[k][n] +
             deltaT * (1. + 0.5 * a * deltaT) * ra[k][n];
15        r[k][n] = r[k][n] + deltaT * rv[k][n];
      } }
    }
```

The temperature constraint is only preserved to the accuracy of the numerical integration. Any temperature drift must be corrected periodically (at intervals of stepAdjustTemp) by velocity rescaling, although this can now be based on the instantaneous value of E_k rather than on an average over the preceding timesteps as in Chapter 3. The function used for this is

```
    AdjustTemp () {
      real vFac;
      int k, n;
      vvSum = 0.;
5     for (n = 1; n <= nAtom; n ++) {
        for (k = 1; k <= NDIM; k ++)
          vvSum = vvSum + Sqr (rv[k][n]);
      }
      vFac = vMag / sqrt (vvSum / nAtom);
10    for (n = 1; n <= nAtom; n ++) {
        for (k = 1; k <= NDIM; k ++)
          rv[k][n] = rv[k][n] * vFac;
      }
    }
```

In Table 6.2 we make a limited comparison between the constant energy MD results obtained earlier for an $N_a = 500$ LJ system (with $r_c = 2.2$) and the corresponding results obtained using isothermal dynamics. Leapfrog integration is used, with $\Delta t = 0.005$ and $\rho = 0.8$. The agreement is satisfactory, as more careful tests will confirm.

6.3.2 *Constant pressure and temperature*

The idea of using mechanical constraints to fix thermodynamic properties can be extended to include pressure as well [eva84]. The problem is formulated in terms of scaled coordinates, exactly as in the feedback case, and so the unconstrained Lagrangian is

$$\mathscr{L} = \tfrac{1}{2} m V^{2/3} \sum_i \dot{r}_i^2 - \sum_{i<j} u(V^{1/3} r_{ij}) \tag{6.35}$$

Table 6.2. *Comparison between constant-energy and constant-temperature results for an LJ fluid*

	const-E	const-T		const-E		const-T	
T	E	$\langle E \rangle$	$\sigma(E)$	$\langle P \rangle$	$\sigma(P)$	$\langle P \rangle$	$\sigma(P)$
0.8	-3.960	-3.901	0.036	0.899	0.124	1.001	0.177
1.0	-3.472	-3.419	0.041	1.923	0.134	2.028	0.209
1.2	-2.974	-2.957	0.049	2.883	0.161	2.913	0.223
1.4	-2.528	-2.488	0.056	3.756	0.172	3.815	0.253
1.6	-2.063	-2.039	0.058	4.644	0.197	4.662	0.264

Here the kinetic energy is again defined using velocities measured relative to the rate at which the region size changes. The equations for the T and P constraints are

$$\frac{1}{2}mV^{2/3}\sum_i \dot{r}_i^2 = N_a E_k \tag{6.36}$$

$$mV^{2/3}\sum_i \dot{r}_i^2 + V^{1/3}\sum_{i<j} r_{ij} \cdot f_{ij} = 3PV \tag{6.37}$$

The equation of motion is

$$\ddot{r}_i = F_i/mV^{1/3} + (\alpha' - 2\gamma)\dot{r}_i \tag{6.38}$$

where $\gamma = \dot{V}/3V$ is the dilation rate and α' the Lagrange multiplier. From the constant-T condition $\dot{E}_k = 0$ we have

$$\sum_i \dot{r}_i \cdot \ddot{r}_i + \gamma \sum_i \dot{r}_i^2 = 0 \tag{6.39}$$

and it follows that

$$\alpha \equiv \alpha' - \gamma = -\frac{\sum_i \dot{r}_i \cdot F_i}{mV^{1/3}\sum_i \dot{r}_i^2} \tag{6.40}$$

so that (6.38) can be expressed in terms of α rather than α'.

The constant-P condition provides the means for computing γ. Starting with

$$\frac{d}{dt}(PV) = P\dot{V} = 3\gamma PV = \frac{1}{3}\sum_{i<j} \frac{d}{dt}\left(r'_{ij} \cdot f_{ij}\right) \tag{6.41}$$

and noting that for pair potentials (such as LJ) that depend only on distance $r' \cdot f = -r' du(r')/dr'$, we obtain

$$d(r' \cdot f)/dt = -\psi r' \cdot \dot{r}' \tag{6.42}$$

where we have defined

$$\psi(r) = d^2u(r)/dr^2 + r^{-1}du(r)/dr \tag{6.43}$$

Thus,

$$9\gamma PV = -V^{2/3}\sum_{i<j}\psi_{ij}\left(\mathbf{r}_{ij}\cdot\dot{\mathbf{r}}_{ij} + \gamma r_{ij}^2\right) \tag{6.44}$$

where ψ_{ij} denotes $\psi(r'_{ij})$, and for the LJ (or soft-sphere) potential:

$$\psi(r) = 144(4r^{-14} - r^{-8}) \tag{6.45}$$

Rearrangement of (6.44) yields an expression for γ in terms of known quantities, namely,

$$\gamma = -\frac{V^{2/3}\sum_{i<j}\psi_{ij}\mathbf{r}_{ij}\cdot\dot{\mathbf{r}}_{ij}}{9PV + V^{2/3}\sum_{i<j}\psi_{ij}r_{ij}^2} \tag{6.46}$$

In situations where the use of an interaction cutoff has a significant effect on P the estimated corrections can be included in the evaluation of γ.

The equations of motion need to be supplemented by the dilation equation (where $L = V^{1/3}$)

$$\dot{L} = \gamma L \tag{6.47}$$

This equation must be integrated numerically to obtain $L(t)$ at each timestep, given the current value of $\gamma(t)$. The pressure can initially be set to the required value and any subsequent small drift eliminated by solving the nonlinear equation $P(V) - P = 0$ to obtain the appropriate V. The solution is obtained using the Newton–Raphson method [pre92]; here this entails iterating the expression

$$L \leftarrow L\left(1 + \frac{P(V) - P}{3P(V) + \sum_{i<j}\psi_{ij}r_{ij}'^2/3V}\right) \tag{6.48}$$

and recomputing the interactions and pressure at each cycle, until $|P(V) - P|/P < \epsilon$. The equilibrium averages are those of the NPT ensemble [eva84].

Many of the implementation details are very similar to the feedback case treated previously. The cell method is used, and the required additions to ComputeForces (Chapter 3) are

```
    real ... w, wrv;
    ...
    dvirSum1 = dvirSum2 = 0.;
    ...
5   w = 144. * rri3 * (4. * rri3 - 1.) * rri;
```

```
   wrv = 0.;
   for (k = 1; k <= NDIM; k ++)
      wrv = wrv + dr[k] * (rv[k][j1] - rv[k][j2]);
   dvirSum1 = dvirSum1 + w * wrv;
10 dvirSum2 = dvirSum2 + w * rr;
```

The function `ApplyThermostat` needs the following change:

```
   for (k = 1; k <= NDIM; k ++)
      ra[k][n] = ra[k][n] / varL + (vFac / varL -
         dilateRate) * rv[k][n];
```

and `EvalProps` the addition

```
   vvSum = vvSum * Sqr (varL);
```

where varL is the value of L. The new variables used in these computations are

```
   real dilateRate, dilateRate1, dilateRate2, dvirSum1,
      dvirSum2, extPressure, tolPressure, varL, varLo,
      varLv, varLv1, varLv2;
   int maxEdgeCells, nPressCycle, stepAdjustPress;
```

and the inputs are

```
   RNAME (extPressure),
   RNAME (tolPressure),
   INAME (stepAdjustPress),
```

A new function is needed to evaluate the dilation rate γ:

```
   ApplyBarostat () {
     real vvS;
     int k, n;
     vvS = 0.;
5    for (n = 1; n <= nAtom; n ++) {
       for (k = 1; k <= NDIM; k ++)
          vvS = vvS + Sqr (rv[k][n]);
     }
     dilateRate = - dvirSum1 * varL / (3. * (vvS *
10      Sqr (varL) + virSum) + dvirSum2);
   }
```

The dilation equation (6.47) is solved by the PC method using functions for first-order equations based on the description in Chapter 3 (with $k = 3$):

```
   PredictorStepBox () {
     real c[] = {23.,-16.,5.}, div = 12.;
     int k;
     varLv = dilateRate * varL;    varLo = varL;
5    varL = varL + (deltaT / div) * (c[0] * varLv +
```

```
              c[1] * varLv1 + c[2] * varLv2);
        varLv2 = varLv1;    varLv1 = varLv;
        dilateRate2 = dilateRate1;    dilateRate1 = dilateRate;
        for (k = 1; k <= NDIM; k ++) {
10        region[k] = varL;    regionH[k] = ... ;
        }
    }

    CorrectorStepBox () {
        real c[] = {5.,8.,-1.}, div = 12.;
        int k;
        varLv = dilateRate * varL;
5       varL = varLo + (deltaT / div) * (c[0] * varLv +
              c[1] * varLv1 + c[2] * varLv2);
        for (k = 1; k <= NDIM; k ++) {
          region[k] = varL;  regionH[k] = ... ;
        }
10  }
```

The changes that must be made to SingleStep to accommodate the extra computations are

```
    PredictorStepBox ();
    ApplyBoundaryCond ();
    UpdateCellSize ();
    UnscaleCoords ();
5   ...
    ApplyBarostat ();
    ApplyThermostat ();
    ...
    CorrectorStepBox ();
10  ...
    nPressCycle = 0;
    if (stepCount % stepAdjustPress == 0) AdjustPressure ();
    if (stepCount % stepAdjustPress == 10) AdjustTemp ();
```

and in EvalProps:

```
    pressure = (vvSum + virSum) / (3. * pow (varL, 3.));
```

The reason for separating the pressure and temperature adjustments by several timesteps is to allow the system to settle down after the volume change; the effect of the delay can be seen in the results.

Pressure adjustments employ a Newton–Raphson procedure to modify the region size, as discussed earlier:

```
    AdjustPressure () {
        real rFac, w;
        int k, maxPressCycle, n;
        maxPressCycle = 20;
```

```
 5    if (fabs (pressure - extPressure) > tolPressure *
         extPressure) {
       UnscaleCoords ();
       vvSum = vvSum / Sqr (varL);
       for (nPressCycle = 0; nPressCycle < maxPressCycle;
10        nPressCycle ++) {
         UpdateCellSize ();
         ComputeForces ();
         w = 3. * pow (varL, 3.);
         pressure = (vvSum * Sqr (varL) + virSum) / w;
15        rFac = 1. + (pressure - extPressure) / (3. *
             pressure + dvirSum2 / w);
         for (n = 1; n <= nAtom; n ++) {
           for (k = 1; k <= NDIM; k ++)
             r[k][n] = r[k][n] * rFac;
20        }
         for (k = 1; k <= NDIM; k ++) {
           region[k] = region[k] * rFac;    regionH[k] = ... ;
         }
         varL = varL * rFac;
25        if (fabs (pressure - extPressure) < tolPressure *
             extPressure) break;
       }
       ScaleCoords ();
       vvSum = vvSum * Sqr (varL);
30   }
     }
```

The variable `maxPressCycle` is provided as a safety measure in the un-likely event of the method failing to converge. The counter `nPressCycle` is globally defined to allow its inclusion in the output. (The previous acceleration values used by the PC method are not modified following the volume change; since this change ought to be small the consequences of this omission should be negligible.)

The function `AdjustTemp` shown earlier requires the addition

```
vvSum = vvSum * Sqr (varL);
vFac = ...
```

Initialization now includes

```
InitBoxVars ();
ScaleCoords ();
ScaleVels ();
```

and the variables associated with the region size are initialized by

```
InitBoxVars () {
   varL = region[1];
```

Table 6.3. *Results from constant-PT run*

step	$\langle E \rangle$	$\langle E_k \rangle$	$\langle P \rangle$	L
1000	1.6359	1.4970	0.8207	11.2606
2000	1.6434	1.4971	0.8347	9.9923
4000	2.3309	1.4987	6.7297	8.5209
8000	2.3147	1.4979	6.5308	8.5642
12000	2.3139	1.4979	6.5351	8.5720
16000	2.3176	1.4979	6.5400	8.5819
20000	2.3212	1.4979	6.5437	8.6035

```
    varLv = varLv1 = varLv2 = 0.;
    dilateRate1 = dilateRate2 = 0.;
5 }
```

The variable `maxEdgeCells` is used as in the feedback case to allow for extra cells. Note that `varL` (the current region edge) is used instead of `varV` (the region volume in the feedback program) in converting between real and scaled coordinates.

The demonstration of this method uses a soft-sphere system with $N_a = 500$, $T = 1$, $\rho = 0.8$, $\Delta t = 0.002$, and other input data:

```
extPressure      6.5
stepAdjustPress  2000
stepAvg          500
tolPressure      0.001
```

Edited output of this run is shown in Table 6.3; the results include the current value of the region edge L. The pressure is adjusted every 2000 timesteps, but the drift is sufficiently small that only two cycles of the correction process are required (except on the very first call where 12 are needed). The typical value of $\sigma(P)$ for this run is 0.002.

6.4 Further work

6.1 By examining the relevant partition functions confirm that the equilibrium properties of these methods are those of the NVT and NPT ensembles.

6.2 If scaled coordinates are not used in the constrained pressure method examine the implications for periodic boundary processing.

6.3 Make a careful comparison of $E(T)$ and $P(T)$ measurements using feedback and basic MD methods.

6.4 Implement the pressure feedback simulation for the case of variable region shape [nos83].

6.5 A further extension of the feedback approach is to the case of fixed external stress [par81]; investigate the applications of this method.

6.6 Study the soft-sphere and LJ melting transitions at constant pressure; when the fluid freezes what is the crystal structure?

7

Nonequilibrium dynamics

7.1 Introduction

In the study of equilibrium behavior, MD is used to probe systems that, at least in principle, are amenable to treatment by statistical mechanics. The fact that statistical mechanics is generally unable to make much headway without resorting to simplification and approximation is merely a practical matter; the concepts and general relationships are extremely important even in the absence of closed-form solutions. When one departs from equilibrium very little theoretical guidance is available, and here MD really begins to fill the role of an experimental tool.

There are many nonequilibrium phenomena worthy of study, but MD applications have so far tended to concentrate on relatively simple systems, and the case studies in this chapter will focus on the simplest of problems. To be more specific, we will demonstrate two very different approaches to questions related to fluid transport. The first approach uses genuine Newtonian dynamics applied to spatially inhomogeneous systems, in which the boundaries play an essential role: simulations of fluids partly constrained by hard walls will be used to determine both shear viscosity and thermal conductivity. The second approach is based on a combination of modified equations of motion and fully homogeneous systems: the same transport coefficients will be measured, but since there are no explicit boundaries the dynamics must be altered in very specific ways to compensate for their absence.

7.2 Homogeneous and inhomogeneous systems

As computational tools both homogeneous and inhomogeneous nonequilibrium methods have their strengths and weaknesses. Before delving into the case studies, which include a sampling of both approaches, it is appropriate to point out the benefits and limitations of the different methods.

The reason for preferring homogeneous systems is that if physical walls can be eliminated (and replaced by periodic boundaries) all atoms perceive a similar environment. Inhomogeneous systems, on the other hand, must allow for perturbations to the structure and dynamics due to the presence of the walls. Furthermore, inhomogeneous systems may not exist at a uniquely defined temperature or density – essential if any comparison with experiment is to be made – as a consequence of the relatively large force needed to drive the mass or heat flow combined with the small system size. Not all problems offer the homogeneous alternative, although the more familiar transport coefficients can indeed be studied in this way.

The disadvantage of homogeneous nonequilibrium systems in general is the unphysical nature of the dynamics, for not only are the equations of motion modified in such a way that the desired transport coefficient emerges directly from linear response theory (actually a version of the theory extended to handle isothermal systems), they are also altered to mechanically suppress the heat generated by the applied force [eva84, eva90]. The method is therefore best regarded as a computational technique whose results are valid in the limit of zero applied force. The fact that each transport property requires a separate simulation because of the differing dynamical requirements leads to the question: if homogeneous systems are already being used why isn't it better to follow the more straightforward approach with Newtonian trajectories and autocorrelation functions as described in Chapter 5, where all the transport coefficients can be computed together? The historical answer focuses on the accuracy of results obtained for comparatively small systems, with non-Newtonian methods having a clear advantage. Whether this advantage remains, now that much more extensive simulation is possible, remains to be seen.

7.3 Direct measurement

7.3.1 Viscous flow

Two of the more elementary exercises in fluid dynamics, both with closed-form solutions, are Couette flow and Poiseuille flow. In planar Couette flow the fluid is confined between two parallel walls that slide relative to one another at a constant rate. An example of Poiseuille flow is a fluid forced to flow between two fixed walls. The walls are rough, so that a thin, stationary (relative to the wall) layer of fluid exists close to each wall. In each of these flow problems we can assume the system to be unbounded (or, for MD purposes, periodic) in the remaining two directions.

The viscous nature of the fluid requires sustained work to maintain motion. For Couette flow a force must be applied to keep the walls moving relative to one another, whereas for Poiseuille flow a pressure head or gravity-like force acts in the flow direction. This work is converted to heat that must be removed from the fluid through the walls to limit the temperature rise. Temperature will vary in the direction perpendicular to the walls, a reflection of the fact that heat generated in the interior must be transported to the walls. This is true in both experiment and simulation.

Once walls have been introduced explicitly into the problem, the question arises as to how realistically they have to be modeled. Real walls are complicated and can only be represented in an average sense because roughness is essentially a statistical notion; this is of little help when trying to develop a detailed microscopic simulation. All we require are walls that are sufficiently rough to ensure nonslip flow, but the precise way by which this is achieved is unlikely to affect the overall flow. One could, for example, use a layer of either fixed or tethered atoms that mimic the effect of a rough wall [ash75]; by adjusting the way the wall atoms are arranged the roughness can even be varied to a certain extent. While this scheme offers a semblance of reality, it presents a problem if the walls are also required to transfer heat in and out of the fluid; by using a thermostat applied to the tethered wall atoms this issue can also be resolved, but the question is whether such an intricate scheme is really necessary.

At the opposite extreme there are 'stochastic' walls [tro84]. Whenever an atom attempts to cross a wall it is reflected back into the interior; the effects of wall roughness and temperature are achieved by randomizing the direction of the reflected velocity and scaling its magnitude to match the wall temperature. The approach may appear rather simplistic and, if not used carefully, could interfere with the integration of the equations of motion, especially if the region near the wall is at relatively high density and temperature. Whether such boundary conditions actually work (they do) can only be established by trying them out.

This case study deals with Poiseuille flow; the Couette problem will be discussed in Section 7.4 in the context of homogeneous systems. The analytic results for (incompressible) Poiseuille flow are as follows. Assume that the two fixed walls lie in the xy-plane and that flow is in the x-direction. In terms of the normalized cross-stream coordinate z, where $0 \le z \le 1$, solving the Navier–Stokes and heat conduction equations [lan59] leads to polynomial velocity and temperature profiles:

$$v_x(z) = \frac{\rho g L_z^2}{2\eta} \left[\tfrac{1}{4} - (z - \tfrac{1}{2})^2 \right] \tag{7.1}$$

$$T(z) = T_w + \frac{\rho^2 g^2 L_z^4}{12\lambda\eta} \left[\tfrac{1}{16} - (z - \tfrac{1}{2})^4 \right] \tag{7.2}$$

where L_z is the channel width, T_w the wall temperature, and g the external field driving the flow. By dividing the simulation region into slices parallel to the walls and computing the mean flow velocity and temperature in each slice, the shear viscosity η and thermal conductivity λ can be determined by fitting second- and fourth-degree polynomials to the results.

In order to carry out this simulation we must first modify the neighbor-list generation function to allow for the fact that the z-boundaries are no longer periodic. In BuildNebrList (Chapter 3) the line

```
if (m2Z == 0 || m2Z > cells[3]) continue;
```

replaces the entire block

```
if (m2Z > cells[3]) ... else ...
```

This removes all reference to nonexistent cells beyond the walls during the search for interaction partners.

Flow rate depends directly on the force used to drive the fluid (experimentally the flow is more likely to be due to a pressure difference between the pipe inlet and outlet); the value needed for a given flow rate will have to be determined by experiment because it depends on both ρ and L_z. The variable corresponding to g is defined as

```
real gravField;
```

and its value is included with the input data:

```
RNAME (gravField),
```

As part of the interaction computation the following function must be called:

```
ComputeExternalForce () {
    int n;
    for (n = 1; n <= nAtom; n ++)
        ra[1][n] = ra[1][n] + gravField;
5 }
```

The presence of hard walls calls for a change in the usual boundary processing in the z-direction. Whenever an atom attempts to cross one of these walls it is reflected back into the interior; the magnitude of the new velocity is set to a fixed value corresponding to the wall temperature

and the direction is randomized. In addition, because the atom will
have slightly overshot the wall during the current timestep, it is moved
back inside (and away from the wall by a minute amount to avoid any
numerical problems). The revised version of the function is

```
ApplyBoundaryCond () {
  real e[NDIM + 1];
  int k, n, vSign;
  for (n = 1; n <= nAtom; n ++) {
5   for (k = 1; k <= NDIM - 1; k ++) {
      if (r[k][n] >= regionH[k])
          r[k][n] = r[k][n] - region[k];
      else if (r[k][n] < - regionH[k])
          r[k][n] = r[k][n] + region[k];
10  }
    vSign = 0;
    if (r[NDIM][n] >= regionH[NDIM]) {
      r[NDIM][n] = regionH[NDIM] * 0.9999;   vSign = -1;
    } else if (r[NDIM][n] < - regionH[NDIM]) {
15    r[NDIM][n] = - regionH[NDIM] * 0.9999;   vSign = 1;
    }
    if (vSign) {
      RandVec3 (e, &randSeed);
      for (k = 1; k <= NDIM; k ++) rv[k][n] = vMag * e[k];
20    if (rv[NDIM][n] * vSign < 0.)
        rv[NDIM][n] = - rv[NDIM][n];
  } }
}
```

In order to use the function for a two-dimensional version of this problem,
where it is the y-direction that is nonperiodic, the only change needed is
in the way random velocities are generated – the function InitVels in
Chapter 2 shows how this is done.

Analysis of the flow requires the construction of cross-stream v_x-
and T-profiles based on slices in the xy-plane; in the case of T each
profile value must be computed in a frame of reference moving with the
mean flow in that slice. However, rather than show the simple profile
computation we will introduce a more general scheme for computing
properties based on a two-dimensional grid subdivision of the simulation
region, with profiles being produced as a byproduct. This method will
prove useful in later work.

The following function performs the grid averaging using cells based on
the x- and z-coordinates (the y-coordinate is not involved). Depending on
the value of the argument opCode, the function initializes the arrays used
for collecting the results, accumulates the results for a single measurement,

or computes the final averages. The five quantities collected for each cell (the parameter NHIST – used for flexibility – is equal to 5) are the occupation count, the sums over the squared atomic velocities, and the sums of each of the velocity components. The final averaging produces the cell-averaged densities, temperatures, and velocities:

```
   GridAverage (int opCode) {
      real pSum;
      int c, hSize, j, n;
      hSize = sizeHistGrid[1] * sizeHistGrid[2];
5     if (opCode == 0) {
         for (j = 1; j <= NHIST; j ++) {
            for (n = 1; n <= hSize; n ++) histGrid[j][n] = 0.;
         }
      } else if (opCode == 1) {
10       for (n = 1; n <= nAtom; n ++) {
            c = (int) ((r[3][n] + regionH[3]) *
               sizeHistGrid[2] / region[3]) * sizeHistGrid[1] +
               (int) ((r[1][n] + regionH[1]) *
               sizeHistGrid[1] / region[1]) + 1;
15          histGrid[1][c] = histGrid[1][c] + 1.;
            histGrid[2][c] = histGrid[2][c] +
               Sqr (rv[1][n]) + Sqr (rv[2][n]) + Sqr (rv[3][n]);
            histGrid[3][c] = histGrid[3][c] + rv[1][n];
            histGrid[4][c] = histGrid[4][c] + rv[2][n];
20          histGrid[5][c] = histGrid[5][c] + rv[3][n];
         }
      } else if (opCode == 2) {
         pSum = 0.;
         for (n = 1; n <= hSize; n ++) {
25          if (histGrid[1][n] > 0.) {
               for (j = 2; j <= NHIST; j ++) histGrid[j][n] =
                  histGrid[j][n] / histGrid[1][n];
               histGrid[2][n] = (histGrid[2][n] -
                  (Sqr (histGrid[3][n]) + Sqr (histGrid[4][n]) +
30                Sqr (histGrid[5][n]))) / 3.;
               pSum = pSum + histGrid[1][n];
            } }
         pSum = pSum / hSize;
         for (n = 1; n <= hSize; n ++)
35          histGrid[1][n] = histGrid[1][n] / pSum;
      }
   }
```

The grid computation requires several additional variables, namely,

```
   real **histGrid;
   int sizeHistGrid[3], countGrid, limitGrid, stepGrid;
```

extra input data:

```
INAME (limitGrid),
INAME (sizeHistGrid),
INAME (stepGrid),
```

and array allocation (in `AllocArrays`):

```
histGrid = AllocMatR (NHIST, sizeHistGrid[1] *
   sizeHistGrid[2]);
```

The following code is added to `SingleStep`:

```
   if (stepCount >= stepEquil && (stepCount - stepEquil)
      stepGrid == 0) {
      countGrid = countGrid + 1;
      GridAverage (1);
5     if (countGrid % limitGrid == 0) {
         GridAverage (2);
         EvalProfile ();
         PrintProfile (stdout);
         GridAverage (0);
10    }
   }
```

and to `SetupJob` for initialization:

```
GridAverage (0);
countGrid = 0;
```

In this case study the grid results will be used to compute profiles, but other kinds of processing could be carried out that utilize the two-dimensional nature of the data, including graphics (Chapter 13).

The functions that extract the profiles from the grid data and output the results are as follows. An array `profileV` is used, allocated by

```
profileV = AllocVecR (sizeHistGrid[2]);
```

Here just v_x is shown, but T is treated similarly:

```
   EvalProfile () {
      int k, n;
      for (n = 1; n <= sizeHistGrid[2]; n ++) profileV[n] = 0.;
      for (n = 1; n <= sizeHistGrid[1] * sizeHistGrid[2];
5        n ++) {
         k = (n - 1) / sizeHistGrid[1] + 1;
         profileV[k] = profileV[k] + histGrid[3][n];
      }
      for (n = 1; n <= sizeHistGrid[2]; n ++)
10       profileV[n] = profileV[n] / sizeHistGrid[1];
   }
```

```
  PrintProfile (FILE *fp) {
    real zVal;
    int n;
    fprintf (fp, "v profile\n");
5   for (n = 1; n <= sizeHistGrid[2]; n ++) {
      zVal = (n - 0.5) / sizeHistGrid[2];
      fprintf (fp, "%.2f %.3f\n", zVal, profileV[n]);
    }
  }
```

A three-dimensional soft-sphere system is used in this study, but the simulation region is not cubic in shape. Since the sheared flow develops in the xz-plane the two longer region edges are assigned to the x- and z-directions; in this way we achieve a relatively large area for examining the flow details while retaining the three-dimensional nature of the system. We use $\rho = 0.8$, $T = 1$, $\Delta t = 0.005$, and additional input data that includes

initUcell	20 5 20
gravField	0.1
stepAvg	2000
stepEquil	1000
stepLimit	31000
limitGrid	100
sizeHistGrid	1 50
stepGrid	50

The parameter gravField takes values between 0.1 and 0.4. The initial state is a simple cubic lattice, so that $N_a = 2000$; at the start of the run there is no flow.

A certain amount of time is required for the system to achieve steady flow. Table 7.1 shows how the mean kinetic energy initially increases with time, eventually reaching a steady value after about 15 000 timesteps (there is even a suggestion of overshoot). The flow velocity and temperature measurements made over the last 15 000 steps of each run appear in Figure 7.1. Least-square polynomial fits based on (7.1) and (7.2) are also shown. Estimates of η and λ derived from the fits are listed in Table 7.2.

The fits are not forced to comply with the boundary conditions, namely, $v_x = 0$, $T = 1$, because of the limited spatial resolution of the coarse-grained measurements. Other examples of problems that might complicate the analysis include a small amount of slip at the walls and density variations across the flow. The fact that the transport coefficients can depend on ρ, T, and even on the local shear rate, contributes to the error when trying to fit to analytic results that assume that η and λ are

Table 7.1. *Kinetic energy during the runs*

step	$g = 0.1$	0.2	0.3	0.4
3000	1.39	2.35	4.09	6.46
7000	2.72	7.71	16.21	26.99
11000	3.88	11.77	22.89	35.33
15000	4.42	12.91	23.24	34.24
19000	4.63	12.53	22.30	33.65
23000	4.66	12.13	22.28	33.25
27000	4.60	12.36	21.75	33.41
31000	4.57	12.34	22.16	33.22

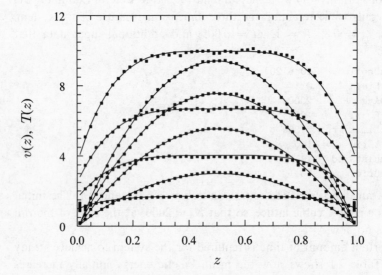

Fig. 7.1 Plots of flow velocity and temperature for field values $g = 0.1, 0.2, 0.3,$ and 0.4; polynomial fits are included.

Table 7.2. *Transport coefficients*

g	η	λ
0.1	1.14	5.0
0.2	1.22	5.3
0.3	1.34	6.3
0.4	1.43	6.9

constant. Despite all these reservations the fits obtained here appear remarkably good. On the other hand, there are too many variables involved to determine the reason why η and λ vary with flow rate.

7.3.2 *Heat transport*

We now turn to another example of the use of MD to mimic a real experiment, in this case heat flow between two parallel walls maintained at different temperatures [ten82]. If heat is transferred only by conduction, from Fourier's law [mcq76] the thermal conductivity is the ratio of the rate of heat (kinetic energy) transfer across the system to the temperature gradient:

$$\lambda = \frac{\Delta E_{k,in} + \Delta E_{k,out}}{2t_m A \Delta T / L_z} \tag{7.3}$$

where t_m is the measurement interval, ΔT the temperature difference, and $A = L_x L_y$ the wall area.

The program is similar to the previous one, differing only in that the external force is absent and the walls are maintained at different temperatures. In the modified version of ApplyBoundaryCond shown below, the z-boundary processing has been altered to handle two distinct wall temperatures and evaluate the total heat transferred in and out of the system as the impacting atoms exchange energy with the constant-temperature walls:

```
    ApplyBoundaryCond () {
       real e[NDIM + 1], vNew, vvNew, vvOld;
       int k, n, vSign;
       for (n = 1; n <= nAtom; n ++) {
5          ... (as before) ...
          if (vSign) {
             vvOld = 0.;
             for (k = 1; k <= NDIM; k ++)
                vvOld = vvOld + Sqr (rv[k][n]);
10           if (vSign > 0) vNew = sqrt (NDIM * wallTempHi);
             else vNew = sqrt (NDIM * wallTempLo);
             RandVec3 (e, &randSeed);
             for (k = 1; k <= NDIM; k ++) rv[k][n] = vNew * e[k];
             vvNew = 0.;
15           for (k = 1; k <= NDIM; k ++)
                vvNew = vvNew + Sqr (rv[k][n]);
             sEnergyTrans = sEnergyTrans + 0.5 * vSign *
                (vvNew - vvOld);
             if (rv[NDIM][n] * vSign < 0.)
```

```
20            rv[NDIM][n] = - rv[NDIM][n];
   } }
 }
```

New variables are

```
real sEnergyTrans, wallTempHi, wallTempLo;
```

and the extra input values are

```
RNAME (wallTempHi),
RNAME (wallTempLo),
```

Heat transfer measurements require additions to `AccumProps`, namely,

```
sEnergyTrans = 0.;
```

when clearing the accumulated values, and

```
sEnergyTrans = 0.5 * sEnergyTrans / (stepAvg * deltaT *
   region[1] * region[2] * ((wallTempHi - wallTempLo) /
   region[3]));
```

when computing the averages; this value (output by `PrintSummary`) is the estimate of λ.

The system used is similar to the previous case, with input data

```
wallTempHi      1.5
wallTempLo      1.0
sizeHistGrid    1 25
```

and the parameter `wallTempHi` varying between 1.5 and 3.0. The simulation is started with the fluid at uniform temperature. Achieving a steady state once again requires a certain amount of time, after which measurement can begin.

The temperature measurements made over the latter 15 000 steps of each run appear in Figure 7.2 (the z-coordinate is normalized). Linear least-square fits are shown; the fits ignore the three values closest to each wall where the deviations from linearity are concentrated. Estimates of λ derived from the fits and the measured heat flow are 5.60, 5.88, 5.45, and 5.54 for $\Delta T = 0.5$, 1.0, 1.5, and 2.0 respectively; these may be compared with the earlier Poiseuille flow values.

As in the case of Poiseuille flow, the fact that the system is inhomogeneous can create problems when trying to interpret the results. Here a kind of thermal boundary layer exists close to each wall where the fluid temperature varies more rapidly than in the bulk, so that the effective thermal gradient is overestimated. Failure to account for this leads to λ-estimates that are some 30% too small. Temperature-dependent density variations may also create problems.

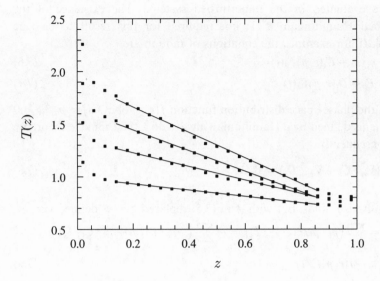

Fig. 7.2 Temperature profiles for ΔT between 0.5 and 2.0; linear fits to the interior values are included.

7.4 Modified dynamics

7.4.1 Linear response theory

The question addressed by linear response theory [han86b, eva90] can be stated in the following way. Given a system with Hamiltonian $\mathscr{H}_0$, evaluate the change in some dynamical variable $B(t)$ caused by an external field $F_e(t)$ applied starting at $t = 0$, whose effect on the system can be expressed schematically in terms of atomic coordinates and momenta as

$$\mathscr{H} = \mathscr{H}_0 - A(\boldsymbol{r}, \boldsymbol{p})F_e(t) \tag{7.4}$$

The actual coupling of F_e to the system may be more general than the form suggested here, with vector or tensor products being involved. The step function is just one possible form of perturbation, and sinusoidal and delta functions are also of interest.

Assuming that the effect of F_e is small enough to permit a linear perturbation treatment, an analysis based on the Liouville equation then leads to

$$\langle B(t)\rangle - \langle B(0)\rangle = \frac{1}{k_{\mathrm{B}}T} \int_0^t \langle B(t - t')\dot{A}(0)\rangle_0 F_e(t')dt' \tag{7.5}$$

This result is valid in the limit $F_e \to 0$, with $\langle \ldots \rangle_0$ denoting an equilibrium average evaluated in the unperturbed system. The existence of the perturbed Hamiltonian $\mathscr{H}$ is not required for this result to be true [eva84]. If, for example, the equations of motion are

$$\dot{r}_i = p_i/m + C_i(r,p)F_e(t) \tag{7.6}$$

$$\dot{p}_i = F_i + D_i(r,p)F_e(t) \tag{7.7}$$

then if the phase-space distribution function $f(r,p)$ obeys $df(r,p)/dt = 0$ (as systems defined by a Hamiltonian always do – the Liouville theorem), and consequently,

$$\sum_i \left(\nabla_{r_i} \cdot C_i + \nabla_{p_i} \cdot D_i \right) = 0 \tag{7.8}$$

the result also holds, but with $\dot{A}$ in (7.5) replaced by $-J$ defined via

$$\dot{\mathscr{H}}_0 = \sum_i \left(\dot{p}_i \cdot p_i/m - \dot{r}_i \cdot F_i \right) = -\sum_i \left(-p_i \cdot D_i/m + F_i \cdot C_i \right) F_e(t)$$

$$\equiv -J(r,p)F_e(t) \tag{7.9}$$

Since the applied force F_e performs mechanical work on the system the temperature rises, and equilibrium is never attained. To eliminate this problem [mor85] a thermostat is included in the dynamics (as in Chapter 6) by adding a term αp_i to (7.7); constant kinetic energy is assured if the value of the Lagrange multiplier is

$$\alpha = -\frac{\sum_i (F_i + D_i F_e) \cdot p_i}{\sum_i p_i^2} \tag{7.10}$$

(Because this analysis uses Hamilton's equations of motion, the momenta play a prominent role in the analysis, but we will dispense with them shortly.)

Transport coefficients can be evaluated by applying the appropriate force F_e and examining the behavior in the limit $F_e \to 0$. If the transport coefficient can be expressed as the integrated autocorrelation function of some dynamical quantity J:

$$Q = \frac{1}{k_B T} \int_0^\infty \langle J(t)J(0) \rangle dt \tag{7.11}$$

then, provided the perturbation is designed so that linear response theory yields an expression that is formally identical to the definition of the transport coefficient (7.11) – in other words $B = J$ – we obtain

$$Q = \lim_{F_e \to 0} \lim_{t \to \infty} \langle J(t) \rangle / F_e \tag{7.12}$$

The order of the limits is important, with the large-t results obtained at finite F_e being extrapolated to $F_e = 0$. Since the goal is to use this definition for Q in constant-temperature simulations, but the usual formulation of linear response theory assumes constant energy (Newtonian) dynamics, the theory must be extended to cover this situation. When this is done [mor85, eva90] the same expressions appear, but with the averages now evaluated at constant temperature. In the remainder of this section we will use this approach to evaluate the two transport coefficients considered previously – shear viscosity and thermal conductivity.

7.4.2 Shear viscosity

We consider the case of Couette flow in which the fluid undergoes sheared flow due to boundary walls that are in relative motion. The equations of motion used in this shear viscosity study are based on the constant-temperature dynamics described in Chapter 6. A small but significant change is required in order to ensure that temperature is correctly defined in the presence of flow, and this affects the velocity terms in the Lagrange multiplier used for the thermostat.

The usual microscopic definition of temperature in terms of mean-square velocity assumes that there is no overall motion; any local flow must be subtracted from the velocities before using them to evaluate temperature (similar situations have been encountered in earlier case studies). The same holds true for the velocities used in the thermostat. However, knowing the bulk flow to an accuracy suitable for use in the equations of motion implies that the problem has already been solved; this circularity can be removed by assuming the nature of the flow, and only later checking to see whether consistent results are obtained. A less reliable alternative is to evaluate local flow by means of coarse-grained averaging, and then use the results in the equations of motion; such an approach is unstable to any fluctuations in the flow because these variations are interpreted by the equations of motion as temperature fluctuations that must be suppressed.

In this study we impose the reasonable requirement that the MD flow obeys the linear velocity profile known from the exact solution of the continuum problem. Assuming it is the z-boundaries that are in motion, then if the relative velocity of the walls is γL_z, the shear rate dv_x/dz has the constant value γ. The thermostated equation of motion is then [eva84]

$$\ddot{\mathbf{r}}_i = \mathbf{F}_i/m + \alpha(\dot{\mathbf{r}}_i - \gamma r_{zi}\mathbf{e}_x) \tag{7.13}$$

where e_x is a unit vector in the x-direction. The value of the Lagrange multiplier α follows from the constant temperature constraint:

$$\alpha = -\frac{\sum_i (\dot{r}_i - \gamma r_{zi} e_x) \cdot (F_i/m - \gamma \dot{r}_{zi} e_x)}{\sum_i (\dot{r}_i - \gamma r_{zi} e_x)^2} \tag{7.14}$$

The equation of motion (7.13) assumes that the linear velocity profile has already been established; creating the initial sheared flow is most readily done as part of the initial conditions, and from the more formal point of view this amounts to applying an impulse of the correct size and direction to each atom at $t = 0$. The sliding boundaries, in the form of a special type of boundary condition (see below), maintain the constant shear rate. The constant-temperature version of linear response theory for this problem provides an expression for η based on the pressure tensor

$$\eta = -\lim_{\gamma \to 0} \lim_{t \to \infty} \langle P_{xz} \rangle / \gamma \tag{7.15}$$

To show that (7.13) corresponds to the more general form given by (7.6) and (7.7) we define the momentum measured relative to the local flow $p_i/m = \dot{r}_i - \gamma r_{zi} e_x$; the first-order equations are then

$$\dot{r}_i = p_i/m + \gamma r_{zi} e_x \tag{7.16}$$

$$\dot{p}_i = F_i - \gamma p_{zi} e_x + \alpha p_i \tag{7.17}$$

exactly as required (if the thermostat is ignored).

The boundaries are periodic, but of a special form to accommodate the uniformly sheared flow [lee72]. The idea is to replace sliding walls by sliding replica systems: layers of replicas that are adjacent in the z-direction move with a relative velocity $\gamma L_z e_x$, an arrangement designed to ensure periodicity at shear rate γ. An atom crossing a z-boundary requires special treatment because the x-components of position and velocity are both discontinuous (not for the replica system just entered but relative to the opposite side of the region itself into which the atom is actually inserted). The velocity change whenever a $\pm z$-boundary is crossed is $\mp \gamma L_z e_x$, and the coordinate change is $\mp d_x e_x$, where the total relative displacement of the neighboring replicas – only meaningful over the range $(-L_x/2, L_x/2)$ – is given by

$$d_x = (\gamma L_z t + L_x/2) \bmod L_x - L_x/2 \tag{7.18}$$

Note that because the x-coordinate changes when a z-boundary is crossed, an additional correction for periodic wraparound in the x-direction may be needed. Interactions that occur between atoms separated by the z-

boundary require an offset value $-d_x$ to be included in the distance computation.

When using the cell method for the interaction calculation, the range of neighbor cells in the x-direction for adjacent cells on opposite sides of the z-boundary must extend over four cells, rather than the usual three, to allow for the fact that the cell x-edges of the sliding replicas are not usually aligned. If there are M_x cells on an edge the additional cell offset across the $+z$-boundary is

$$\Delta m_x = \left\lfloor M_x \left(1 - d_x/L_x\right)\right\rfloor - M_x \qquad (7.19)$$

Taking these considerations into account, the modified form of the interaction calculation is as follows:

```
ComputeForces () {
  ...
  int ... bdyOffsetX, offsetHi,
    iofX[] = {0,0,1,1,0,-1,0,1,1,0,-1,-1,-1,0,1,2,2,2},
5   iofY[] = {0,0,0,1,1,1,0,0,1,1,1,0,-1,-1,-1,-1,0,1},
    iofZ[] = {0,0,0,0,0,0,0,1,1,1,1,1,1,1,1,1,1,1,1};
  ...
  bdyOffsetX = (int) (cells[1] * (1. - boundarySlide /
    region[1])) - cells[1];
10  pressureTensorXZ = 0.;
  for (m1Z = ...
    ...
    offsetHi = 14;
    if (m1Z == cells[3]) offsetHi = 17;
15  for (offset = 1; offset <= offsetHi; offset ++) {
      m2X = m1X + iofX[offset];
      if (m1Z == cells[3] && iofZ[offset] == 1) {
        m2X = m2X + bdyOffsetX;
        shift[1] = boundarySlide;
20      if (m2X > cells[1]) {
          m2X = m2X - cells[1];
          shift[1] = shift[1] + region[1];
        } else if (m2X <= 0) {
          m2X = m2X + cells[1];
25        shift[1] = shift[1] - region[1];
        }
      } else {
        shift[1] = 0.;
        if (m2X > cells[1]) {
30          ... (regular treatment) ...
        }
      }
```

```
      m2Y = ...
      . . .
35    pressureTensorXZ = pressureTensorXZ +
         fcVal * dr[1] * dr[3];
      . . .
```

The quantity boundarySlide – corresponding to d_x – is computed at the start of SingleStep:

```
boundarySlide = shearRate * region[3] * timeNow +
   regionH[1];
boundarySlide = boundarySlide - (int) (boundarySlide /
   region[1]) * region[1] - regionH[1];
```

The sliding boundary conditions are handled by a modified version of ApplyBoundaryCond that treats z-boundary crossings in the required way:

```
ApplyBoundaryCond () {
  int k, n;
  for (n = 1; n <= nAtom; n ++) {
    for (k = 1; k <= NDIM - 1; k ++) {
5       ... (usual treatment) ...
    }
    if (r[NDIM][n] >= regionH[NDIM]) {
      r[1][n] = r[1][n] - boundarySlide;
      if (r[1][n] < regionH[1])
10         r[1][n] = r[1][n] + region[1];
      rv[1][n] = rv[1][n] - shearRate * region[NDIM];
      r[NDIM][n] = r[NDIM][n] - region[NDIM];
    } else if (r[NDIM][n] < - regionH[NDIM]) {
      r[1][n] = r[1][n] + boundarySlide;
15       if (r[1][n] >= regionH[1])
        r[1][n] = r[1][n] - region[1];
      rv[1][n] = rv[1][n] + shearRate * region[NDIM];
      r[NDIM][n] = r[NDIM][n] + region[NDIM];
  } } }
```

New variables introduced are

```
real boundarySlide, pressureTensorXZ, shearRate, shearVisc,
   sShearVisc, ssShearVisc, vvSumXZ;
```

and there is an additional input data item:

```
RNAME (shearRate),
```

The thermostat plays a key role in this approach, and temperature must be evaluated with respect to the local flow to adhere to the correct definition. We assume that local flow is determined by the constant shear rate γ. The changes to `ApplyThermostat` – PC integration is used – involve the x-component of velocity being singled out for special treatment as in (7.14):

```
s1 = s1 + (rv[1][n] - shearRate * r[NDIM][n]) *
    (ra[1][n] - shearRate * rv[NDIM][n]);
s2 = s2 + Sqr (rv[1][n] - shearRate * r[NDIM][n]);
...
ra[1][n] = ra[1][n] + vFac * (rv[1][n] -
    shearRate * r[NDIM][n]);
```

Similar modification is required to `rv[1][n]` in `AdjustTemp`. The initial velocities include the uniform shear flow; the change to `InitVels` is

```
rv[1][n] = rv[1][n] + shearRate * r[NDIM][n];
```

The value of $\eta(\gamma)$ is computed by

```
EvalShearVisc () {
  pressureTensorXZ = (pressureTensorXZ + vvSumXZ) *
      density / nAtom;
  shearVisc = - pressureTensorXZ / shearRate;
}
```

with the quantity `vvSumXZ` evaluated in `EvalProps`:

```
vvSumXZ = 0.;
for (n = 1; n <= nAtom; n ++) {
  ...
  vvSumXZ = vvSumXZ + (rv[1][n] - shearRate * r[NDIM][n]) *
      rv[NDIM][n];
```

The computation of `vvSum` in `EvalProps` also requires a change:

```
if (k == 1) v = rv[1][n] - shearRate * r[NDIM][n];
else v = rv[k][n];
```

The averaging of `shearVisc` is included in `AccumProps`:

```
sShearVisc = ssShearVisc = 0.;
...
sShearVisc = sShearVisc + shearVisc;
ssShearVisc = ssShearVisc + Sqr (shearVisc);
...
sShearVisc = sShearVisc / stepAvg;
ssShearVisc = sqrt (ssShearVisc / stepAvg -
    Sqr (sShearVisc));
```

and the appropriate output added to `PrintSummary`.

The runs used for the shear viscosity measurements include the following input data:

```
initUcell    4 4 4
```

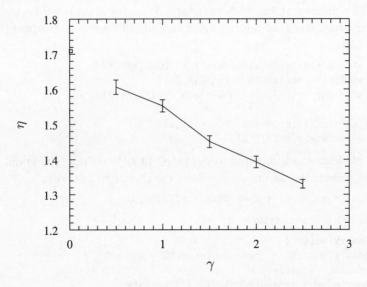

Fig. 7.3 Shear viscosity for various shear rates γ; the equilibrium value appears at $\gamma = 0$.

```
shearRate        0.5
stepAdjustTemp   5000
stepAvg          2000
stepEquil        2000
stepLimit        22000
```

with the value of `shearRate` ranging between 0.5 and 2.5. An FCC initial state is used, so that $N_a = 256$. In addition, $\rho = 0.8$, $T = 1$, and $\Delta t = 0.002$.

Estimates of $\eta(\gamma)$ are shown in Figure 7.3; the error bars show the standard deviation of the mean $\sigma(\langle \eta \rangle)$ computed from the block averages that are produced every 2000 steps. Taking into account the more extensive computation required in the equilibrium case (Chapter 5) and the larger uncertainty in the final estimate, the advantage of the nonequilibrium approach here is obvious. Further discussion appears in [eva90, fer91, lie92].

7.4.3 Thermal conductivity

The thermal conductivity is another example of a transport coefficient that can be measured by a similar approach, assuming a suitable equation of motion can be synthesized. In this case a fictitious external field F_e of a rather unusual kind is introduced [eva82, gil83]: it has the effect of driving atoms with a higher than average energy in the direction of F_e, while those with a lower energy are driven in the opposite direction; in other words F_e generates heat flow and so, at least for small values of the field, produces the effect of an imposed temperature difference.

The additional force acting on each atom is defined as

$$F'_i = e_i F_e + \tfrac{1}{2} \sum_{j(\neq i)} f_{ij}(r_{ij} \cdot F_e) - \frac{1}{2N_a} \sum_{j \neq k} f_{jk}(r_{jk} \cdot F_e) \qquad (7.20)$$

where e_i is the excess energy of atom i (the signs differ from [eva82] because of the way r_{ij} is defined). Here F'_i has been chosen so that in terms of the heat current S, defined in Chapter 5,

$$\sum_i \dot{r}_i \cdot F'_i = V S \cdot F_e \qquad (7.21)$$

The force conserves total momentum because $\sum F'_i = 0$. Since only relative distances occur in F'_i, and assuming the force is sufficiently weak that the system remains homogeneous, there is nothing to prevent the use of periodic boundary conditions – exactly the motivation for devising methods of this kind. If $J = S_z$ and $F_e = F_e e_z$, then the constant-temperature version of (7.12) leads to the result

$$\lambda = \lim_{F_e \to 0} \lim_{t \to \infty} \frac{\langle S_z \rangle}{F_e T} \qquad (7.22)$$

The thermostat is the usual one, based on the total force, so that the equations of motion are simply $\ddot{r}_i = F_i + F'_i + \alpha \dot{r}_i$. For computational convenience we introduce a matrix with elements

$$B_{xyi} = \sum_{j(\neq i)} f_{xij} r_{yij} \qquad (7.23)$$

so that the components of (7.20) can be written

$$F'_{ix} = e_i \delta_{xz} F_e + \tfrac{1}{2}(B_{xzi} - \langle B_{xz} \rangle) F_e \qquad (7.24)$$

where $\langle B_{xz} \rangle$ is just the mean of B_{xzi}

New variables used in this calculation are

```
real **rfAtom, *enAtom, heatForce, thermalCond,
    sThermalCond, ssThermalCond;
```

with rfAtom and enAtom defined and evaluated exactly as in the equilibrium case (Chapter 5). The only new input item is the thermal driving force

```
RNAME (heatForce),
```

Evaluating the right-hand side of the equation of motion, again assuming the use of PC integration, requires the additional function

```
   ComputeThermalForce () {
     real rfAtomSumZ[NDIM + 1], enAtomSum, vv;
     int k, n;
     for (k = 1; k <= NDIM; k ++) rfAtomSumZ[k] = 0.;
5    enAtomSum = 0.;
     for (n = 1; n <= nAtom; n ++) {
       vv = 0.;
       for (k = 1; k <= NDIM; k ++) {
         vv = vv + Sqr (rv[k][n]);
10       ra[k][n] = ra[k][n] +
             0.5 * rfAtom[NDIM * 2 + k][n] * heatForce;
         rfAtomSumZ[k] = rfAtomSumZ[k] +
             rfAtom[NDIM * 2 + k][n];
       }
15     ra[3][n] = ra[3][n] + 0.5 * (enAtom[n] + vv) *
           heatForce;
       enAtomSum = enAtomSum + enAtom[n] + vv;
     }
     for (n = 1; n <= nAtom; n ++) {
20     for (k = 1; k <= NDIM; k ++)
         ra[k][n] = ra[k][n] - 0.5 * rfAtomSumZ[k] *
             heatForce / nAtom;
       ra[3][n] = ra[3][n] - 0.5 * enAtomSum *
           heatForce / nAtom;
25   }
   }
```

The value of $\lambda(F_e)$ is computed by

```
   EvalThermalCond () {
     real thermVecZ;
     int k, n;
     thermVecZ = 0.;
5    for (n = 1; n <= nAtom; n ++) {
       for (k = 1; k <= NDIM; k ++)
         enAtom[n] = enAtom[n] + Sqr (rv[k][n]);
       thermVecZ = thermVecZ + 0.5 * rv[3][n] * enAtom[n];
       for (k = 1; k <= NDIM; k ++)
10       thermVecZ = thermVecZ + 0.5 * rv[k][n] *
             rfAtom[NDIM * 2 + k][n];
     }
```

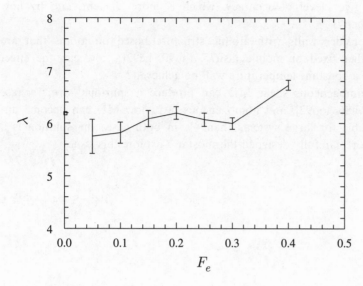

Fig. 7.4 Thermal conductivity for different values of applied heat force F_e; the equilibrium result appears at $F_e = 0$.

```
      thermalCond = thermVecZ * density / (temperature *
         heatForce * nAtom);
15 }
```

Final averages are processed by AccumProps in exactly the same way as the viscosity – simply change the variable names to those used here.

The results of this case study are based on the same input data as previously, except that the shear rate is replaced by the parameter heatForce with values in the range 0.05 to 0.4. For the three smallest values of heatForce the run length is extended to 62 000 steps to reduce the error estimates. The results appear in Figure 7.4 (error bars are computed as before); the benefits of the nonequilibrium approach are not as pronounced here as in the case of shear viscosity, although the fact that these runs are almost an order of magnitude shorter than the run of Chapter 5 should not be forgotten.

7.5 Further work

7.1 Examine the Couette flow problem when hard sliding walls are included [tro84].

7.2 Compare the transport coefficients obtained by nonequilibrium methods with those from the autocorrelation integrals [erp77]; for a given level of accuracy, which is more efficient, and by how much?

7.3 Consider walls with atomic structure based on atoms that are either fixed or mobile [ash75, tho90, lie92]; how can the effect of a constant-temperature wall be achieved?

7.4 Homogeneous shear MD can produce a spurious 'string' phase [eva86, loo92], and homogeneous heat flow MD can become unstable for large systems [han94]; in both cases the solution is a more carefully designed thermostat – explore this issue.

8

Rigid molecules

8.1 Introduction

The elementary constituents of most substances are structured molecules, rather than the spherically symmetric atoms treated in previous chapters. The emphasis on simple monatomic models is justified for a number of reasons: the dynamics are simpler, thereby making life easier for newcomers; it reflects the historical development of the field, since the original work establishing the viability of the MD approach as a quantitative tool dealt with liquid argon [rah64]; and once the basic techniques have been mastered they can be extended to a variety of more complex situations. In this chapter we discuss the first of these excursions – to molecules constructed from a rigidly linked atomic framework. This approach is suitable for small, relatively compact molecules, where rigidity seems a reasonable assumption, but if this is not true then motion within the molecule must also be taken into account, as we will see in later chapters. There is really no such thing as a rigid molecule, but from the practical point of view it is a very effective simplification of the underlying quantum problem; the model also does not account for chemical processes – no mechanism is provided for molecular formation and dissociation.

The chapter begins with a summary of rigid-body dynamics, but with a rather unfamiliar emphasis. In treatises on classical mechanics Euler angles play a central part [gol80]; while they provide the most intuitive means for describing the orientation of a rigid body and are helpful for analyzing certain exactly soluble problems, in numerical applications they actually represent a very poor choice. Quaternions, originally a purely theoretical development due to Hamilton, turn out to be the preferred method, and the dynamics will be described using such quantities. Linear molecules, with only two rotational degrees of freedom rather than three, are treated separately. The case study in this chapter addresses a simple but useful model for liquid water; the results derived for linear molecules will be used in Chapter 11. The methods shown here also apply when there is no translational motion and the molecules merely rotate about fixed lattice sites (as in molecular crystals [kle86]); the force computations are simpler because the neighbors within interaction range never change.

8.2 Dynamics

8.2.1 Coordinates

Rigid-body motion can be decomposed into two completely independent parts, translational motion of the center of mass and rotation about the center of mass. A basic result of classical mechanics is that the former is governed by the total force acting on the body, whereas the latter depends on the total applied torque. Thus translation can be treated as before, and we need only consider the dynamics of rotation.

Fully rigid molecules come in two flavors, linear and nonlinear, with each molecule having two or three rotational degrees of freedom respectively. The orientation of a rod-like linear molecule can be specified using two angular coordinates, whereas the more general case requires three, and it is for the latter that Euler angles are usually introduced as a particularly simple way of describing orientation. We begin with the nonlinear case [gol80].

The Euler angles are defined in terms of a sequence of rotations of a set of Cartesian coordinate axes about the origin. The first rotation is through an angle (measured counterclockwise) ϕ about the z-axis; this is followed by a rotation θ about the new x-axis; the final rotation is through an angle ψ about the new z-axis. The full rotation matrix $\mathbf{A}$ is the product of the individual rotation matrices, namely,

$$\mathbf{A} = \mathbf{R}(\psi)\mathbf{R}(\theta)\mathbf{R}(\phi) \tag{8.1}$$

and, if required, the elements of $\mathbf{A}$ can be expressed in terms of the Euler angles. There are two ways of interpreting the rotation expressed by $\mathbf{A}$. One is to consider a vector r' and use $\mathbf{A}$ to obtain its components in the rotated coordinate system: $r = \mathbf{A}r'$. The other is to rotate the vector itself, beginning with r and applying the opposite rotations in reversed order by means of the transpose of $\mathbf{A}$, in which case the result is the rotated vector $r' = \mathbf{A}^{\mathrm{T}}r$.

8.2.2 Quaternions

There are other ways of describing rotations, perhaps not as intuitively obvious, but often more convenient for numerical problem solving. Here we consider a particularly useful method – Hamilton's quaternions [gol80].

We begin by specifying the components of a quaternion in terms of the Euler angles:

$$\begin{aligned}
q_1 &= \sin(\theta/2)\cos((\phi - \psi)/2) \\
q_2 &= \sin(\theta/2)\sin((\phi - \psi)/2) \\
q_3 &= \cos(\theta/2)\sin((\phi + \psi)/2) \\
q_4 &= \cos(\theta/2)\cos((\phi + \psi)/2)
\end{aligned} \tag{8.2}$$

(with a minor change of indices in which q_4 replaces q_0). The components are normalized, with $\sum_m q_m^2 = 1$. The inverse relations are

$$\begin{aligned}
\sin\theta &= 2\sqrt{(q_1^2 + q_2^2)(1 - q_1^2 - q_2^2)} \\
\cos\theta &= 1 - 2(q_1^2 + q_2^2) \\
\sin\phi &= 2(q_1 q_3 + q_2 q_4)/\sin\theta \\
\cos\phi &= 2(q_1 q_4 - q_2 q_3)/\sin\theta \\
\sin\psi &= 2(q_1 q_3 - q_2 q_4)/\sin\theta \\
\cos\psi &= 2(q_1 q_4 + q_2 q_3)/\sin\theta
\end{aligned} \tag{8.3}$$

These results break down when $\theta = 0$ or π, corresponding to the coincidence of two of the rotation axes; since ϕ and ψ cannot be identified separately in this case we (arbitrarily) set $\psi = 0$ to remove any ambiguity.

An alternative definition is motivated by the fact that any rotation about a fixed point can be expressed in the form

$$r' = r\cos\zeta + (c \cdot r)c(1 - \cos\zeta) + (c \times r)\sin\zeta \tag{8.4}$$

where c is a unit vector specifying the axis of rotation and ζ is the rotation angle. If we define $q_m = c_m \sin\zeta/2$ for $m = 1, 2, 3$, and $q_4 = \cos\zeta/2$, then

$$r' = (2q_4^2 - 1)r + 2(q \cdot r)q + 2q_4 q \times r \tag{8.5}$$

While the definition based on Euler angles (8.2) is useful for converting to and from the quaternion representation, the completely equivalent result (8.5) leads directly to the rotation matrix

$$\mathbf{A} = 2\begin{pmatrix}
q_1^2 + q_4^2 - \frac{1}{2} & q_1 q_2 + q_3 q_4 & q_1 q_3 - q_2 q_4 \\
q_1 q_2 - q_3 q_4 & q_2^2 + q_4^2 - \frac{1}{2} & q_2 q_3 + q_1 q_4 \\
q_1 q_3 + q_2 q_4 & q_2 q_3 - q_1 q_4 & q_3^2 + q_4^2 - \frac{1}{2}
\end{pmatrix} \tag{8.6}$$

One of the benefits of quaternions is obvious – no trigonometric functions are required in evaluating $\mathbf{A}$. However, there is a more important advantage that will become apparent shortly.

A more formal treatment of quaternions [gol80] can be summarized as follows. Define a quaternion as the complex sum of a scalar and a vector:

$$\tilde{q} = q_4 + i\boldsymbol{q} \qquad (8.7)$$

The product of two quaternions is then

$$\tilde{q}\tilde{q}' = q_4 q_4' - \boldsymbol{q} \cdot \boldsymbol{q}' + i(q_4 \boldsymbol{q}' + \boldsymbol{q} q_4' + \boldsymbol{q} \times \boldsymbol{q}') \qquad (8.8)$$

itself also a quaternion. The complex conjugate of $\tilde{q}$ is $\tilde{q}^* = q_4 - i\boldsymbol{q}$ so that normalization implies $\tilde{q}\tilde{q}^* = 1$. The connection with rotation is made by choosing a vector $\boldsymbol{r}$, defining two quaternions $\tilde{r} = 0 + i\boldsymbol{r}$ and $\tilde{r}' = 0 + i\boldsymbol{r}'$, where $\boldsymbol{r}$ and $\boldsymbol{r}'$ are related as above, and with a little algebra arriving at the result $\tilde{r}' = \tilde{q}\tilde{r}\tilde{q}^*$. In this way quaternions are seen to provide the correct answers.

There are other ways of describing rotations, and even the quaternions can be expressed in an alternative fashion as complex 2×2 matrices, but our interest is confined to the set of real numbers $\{q_m\}$. The next step is to demonstrate the important role of quaternions in the dynamics of rigid bodies.

8.2.3 *Equations of motion for nonlinear molecules*

Rigid-body dynamics deals with two coordinate frames, one fixed in space, the other attached to the principal axes of the rotating body. The expression for the angular velocity ω' measured in the body-fixed frame in terms of Euler angles is a familiar one [gol80]:

$$\begin{pmatrix} \omega_x' \\ \omega_y' \\ \omega_z' \end{pmatrix} = \begin{pmatrix} \sin\theta\sin\psi & \cos\psi & 0 \\ \sin\theta\cos\psi & -\sin\psi & 0 \\ \cos\theta & 0 & 1 \end{pmatrix} \begin{pmatrix} \dot{\phi} \\ \dot{\theta} \\ \dot{\psi} \end{pmatrix} \qquad (8.9)$$

Mechanics texts tend to ignore the fact that the matrix in this equation is singular when $\sin\theta = 0$. Since the inverse of the matrix appears in the equations of motion, the numerical treatment will become unstable whenever θ even approaches 0 or π. The simplest and most elegant way to avoid this inconvenience is to abandon Euler angles and use quaternions instead [eva77a, eva77b]; this eliminates the problem of singular matrices.

The angular velocity ω' is related to $\dot{q}$ by

$$\begin{pmatrix} \omega'_x \\ \omega'_y \\ \omega'_z \\ 0 \end{pmatrix} = 2\mathbf{W} \begin{pmatrix} \dot{q}_1 \\ \dot{q}_2 \\ \dot{q}_3 \\ \dot{q}_4 \end{pmatrix} \tag{8.10}$$

where $\mathbf{W}$ is the orthogonal matrix

$$\mathbf{W} = \begin{pmatrix} q_4 & q_3 & -q_2 & -q_1 \\ -q_3 & q_4 & q_1 & -q_2 \\ q_2 & -q_1 & q_4 & -q_3 \\ q_1 & q_2 & q_3 & q_4 \end{pmatrix} \tag{8.11}$$

This result [cor60, eva77a] follows from the fact that if $\tilde{r}(t) = \tilde{q}(t)\tilde{r}(0)\tilde{q}^*(t)$, then the time derivative is $\dot{\tilde{r}}(t) = \dot{\tilde{q}}(t)\tilde{r}(0)\tilde{q}^*(t) + \tilde{q}(t)\tilde{r}(0)\dot{\tilde{q}}^*(t)$. Now $\tilde{r}(0) = \tilde{q}^*(t)\tilde{r}(t)\tilde{q}(t)$ and $\dot{\tilde{q}}\tilde{q}^* = -\tilde{q}\dot{\tilde{q}}^*$ (here we drop the explicit t-dependence), so that $\dot{\tilde{r}} = \dot{\tilde{q}}\tilde{q}^*\tilde{r} - \tilde{r}\dot{\tilde{q}}\tilde{q}^*$. Thus $\dot{r} = u \times r - r \times u = 2u \times r$, where u is the vector part of the quaternion $\tilde{u} = \dot{\tilde{q}}\tilde{q}^*$ (the scalar part is zero). In other words, $\omega = 2u$ (since $\dot{r} = \omega \times r$); (8.10) follows immediately because $\tilde{u}' = \tilde{q}^*\tilde{u}\tilde{q} = \tilde{q}^*\dot{\tilde{q}}$.

In a space-fixed coordinate frame, torque equals the rate of change of angular momentum:

$$\frac{d\mathbf{L}}{dt} = \mathbf{N} \tag{8.12}$$

Given that the general relation between time derivatives in space- and body-fixed coordinate frames is

$$\left(\frac{d\mathbf{L}}{dt}\right)_{\text{space}} = \left(\frac{d\mathbf{L}}{dt}\right)_{\text{body}} + \omega \times \mathbf{L} \tag{8.13}$$

the body-fixed version of (8.12) is

$$\dot{L}_x + \omega'_y L_z - \omega'_z L_y = N_x \tag{8.14}$$

with corresponding expressions for the other two components. Now for the principal axes, $L_x = I_x \omega'_x$ (etc.), and we obtain the Euler equations for rigid body rotation, a typical component of which is

$$I_x \dot{\omega}'_x = N_x + (I_y - I_z)\omega'_y \omega'_z \tag{8.15}$$

The quaternion accelerations are obtained from $\dot{\tilde{u}}' = \dot{\tilde{q}}^*\dot{\tilde{q}} + \tilde{q}^*\ddot{\tilde{q}}$; premultiply by $\tilde{q}$ and rearrange to get $\ddot{\tilde{q}} = \tilde{q}(\dot{\tilde{u}}' - \dot{\tilde{q}}^*\dot{\tilde{q}})$, or, equivalently,

$$\begin{pmatrix} \ddot{q}_1 \\ \ddot{q}_2 \\ \ddot{q}_3 \\ \ddot{q}_4 \end{pmatrix} = \frac{1}{2}\mathbf{W}^{\mathrm{T}} \begin{pmatrix} \dot{\omega}'_x \\ \dot{\omega}'_y \\ \dot{\omega}'_z \\ -2\sum \dot{q}_m^2 \end{pmatrix} \qquad (8.16)$$

We can eliminate $\dot{\omega}'$ from the right-hand side of (8.16) by using the Euler equations, and if the components of ω' that then appear are replaced by linear combinations of the $\dot{q}_i$ from (8.10) the result is a set of equations of motion entirely in terms of quaternions and their derivatives [pow79, rap85]; Euler angles and angular velocities no longer play any part in the calculation.

The function for evaluating the quaternion accelerations incorporates the complete version of the equations of motion:

```
   ComputeAccelsQ () {
     real w[4], s1, s2, s3, s4;
     int n;
     for (n = 1; n <= nMol; n ++) {
5      ComputeAngVel (n, w);
       s1 = (torq[1][n] + (momInertia[2] - momInertia[3]) *
         w[2] * w[3]) / momInertia[1];
       s2 = (torq[2][n] + (momInertia[3] - momInertia[1]) *
         w[3] * w[1]) / momInertia[2];
10     s3 = (torq[3][n] + (momInertia[1] - momInertia[2]) *
         w[1] * w[2]) / momInertia[3];
       s4 = -2. * (Sqr (qv[1][n]) + Sqr (qv[2][n]) +
         Sqr (qv[3][n]) + Sqr (qv[4][n]));
       qa[1][n] = 0.5 * (  q[4][n] * s1 - q[3][n] * s2 +
15       q[2][n] * s3 + q[1][n] * s4);
       qa[2][n] = 0.5 * (  q[3][n] * s1 + q[4][n] * s2 -
         q[1][n] * s3 + q[2][n] * s4);
       qa[3][n] = 0.5 * (- q[2][n] * s1 + q[1][n] * s2 +
         q[4][n] * s3 + q[3][n] * s4);
20     qa[4][n] = 0.5 * (- q[1][n] * s1 - q[2][n] * s2 -
         q[3][n] * s3 + q[4][n] * s4);
     }
   }
```

New variables appearing here are as follows. The quantity nMol is the number of molecules in the system (replacing nAtom). The quaternion components and their first and second time derivatives are denoted by q, qv, and qa – the variables are named to correspond with their translational counterparts. The components of the vector torq are expressed in the body-fixed coordinate frame.

The angular velocities w required here are computed by a separate function that is called for each molecule n:

```
ComputeAngVel (int n, real *w) {
    w[1] = 2. * (   q[4][n] * qv[1][n] + q[3][n] * qv[2][n] -
                    q[2][n] * qv[3][n] - q[1][n] * qv[4][n]);
    w[2] = 2. * (- q[3][n] * qv[1][n] + q[4][n] * qv[2][n] +
5                  q[1][n] * qv[3][n] - q[2][n] * qv[4][n]);
    w[3] = 2. * (   q[2][n] * qv[1][n] - q[1][n] * qv[2][n] +
                    q[4][n] * qv[3][n] - q[3][n] * qv[4][n]);
}
```

The interactions between rigid molecules are usually expressed as sums of contributions from pairs of 'interaction sites' on different molecules. Once the necessary details have been dealt with – the subject is discussed in Section 8.3 – it is sufficient to know the center of mass separation of two molecules and their orientations in order to be able to compute the interactions between all pairs of sites. Assuming that the site forces have already been computed, the forces and torques acting on the molecules as a whole are evaluated by the function given below. If μ labels the interaction sites and r_μ is the location of a site relative to the center of mass of a given molecule, then the total torque acting on the molecule is $\sum_\mu r_\mu \times F_\mu$. The variable sites-Mol is the number of interaction sites in each molecule (all molecules are assumed identical for simplicity), rSite contains the current coordinates of the sites, and fSite the results of the force calculation. The torques are evaluated in the space-fixed coordinate frame and then transformed to the body-fixed frame as required by the equations of motion:

```
ComputeTorqs () {
    real dr[4], torqS[4];
    int j, k, kk, n, nj;
    for (n = 1; n <= nMol; n ++) {
5       for (k = 1; k <= 3; k ++) {
            ra[k][n] = 0.;    torqS[k] = 0.;
        }
        for (j = 1; j <= sitesMol; j ++) {
            nj = (n - 1) * sitesMol + j;
10          for (k = 1; k <= 3; k ++) {
                ra[k][n] = ra[k][n] + fSite[k][nj];
                dr[k] = rSite[k][nj] - r[k][n];
            }
            torqS[1] = torqS[1] + dr[2] * fSite[3][nj] -
15              dr[3] * fSite[2][nj];
```

```
          torqS[2] = torqS[2] + dr[3] * fSite[1][nj] -
             dr[1] * fSite[3][nj];
          torqS[3] = torqS[3] + dr[1] * fSite[2][nj] -
             dr[2] * fSite[1][nj];
20     }
       BuildRotMatrix (n);
       kk = 1;
       for (k = 1; k <= 3; k ++) {
          torq[k][n] = rMat[kk] * torqS[1] + rMat[kk + 1] *
25           torqS[2] + rMat[kk + 2] * torqS[3];
          kk = kk + 3;
       } }
    }
```

The rotation matrix for each molecule is constructed from the quaternion components and stored as a linear array:

```
    BuildRotMatrix (int n) {
       real p[11];
       int k, k1, k2;
       k = 0;
5      for (k2 = 1; k2 <= 4; k2 ++) {
          for (k1 = k2; k1 <= 4; k1 ++) {
             k = k + 1;     p[k] = 2. * q[k1][n] * q[k2][n];
       } }
       rMat[1] = p[1] + p[10] - 1.;    rMat[2] = p[2] + p[9];
10     rMat[3] = p[3] - p[7];          rMat[4] = p[2] - p[9];
       rMat[5] = p[5] + p[10] - 1.;    rMat[6] = p[6] + p[4];
       rMat[7] = p[3] + p[7];          rMat[8] = p[6] - p[4];
       rMat[9] = p[8] + p[10] - 1.;
    }
```

Computation of the interaction-site coordinates in preparation for the force calculation is as follows:

```
    GenSiteCoords () {
       int j, k, kk, n, nj;
       for (n = 1; n <= nMol; n ++) {
          BuildRotMatrixT (n);
5         for (j = 1; j <= sitesMol; j ++) {
             nj = sitesMol * (n - 1) + j;     kk = 1;
             for (k = 1; k <= 3; k ++) {
                rSite[k][nj] = r[k][n] + rMat[kk] *
                   molSite[1][j] + rMat[kk + 1] * molSite[2][j] +
10                 rMat[kk + 2] * molSite[3][j];
                kk = kk + 3;
       } } }
    }
```

The function BuildRotMatrixT is similar to BuildRotMatrix but generates the transposed matrix; this is used in rotating the molecule from a predefined reference orientation – in which the site coordinates are denoted by molSite – to the current state.

Numerical integration of these second-order equations uses the same PC method as the translational equations. The integration functions, named PredictorStepQ and CorrectorStepQ, are based on the translational functions (Chapter 3), and only differ in the names of the variables processed and the presence of four rather than NDIM components; for this reason the functions are omitted. Normalization of the quaternions must be enforced separately to prevent gradual accumulation of numerical error (the error over a single timestep is very small); the adjustments can be carried out after each integration step:

```
   AdjustQuat () {
     real qi, qq;
     int k, n;
     for (n = 1; n <= nMol; n ++) {
5      qq = 0.;
       for (k = 1; k <= 4; k ++) qq = qq + Sqr (q[k][n]);
       qi = 1. / sqrt (qq);
       for (k = 1; k <= 4; k ++) q[k][n] = q[k][n] * qi;
     }
10 }
```

The contribution of the rotational motion to the kinetic energy is computed by an addition to EvalProps:

```
   real ... w[4];
   ...
   vvqSum = 0.;
   for (n = 1; n <= nMol; n ++) {
5  ComputeAngVel (n, w);
     for (k = 1; k <= 3; k ++)
        vvqSum = vvqSum + momInertia[k] * Sqr (w[k]);
   }
   vvSum = vvSum + vvqSum;
```

Tests based on momentum and energy conservation serve as partial checks on the correctness of the calculation. Angular momentum is not conserved however; this is a consequence both of the abrupt changes in angular momentum whenever a molecule crosses a periodic boundary and interaction wraparound. In order to verify angular momentum conservation, an isolated cluster of molecules must be simulated in a region that is nominally unbounded, thus eliminating the effects of periodicity.

8.2.4 Equations of motion for linear molecules

Linear rigid bodies are treated in a different way, since there are only two rotational degrees of freedom rather than three (quaternions are another possibility). The torque on a linear molecule can be written as a sum over interaction sites:

$$N = \sum_\mu r_\mu \times F_\mu = s \times \sum_\mu d_\mu F_\mu = s \times G \qquad (8.17)$$

where the orientation is defined by s, the unit vector along the molecular axis, and where d_μ is the distance of each interaction site from the center of mass. In the linear case angular momentum is simply $L = I\omega$, so that the equations of motion are

$$I\dot{\omega} = s \times G \qquad (8.18)$$

$$\dot{s} = \omega \times s \qquad (8.19)$$

There is also a two-dimensional version of this system, in which s is confined to the xy-plane and the only nonzero component of ω is ω_z.

We have a choice of either using this pair of first-order equations, or eliminating ω to obtain a single second-order equation

$$\ddot{s} = \dot{\omega} \times s + \omega \times \dot{s} = I^{-1}(s \times G) \times s + \omega \times (\omega \times s)$$
$$= I^{-1}G - \left(I^{-1}(s \cdot G) + \dot{s}^2\right)s \qquad (8.20)$$

where we have used the results $\omega \cdot s = 0$ – a consequence of (8.18) – and $\dot{s}^2 = \omega^2$. Here it is important that the initial state be defined consistently to ensure that (8.19) is satisfied. In both cases the length of s must be adjusted at regular intervals (not necessarily at every step, although this causes no harm) to avoid any gradual buildup of error.

The PC integration functions for the first-order equations follow (they are not required for the case study in Chapter 11 where second-order equations are used instead). Here, sv and sa denote ω and $\dot{\omega}$ (for the second-order equations they stand for $\dot{s}$ and $\ddot{s}$). Additional arrays – svxs, svxs1, and svxs2 – are used to hold the current and previous values of $\omega \times s$ appearing on the right-hand side of (8.19):

```
    PredictorStepF () {
      real c[] = {23.,-16.,5.}, div = 12., w;
      int k, n;
      w = deltaT / div;
5     for (n = 1; n <= nMol; n ++) {
        for (k = 1; k <= 3; k ++) {
          so[k][n] = s[k][n];    svo[k][n] = sv[k][n];
```

```
        s[k][n] = s[k][n] + w * (c[0] * svxs[k][n] +
            c[1] * svxs1[k][n] + c[2] * svxs2[k][n]);
10      sv[k][n] = sv[k][n] + w * (c[0] * sa[k][n] +
            c[1] * sa1[k][n] + c[2] * sa2[k][n]);
        sa2[k][n] = sa1[k][n];    sa1[k][n] = sa[k][n];
        svxs2[k][n] = svxs1[k][n];
        svxs1[k][n] = svxs[k][n];
15      }
        svxs[1][n] = sv[2][n] * s[3][n] - sv[3][n] * s[2][n];
        svxs[2][n] = sv[3][n] * s[1][n] - sv[1][n] * s[3][n];
        svxs[3][n] = sv[1][n] * s[2][n] - sv[2][n] * s[1][n];
    }
20 }

CorrectorStepF () {
    real c[] = {5.,8.,-1.}, div = 12., w;
    int k, n;
    w = deltaT / div;
5   for (n = 1; n <= nMol; n ++) {
      for (k = 1; k <= 3; k ++) {
        s[k][n] = so[k][n] + ... (as above) ...
        sv[k][n] = svo[k][n] + ...
      }
10      svxs[1][n] = ...
        ...
    }
  }
```

The contribution of the rotational motion to the kinetic energy is once again computed by code added to EvalProps:

```
vvsSum = 0.;
for (n = 1; n <= nMol; n ++) {
  for (k = 1; k <= 3; k ++)
      vvsSum = vvsSum + momInertia * Sqr (sv[k][n]);
5 }
vvSum = vvSum + vvsSum;
```

8.2.5 Temperature control

In the same way that the constant-temperature constraint was applied to simple atoms (Chapter 6), it can also be applied to nonlinear rigid molecules, but now it must be based on the combined translational and rotational kinetic energy. For each molecule we include a Lagrange multiplier term in the translational equations as before and a term of the general form $\alpha I_x \omega'_x$ must be added to each Euler equation (8.15). Since the total kinetic energy is

$$N_m E_k = \tfrac{1}{2} m \sum_i \dot{r}_i^2 + \tfrac{1}{2} \sum_x I_x \sum_i \omega_{xi}'^2 \tag{8.21}$$

with $\sum_x$ denoting a sum over components and N_m the number of molecules, by setting $\dot{E}_k = 0$ we obtain

$$\alpha = -\frac{\sum_i \dot{r}_i \cdot F_i + \sum_i \omega_i' \cdot N_i}{m \sum_i \dot{r}_i^2 + \sum_x I_x \sum_i \omega_{xi}'^2} \tag{8.22}$$

where F_i and N_i are the total force and torque on molecule i. When using quaternions, the kth component of the right-hand side of the equation of motion (8.16) gains an extra term (we omit the molecule index) $+\alpha \dot{q}_k$.

A similar expression for the Lagrange multiplier also applies in the case of linear molecules. The equation for α is similar to (8.22) but involves sums over either $\omega_i \cdot N_i$ and $I\omega_i^2$, or $\dot{s}_i \cdot [G_i - (s_i \cdot G_i + I\dot{s}_i^2)s_i]$ and $I\dot{s}_i^2$, depending on which form of the equation of motion is used.

The additions to the thermostat function `ApplyThermostat` needed for nonlinear rigid molecules are

```
    real ... w[4];
    ...
    s1 = s2 = 0.;
    ...
5   for (n = 1; n <= nMol; n ++) {
      ComputeAngVel (n, w);
      for (k = 1; k <= 3; k ++) {
        s1 = s1 + w[k] * torq[k][n];
        s2 = s2 + momInertia[k] * Sqr (w[k]);
10  } }
    vFac = - s1 / s2;
    ...
    for (n = 1; n <= nMol; n ++) {
      for (k = 1; k <= 4; k ++)
15      qa[k][n] = qa[k][n] + vFac * qv[k][n];
    }
```

In the linear case the additions to `ApplyThermostat` are the following:

```
    s1 = s2 = 0.;
    ...
    for (n = 1; n <= nMol; n ++) {
      for (k = 1; k <= 3; k ++) {
5       s1 = s1 + momInertia * sv[k][n] * sa[k][n];
        s2 = s2 + momInertia * Sqr (sv[k][n]);
    } }
```

```
   vFac = - s1 / s2;
   ...
10 for (n = 1; n <= nMol; n ++) {
      for (k = 1; k <= 3; k ++)
         sa[k][n] = sa[k][n] + vFac * sv[k][n];
   }
```

Temperature adjustment to correct numerical drift is applied separately to the translational and rotational motion. The addition to Adjust-Temp for the nonlinear case, assuming a constant-temperature simulation, is

```
   real ... w[4];
   ...
   vvqSum = 0.;
   for (n = 1; n <= nMol; n ++) {
5     ComputeAngVel (n, w);
      for (k = 1; k <= 3; k ++)
         vvqSum = vvqSum + momInertia[k] * Sqr (w[k]);
   }
   vFac = vMag / sqrt (vvqSum / nMol);
10 for (n = 1; n <= nMol; n ++) {
      for (k = 1; k <= 4; k ++) qv[k][n] = qv[k][n] * vFac;
   }
```

whereas for linear molecules it is

```
   vvsSum = 0.;
   for (n = 1; n <= nMol; n ++) {
      for (k = 1; k <= 3; k ++)
         vvsSum = vvsSum + momInertia * Sqr (sv[k][n]);
5  }
   vFac = vMag / sqrt (1.5 * vvsSum / nMol);
   for (n = 1; n <= nMol; n ++) {
      for (k = 1; k <= 3; k ++) sv[k][n] = sv[k][n] * vFac;
   }
```

In both cases the value of vMag (see below) determines the correct kinetic energy value.

8.2.6 Initial state

The functions listed below are called from SetupJob to handle the initialization of the rotational variables. Here we consider only nonlinear molecules; the linear case will be treated in Chapter 11. Molecular orientation is randomly assigned (atan2 is a standard library function), with each angular velocity having a fixed magnitude based on the temperature

(through the quantity vMag) and a randomly chosen direction. Angular
coordinates and velocities are converted to quaternion form and angular
accelerations used by the PC method are zeroed:

```
    InitAngCoords () {
      real e[4], eulAng[4];
      int n;
      for (n = 1; n <= nMol; n ++) {
5       RandVec3 (e, &randSeed);
        eulAng[1] = atan2 (e[1], e[2]);
        eulAng[2] = acos (e[3]);
        eulAng[3] = 2. * pi * RandR (&randSeed);
        EulerToQuat (n, eulAng[1], eulAng[2], eulAng[3]);
10    }
    }

    InitAngVels () {
      real e[4], w;
      int k, n;
      for (n = 1; n <= nMol; n ++) {
5       RandVec3 (e, &randSeed);
        w = 0.;
        for (k = 1; k <= 3; k ++)
          w = w + momInertia[k] * Sqr (e[k]);
        w = vMag / sqrt (w);
10      qv[1][n] = 0.5 * w * (q[4][n] * e[1] -
          q[3][n] * e[2] + q[2][n] * e[3]);
        qv[2][n] = 0.5 * w * (q[3][n] * e[1] +
          q[4][n] * e[2] - q[1][n] * e[3]);
        qv[3][n] = 0.5 * w * (- q[2][n] * e[1] +
15        q[1][n] * e[2] + q[4][n] * e[3]);
        qv[4][n] = 0.5 * w * (- q[1][n] * e[1] -
          q[2][n] * e[2] - q[3][n] * e[3]);
      }
    }

    InitAngAccels () {
      int k, n;
      for (n = 1; n <= nMol; n ++) {
        for (k = 1; k <= 4; k ++)
5         qa[k][n] = qa1[k][n] = qa2[k][n] = 0.;
      }
    }
```

To convert from Euler angles to quaternion components use

```
    EulerToQuat (int n,
      real eulAng1, real eulAng2, real eulAng3) {
      real a1, a2, a3, c1, s1;
```

```
     a1 = 0.5 * eulAng2;
5    a2 = 0.5 * (eulAng1 - eulAng3);
     a3 = 0.5 * (eulAng1 + eulAng3);
     s1 = sin (a1);     c1 = cos (a1);
     q[1][n] = s1 * cos (a2);     q[2][n] = s1 * sin (a2);
     q[3][n] = c1 * sin (a3);     q[4][n] = c1 * cos (a3);
10 }
```

The translational variables are initialized in exactly the same way as for atomic fluids; the only alteration to the functions of Chapter 3 is to replace nAtom with nMol.

8.3 Molecular construction

8.3.1 General features

Now that we have seen how to formulate and solve the dynamical problem we turn to the details of the molecules themselves. Interactions between rigid molecules are most readily introduced by specifying the locations of the sites in the molecule at which the forces act. The total force between two molecules is then simply the sum of the forces acting between all pairs of interaction sites. The amount of work is proportional to the square of the number of sites, so this should be kept as low as possible. The potential function used for each pair can be defined independently, but molecular symmetry reduces the number of functions needed. Interaction sites may be associated with the positions of the nuclei, but this is not essential and often just serves as the initial version of a model. There is considerable scope for fine-tuning the structure and interactions in this engineering-like approach, with the simulations themselves being used to refine the models. For further details see [gra84, lev92].

Molecular fluids require substantially more computation per molecule than their atomic counterparts because of the need to consider all pairs of interaction sites. Coulomb interactions are usually involved, so the cutoff distance should be as large as possible, again adding to the computational effort; the specialized methods available for such long-range forces are not used here (see Chapter 11). The fact that the interaction range now extends over a substantial fraction of the simulation region can erase the benefits of the cell and neighbor-list methods, so that the all-pairs approach is often adequate.

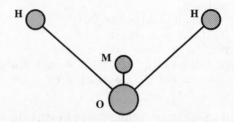

Fig. 8.1 The planar TIP4P water molecule; site coordinates are given in the text.

8.3.2 Model water

The most popular molecular fluid for MD exploration, for obvious reasons, is water. Not only because of its ubiquity and importance, but also for its many unusual features that defy simple explanation, water has long been associated with MD simulation [rah71, sti72, sti74], and numerous models have been proposed to help understand the microscopic mechanisms underlying the behavior.

For our case study we will use one of several currently popular rigid water models – the so-called TIP4P model [jor83]. The model, shown in Figure 8.1, is based on four interaction sites located in a planar configuration, two of which – labeled M and O – are associated with the oxygen nucleus, and two – both labeled H – with the protons. The two distances and one angle required to fully specify the site coordinates are $r_{\mathrm{OH}} = 0.957\,\text{Å}$, $r_{\mathrm{OM}} = 0.15\,\text{Å}$ (M lies on the symmetry axis between O and the line joining the H sites), and $\angle\mathrm{HOH} = 104.5°$.

The potential energy between two molecules i and j consists of a double sum over the interaction sites of both molecules; the terms in the sum allow for Coulomb interactions between the electric charges assigned to the sites as well as an LJ-type contribution:

$$u_{ij} = \sum_{\mu \in i} \sum_{\nu \in j} \left(\frac{q_\mu q_\nu e^2}{r_{i\mu, j\nu}} + \frac{A_{\mu\nu}}{r_{i\mu, j\nu}^{12}} - \frac{C_{\mu\nu}}{r_{i\mu, j\nu}^6} \right) \tag{8.23}$$

The charges associated with the sites, while maintaining some resemblance to the actual molecule, are generally regarded as parameters that can be adjusted to fit known molecular properties, such as the multipole moments. The corresponding force is

$$f_{ij} = \sum_{\mu \in i} \sum_{\nu \in j} \left(\frac{q_\mu q_\nu e^2}{r_{i\mu,j\nu}^3} + \frac{12A_{\mu\nu}}{r_{i\mu,j\nu}^{14}} - \frac{6C_{\mu\nu}}{r_{i\mu,j\nu}^8} \right) r_{i\mu,j\nu} \tag{8.24}$$

The charges appearing in the potential function are $q_H = 0.52\,e$, $q_O = 0$, and $q_M = -2q_H$, where $e = 4.803 \times 10^{-10}$ esu; to convert to units used experimentally note that $e^2 = 331.8\,(\text{kcal/mole})\text{Å}$. As part of the molecular design process, the negative charge has been shifted away from the O site by a small amount to the M site introduced specifically for this purpose. The parameters in the LJ-part of the potential, which only acts between O sites, are $A_{OO} \equiv A = 600 \times 10^3\,(\text{kcal/mole})\text{Å}^{12}$, and $C_{OO} \equiv C = 610\,(\text{kcal/mole})\text{Å}^6$.

We now switch to reduced units. Define the length unit σ to be the value of r for which $A/r^{12} - C/r^6 = 0$, namely, $\sigma = (A/C)^{1/6}$, and the unit of energy $\epsilon = A/4\sigma^{12}$. The mass of the water molecule is 2.987×10^{-23} g. Physical and reduced units are then related by $\sigma = 3.154\,\text{Å}$, $\epsilon = 0.155\,\text{kcal/mole}$ ($= 1.08 \times 10^{-14}$ erg/molecule), and the unit of time is 1.66×10^{-12} s. We also define a reduced unit of charge (e) in terms of which $q_H = 1$ and for convenience we let $b = e^2/\epsilon\sigma$; in reduced units $b = 183.5$.

The coordinates of the interaction sites when the molecule is situated in a reference state in the yz-plane with its center of mass at the origin are (in reduced units)

$$r_O = (0, 0, -0.0206)$$
$$r_M = (0, 0, 0.0274)$$
$$r_H = (0, \pm0.240, 0.165)$$

Masses denoted by m_O and m_H are associated with the O and H sites, and $m_O = 16m_H$; in reduced MD units $m_O + 2m_H = 1$. The principal moments of inertia are

$$I_y = m_O z_O^2 + 2m_H z_H^2 = 0.0034$$
$$I_z = 2m_H y_H^2 = 0.0064$$

and, of course, $I_x = I_y + I_z$.

8.3.3 Interaction calculations

In terms of the reduced units introduced above, the potential energy and force contributions from the different interaction-site pairs of the model, namely, LJ between the O sites and Coulomb between all pairs of charges, are

$$
\begin{aligned}
\text{O}-\text{O}: \quad & u = 4(r^{-12} - r^{-6}) & f = 48(r^{-14} - 0.5r^{-8})r \\
\text{M}-\text{M}: \quad & u = 4b/r & f = (4b/r^3)r \\
\text{M}-\text{H}: \quad & u = -2b/r & f = (-2b/r^3)r \\
\text{H}-\text{H}: \quad & u = b/r & f = (b/r^3)r
\end{aligned}
\tag{8.25}
$$

The function shown below computes these interactions using an all-pairs approach. The different kinds of interaction site are assigned numerical types 1, 2, and 3, corresponding to O, M, and H; these values appear in the array typeSite. The decision as to whether a pair of sites lies within the cutoff range is based on the distance between the centers of mass of the molecules containing the sites, and not on the distance between the sites themselves; not only is this more efficient computationally than testing pairs of sites individually, but it means that there are no partially interacting molecules:

```
    ComputeSiteForces () {
      real dr[NDIM + 1], shift[NDIM + 1], enVal, fcVal, rr,
        rrCut, rri, rri3;
      int i, j1, j2, k, ms1, ms2, m1, m2, typeSum;
5     rrCut = Sqr (rCut);
      for (i = 1; i <= nMol * sitesMol; i ++) {
        for (k = 1; k <= NDIM; k ++) fSite[k][i] = 0.;
      }
      uSum = 0.;
10    for (m1 = 1; m1 <= nMol - 1; m1 ++) {
        for (m2 = m1 + 1; m2 <= nMol; m2 ++) {
          for (k = 1; k <= NDIM; k ++) {
            dr[k] = r[k][m1] - r[k][m2];
            shift[k] = 0.;
15          if (fabs (dr[k]) > regionH[k])
              shift[k] = - SignR (region[k], dr[k]);
            dr[k] = dr[k] + shift[k];
          }
          rr = Sqr (dr[1]) + Sqr (dr[2]) + Sqr (dr[3]);
20        if (rr < rrCut) {
            ms1 = (m1 - 1) * sitesMol;
            ms2 = (m2 - 1) * sitesMol;
            for (j1 = 1; j1 <= sitesMol; j1 ++) {
              for (j2 = 1; j2 <= sitesMol; j2 ++) {
25              typeSum = typeSite[j1] + typeSite[j2];
                if (typeSite[j1] == typeSite[j2] ||
                  typeSum == 5) {
```

```
          for (k = 1; k <= NDIM; k ++)
            dr[k] = rSite[k][ms1 + j1] -
30            rSite[k][ms2 + j2] + shift[k];
          rr = Sqr (dr[1]) + Sqr (dr[2]) + Sqr (dr[3]);
          rri = 1. / rr;
          if (typeSum == 2) {
            rri3 = rri * rri * rri;
35          fcVal = 48. * rri3 * (rri3 - 0.5) * rri;
            enVal = 4. * rri3 * (rri3 - 1.);
          } else if (typeSum == 4) {
            enVal = 4. * bCon * sqrt (rri);
            fcVal = enVal * rri;
40        } else if (typeSum == 5) {
            enVal = -2. * bCon * sqrt (rri);
            fcVal = enVal * rri;
          } else if (typeSum == 6) {
            enVal = bCon * sqrt (rri);
45          fcVal = enVal * rri;
          }
          for (k = 1; k <= NDIM; k ++) {
            fSite[k][ms1 + j1] = fSite[k][ms1 + j1] +
              fcVal * dr[k];
50          fSite[k][ms2 + j2] = fSite[k][ms2 + j2] -
              fcVal * dr[k];
          }
          uSum = uSum + enVal;
    } } } } } }
55 }
```

8.3.4 Further details

New variables appearing in this simulation are

```
real **q, **qv, **qa, **qa1, **qa2, **qo, **qvo, **fSite,
  **molSite, **rSite, **torq, momInertia[4], rMat[10],
  bCon, vvqSum;
int *typeSite, nMol, sitesMol;
```

and in SetParams we set

```
sitesMol = 4;
```

The additional array allocations in AllocArrays are

```
  q = AllocMatR (4, nMol);
  ... (ditto for qv, qa, qa1, qa2, qo, qvo) ...
  fSite = AllocMatR (3, nMol * sitesMol);
  molSite = AllocMatR (3, sitesMol);
5 rSite = AllocMatR (3, nMol * sitesMol);
```

```
torq = AllocMatR (3, nMol);
typeSite = AllocVecI (sitesMol);
```

The details of the molecule itself are specified in a function that is called from SetupJob:

```
DefineMol () {
  int j;
  for (j = 1; j <= sitesMol; j ++)
    molSite[1][j] = molSite[2][j] = molSite[3][j] = 0.;
5   molSite[3][1] = -0.0206;    molSite[3][2] = 0.0274;
  molSite[2][3] = 0.240;      molSite[3][3] = 0.165;
  molSite[2][4] = - molSite[2][3];
  molSite[3][4] = molSite[3][3];
  momInertia[1] = 0.0098;    momInertia[2] = 0.0034;
10  momInertia[3] = 0.0064;
  bCon = 183.5;
  typeSite[1] = 1;    typeSite[2] = 2;
  typeSite[3] = 3;    typeSite[4] = 3;
}
```

The full sequence of calls appearing in SingleStep used for the interaction computations and integration is

```
  PredictorStep ();
  PredictorStepQ ();
  ApplyBoundaryCond ();
  GenSiteCoords ();
5 ComputeSiteForces ();
  ComputeTorqs ();
  ComputeAccelsQ ();
  ApplyThermostat ();
  CorrectorStep ();
10 CorrectorStepQ ();
  AdjustQuat ();
  ApplyBoundaryCond ();
```

A few additional items complete the description: The interaction cutoff is at $r_c = 7.5$ Å, or 2.38 in reduced units; this value of rCut can be added to the input data. A density of $1 \, g/cm^3$ is equivalent to a unit-cell spacing of 3.103 Å for an initial cubic lattice arrangement, or 0.983 in reduced units. Since the unit of energy corresponds to $\epsilon/k_B = 78.2$ K, a typical temperature of 298 K corresponds to 3.8 in reduced units. The timestep is typically $\Delta t = 0.0005$; in real units this amounts to 8×10^{-16} s.

8.4 Measurements

8.4.1 Types of measurement

A model such as the one described here has a variety of properties that are of experimental relevance and others that, although not directly measurable in the laboratory, are able to contribute towards understanding the behavior at the microscopic level. We will consider two examples of the former and one of the latter, all in connection with pure water. (A particularly important use of water models is in the study of solvation of other kinds of molecules, ranging from simple atoms and ions to complex molecules such as biopolymers; we will not attempt to delve into this extensive subject [bro88, lev92].)

The first of the measurements involves the site–site RDFs. Here, rather than simply examining the distribution of center-of-mass separations, it is possible to study RDFs associated with distinct sites on the molecules; together, these RDFs are able to provide clues to local molecular arrangement beyond just the distances themselves. The second measurement deals with rotational diffusion by looking at the rate at which molecules undergo orientational change, an important aspect of certain kinds of spectroscopic study. The final feature examined, the one with no direct laboratory analogy, is the nature of the hydrogen-bond network formed by the fluid.

Other properties, including those of thermodynamic interest, as well as the dielectric constant, can also be measured, although they will not be included here. A quantity such as the pressure, normally expressed in terms of the virial sum, needs to be redefined for use with rigid molecules. There are in fact two ways of dealing with the virial which, for equilibrium systems, are readily shown to be completely equivalent [cic86b]: it can be expressed either as a sum involving just the intermolecular forces and center-of-mass separations, ignoring all the internal details, or as a sum over all pairs of interaction sites in each pair of molecules.

8.4.2 Radial distribution functions

When evaluating the RDF we consider three distinct site–site distribution functions that are accessible experimentally – g_{OO}, g_{OH}, and g_{HH}. For computational purposes we assign numerical labels to the sites to simplify the task of deciding which site pairs contribute to which function. The arrays required are

```
real **histRdf;
int *typeSiteRdf;
```

where (unlike previously) histRdf is a two-dimensional array with pro-vision for several distinct RDF measurements. The necessary array allocations are

```
typeSiteRdf = AllocVecI (sitesMol);
histRdf = AllocMatR (3, sizeHistRdf);
```

and in DefineMol we add

```
typeSiteRdf[1] = 1;    typeSiteRdf[2] = -1;
typeSiteRdf[3] = 2;    typeSiteRdf[4] = 2;
```

The following modifications to EvalRdf (Chapter 4) are required; we skip the obvious changes that initialize and later normalize three sets of RDF data rather than just one:

```
     real ... shift[NDIM + 1];
     int ... ms1, ms2, m1, m2, rdfType, typeSum;
     ...
     for (m1 = 1; m1 <= nMol - 1; m1 ++) {
5      for (m2 = m1 + 1; m2 <= nMol; m2 ++) {
         for (k = 1; k <= NDIM; k ++) {
         ... (as in ComputeSiteForces) ...
         }
         if (rr < rrRange) {
10         ms1 = (m1 - 1) * sitesMol;
           ms2 = (m2 - 1) * sitesMol;
           for (j1 = 1; j1 <= sitesMol; j1 ++) {
             for (j2 = 1; j2 <= sitesMol; j2 ++) {
               typeSum = typeSiteRdf[j1] + typeSiteRdf[j2];
15             if (typeSum >= 2) {
                 for (k = 1; k <= NDIM; k ++)
                   dr[k] = rSite[k][ms1 + j1] -
                     rSite[k][ms2 + j2] + shift[k];
                 rr = Sqr (dr[1]) + Sqr (dr[2]) + Sqr (dr[3]);
20               if (rr < rrRange) {
                   n = (int) (sqrt (rr) / deltaR) + 1;
                   if (typeSum == 2) rdfType = 1;
                   else if (typeSum == 3) rdfType = 2;
                   else rdfType = 3;
25                 histRdf[rdfType][n] =
                     histRdf[rdfType][n] + 1.;
     } } } } } }
```

It is not necessary to recompute the site coordinates after the corrector step, since the values computed for use in the interaction calculations are adequate for this purpose. Because there are two H sites per molecule and we have not allowed for this symmetry in the RDF computation, both g_{OH} and g_{HH} must be divided by four. Other details of the RDF computation resemble the atomic case.

The run used to produce the RDF results is based on the following input data:

```
initUcell        6 6 6
density          0.98
temperature      3.8
deltaT           0.0005
rCut             2.38
stepAdjustTemp   1000
stepAvg          200
stepEquil        1000
limitRdf         100
rangeRdf         2.5
sizeHistRdf      125
stepRdf          50
```

A cubic initial array is used, so that the system contains $N_m = 216$ molecules. The value of Δt is an order of magnitude smaller than that used in the soft-sphere work; this is a result of the higher temperature (in MD units) and the need to allow for the rotational motion of molecules with a low moment of inertia and hence a relatively high angular velocity. Constant-temperature MD is used; with the value of Δt shown, the temperature drift over 1000 steps amounts to about 4%, but if this presents a problem the drift can be reduced by an order of magnitude simply by halving Δt.

The measured g_{OO}, g_{OH}, and g_{HH} are shown in Figure 8.2. The latter two curves are truncated at distances less than `rangeRdf` because the criterion for limiting the distance between sites is applied to the molecular centers of mass (exactly as in the force computation) and not the sites themselves. Without going into detail, the results are consistent with the expected tetrahedral, or ice-like, structural correlations known to occur in liquid water [jor83]. One example of a measurement demonstrating the loose-packed molecular organization of the fluid is the integral of the function $4\pi r^2 g_{OO}(r)$ out to a distance that includes the first peak of g_{OO}; this provides an estimate of the number of molecules that can be regarded as nearest neighbors, and here the value is found to be 4.4.

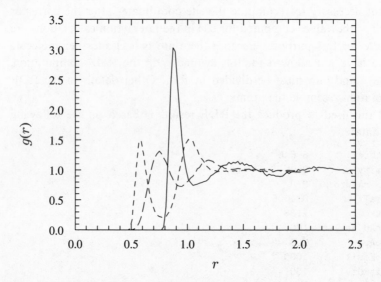

Fig. 8.2 Site–site RDFs for the TIP4P water model (solid curve g_{OO}, short dashes g_{OH}, long dashes g_{HH}).

8.4.3 Rotational diffusion

Rotational diffusion a measure of the rate at which the direction of the molecular dipole changes, is another quantity of experimental significance. The dipole direction appears as the bottom row of the rotation matrix (8.6) and is the unit vector

$$\boldsymbol{\mu} = 2 \begin{pmatrix} q_1 q_3 + q_2 q_4 \\ q_2 q_3 - q_1 q_4 \\ q_3^2 + q_4^2 - \frac{1}{2} \end{pmatrix} \tag{8.26}$$

Rather than measuring the mean-square change in orientation we consider the time-dependence of the dipole autocorrelation function $C(t) = \langle \boldsymbol{\mu}_i(t) \cdot \boldsymbol{\mu}_i(0) \rangle$. Translational diffusion will also be measured, based on the molecular center-of-mass coordinates.

The measurement is organized in the same way as translational diffusion, with the following additions to EvalDiffusion (Chapter 5):

```
real e[4];
  ...
if (indexDiffuse[nb] == 1) {
  for (n = 1; n <= nMol; n ++) {
    ...
    e[1] = 2. * (q[1][n] * q[3][n] + q[2][n] * q[4][n]);
```

```
        e[2] = 2. * (q[2][n] * q[3][n] - q[1][n] * q[4][n]);
        e[3] = sqrt (1. - Sqr (e[1]) - Sqr (e[2]));
        nn = 3 * (n - 1);
10      for (k = 1; k <= 3; k ++)
            aDiffuseOrg[nb][nn + k] = e[k];
    } }
    ni = indexDiffuse[nb];
    ...
15  aDiffuse[nb][ni] = 0.;
    for (n = 1; n <= nMol; n ++) {
        e[1] = 2. * (q[1][n] * q[3][n] + q[2][n] * q[4][n]);
        e[2] = 2. * (q[2][n] * q[3][n] - q[1][n] * q[4][n]);
        e[3] = sqrt (1. - Sqr (e[1]) - Sqr (e[2]));
20      nn = 3 * (n - 1);
        aDiffuse[nb][ni] = aDiffuse[nb][ni] +
            aDiffuseOrg[nb][nn + 1] * e[1] +
            aDiffuseOrg[nb][nn + 2] * e[2] +
            aDiffuseOrg[nb][nn + 3] * e[3];
25  }
```

Additions to AccumDiffusion, in the obvious places, are

```
    for (j = 1; j <= nValDiffuse; j ++)
        aDiffuseAv[j] = aDiffuseAv[j] + aDiffuse[nb][j];
        ...
    fac = 1. / (nMol * limitDiffuseAv);
5   for (j = 1; j <= nValDiffuse; j ++)
        aDiffuseAv[j] = aDiffuseAv[j] * fac;
```

and to ZeroDiffusion

```
    for (j = 1; j <= nValDiffuse; j ++) aDiffuseAv[j] = 0.;
```

In PrintDiffusion the values of aDiffuseAv must be included in the output. New arrays are

```
    real **aDiffuse, **aDiffuseOrg, *aDiffuseAv;
```

and their allocations:

```
    aDiffuse = AllocMatR (nBuffDiffuse, nValDiffuse);
    aDiffuseOrg = AllocMatR (nBuffDiffuse, 3 * nMol);
    aDiffuseAv = AllocVecR (nValDiffuse);
```

The runs used for these measurements are similar to the one described above, but the system size is reduced to $N_m = 125$, and the following additional input data are required:

```
    limitDiffuseAv   10
    nBuffDiffuse     20
    nValDiffuse      200
    stepDiffuse      40
```

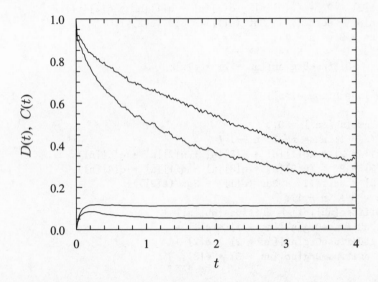

Fig. 8.3 Diffusion coefficients and dipole autocorrelation functions for water at $T = 3.8$ and 4.4.

The translational diffusion coefficients and the dipole autocorrelations at $T = 3.8$ and 4.4 are shown in Figure 8.3. Runs of 13 400 timesteps are used to produce these results.

8.4.4 Hydrogen bonds

The molecular structure of normal ice involves a diamond (or tetrahedral) lattice, and short-range correlations reminiscent of this order persist into the liquid state. The forces responsible for this loosely packed arrangement are attributed to hydrogen bonding, in which each molecule forms four strong and highly directional bonds with its immediate neighbors. One of the basic requirements of any water model is that it should reproduce this behavior. What exactly constitutes a hydrogen bond is not included in the definition of the molecule, since it is a feature whose origin is quantum mechanical, but, for modeling purposes, it is reasonable to assume that the presence of such a bond between two molecules is marked by an interaction energy lying in a particular range and a molecular alignment that satisfies certain conditions insofar as the distance and angles are concerned. Once all the hydrogen bonds have been identified it is possible to study the properties of the network formed by the bonds [rah73, gei79].

Here we will focus on the pair-energy distribution [jor83], to see whether there is anything special about its form to warrant using it in determining where hydrogen bonds have formed. Since this exercise turns out to be successful, we then make use of what has been learned to count the numbers of bonds formed by each molecule.

The first step is to evaluate the interaction energy for each pair of molecules separately and construct an energy histogram of these values. In addition, each pair whose energy lies below a certain threshold is regarded as linked by a hydrogen bond, and bond counts associated with these molecules are incremented. The threshold is determined by a parameter boundPairEng. The following alterations and additions to ComputeSiteForces are required:

```
   real ... uSumPair;
   int ... j;
   ...
   for (i = 1; i <= nMol; i ++) nMolBonds[i] = 0;
5  ...
     if (rr < rrCut) {
       uSumPair = 0.;
       ...
       for (j1 = ...
10       for (j2 = ...
           ...
           uSumPair = uSumPair + enVal;
           ...
       }
15     uSum = uSum + uSumPair;
       j = sizeHistPairEng * (uSumPair - minPairEng) /
         (maxPairEng - minPairEng);
       if (j < 1) j = 1;
       else if (j > sizeHistPairEng) j = sizeHistPairEng;
20     histPairEng[j] = histPairEng[j] + 1.;
       if (uSumPair < boundPairEng) {
         nMolBonds[m1] = nMolBonds[m1] + 1;
         nMolBonds[m2] = nMolBonds[m2] + 1;
       }
25     ...
   for (i = 1; i <= nMol; i ++) {
     j = nMolBonds[i] + 1;
     if (j > sizeHistBondNum) j = sizeHistBondNum;
     histBondNum[j] = histBondNum[j] + 1.;
30 }
```

The data collected is processed by a function called from SingleStep:

```
    if (stepCount >= stepEquil && (stepCount - stepEquil) %
      stepPairEng == 0) EvalPairEng ();
```

This function computes the average pair-energy distribution over a series
of configurations, and also constructs a histogram of the number of
bonds per molecule:

```
    EvalPairEng () {
      real hSum;
      int n;
      countPairEng = countPairEng + 1;
5     if (countPairEng == 1) {
        for (n = 1; n <= sizeHistPairEng; n ++)
          histPairEng[n] = 0.;
        for (n = 1; n <= sizeHistBondNum; n ++)
          histBondNum[n] = 0.;
10    }
      if (countPairEng == limitPairEng) {
        hSum = 0;
        for (n = 1; n <= sizeHistPairEng; n ++)
          hSum = hSum + histPairEng[n];
15      for (n = 1; n <= sizeHistPairEng; n ++)
          histPairEng[n] = histPairEng[n] / hSum;
        ... (also normalize histBondNum) ...
        PrintPairEng (stdout);
        countPairEng = 0;
20    }
    }
```

The new variables used here are as follows:

```
    real *histBondNum, *histPairEng, boundPairEng,
      maxPairEng, minPairEng;
    int *nMolBonds, countPairEng, limitPairEng,
      sizeHistBondNum, sizeHistPairEng, stepPairEng;
```

the additional data to be input are

```
    RNAME (boundPairEng),
    INAME (limitPairEng),
    RNAME (maxPairEng),
    RNAME (minPairEng),
5   INAME (sizeHistBondNum),
    INAME (sizeHistPairEng),
    INAME (stepPairEng),
```

the array allocations (AllocArrays) are

```
    histBondNum = AllocVecR (sizeHistBondNum);
    histPairEng = AllocVecR (sizeHistPairEng);
    nMolBonds = AllocVecI (nMol);
```

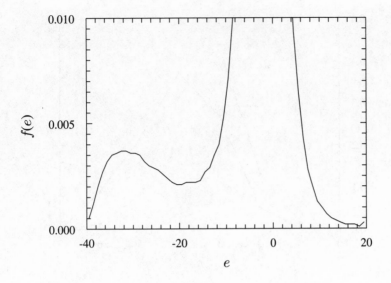

Fig. 8.4 Pair-energy distribution.

and a counter initialization (`SetupJob`):

```
countPairEng = 0;
```

An output function (not shown) must be provided.

To investigate the pair-energy distribution we carry out a run with the following input data:

```
initUcell         6 6 6
deltaT            0.0003
boundPairEng      -8.0
limitPairEng      50
maxPairEng        20.0
minPairEng        -40.0
sizeHistBondNum   8
sizeHistPairEng   60
stepPairEng       20
```

The value of Δt has been reduced to improve the energy measurements. The results shown in Figure 8.4 are obtained by averaging over timesteps 4–5000; the smaller peak corresponds to tightly bound nearest-neighbor molecule pairs.

If we now assume that all pairs of molecules with mutual interaction energy (e) below a certain threshold e_h are hydrogen bonded, we can actually examine the distribution of hydrogen bonds. By way of example

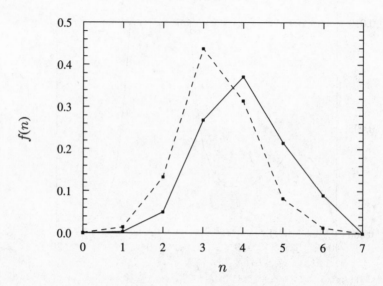

Fig. 8.5 Distribution of mean number of hydrogen bonds per molecule for threshold energy $e_h = -8$ (solid curve) and -10 (dashed).

we use values of boundPairEng (corresponding to e_h) of -8 and -10 to obtain the results shown in Figure 8.5. Although a more detailed analysis taking the relative orientation of the molecules into account is required to ensure a consistent picture, the fact that for physically reasonable values of e_h ($e_h = -10$ corresponds to 1.55 kcal/mole) the average number of hydrogen bonds formed by each molecule is close to four is encouraging. Further analysis appears in [jor83].

8.5 Further work

8.1 Explore the relation between quaternions and other representations of orientation [gol80].

8.2 The leapfrog method can also be used with quaternions [fin93]; examine its effectiveness.

8.3 Because of the unusual properties of water various models have been used in MD studies to account for the experimental observations; investigate the factors contributing to the design of different models.

8.4 Study the relative orientation of neighboring water molecules.

8.5 Constant-pressure techniques can also be applied to rigid bodies; investigate [nos83].

8.6 Study the nature of the hydrogen bond network formed and the bond lifetimes [gei79].

8.7 Compute the dielectric constant [spr91, smi94]; how sensitive is this to the choice of model, and how significant is the effect of truncating the long-range Coulomb forces?

8.8 The rigid molecule used for water ignores important polarizability effects; how can the model be extended [spr88] to incorporate such behavior?

8.9 A subtle but important property of water is the density maximum while still in the liquid state (the reason that ice floats); how successful has MD been in studying this phenomenon [bil94]?

8.10 Study other examples of rigid molecular models – both linear and nonlinear – for real fluids [lev92].

9

Flexible molecules

9.1 Introduction

The rigid molecule approach described in Chapter 8 is limited in its applicability because it is really only appropriate for small compact molecules. Here we consider the opposite extreme – completely flexible molecules of a type used in certain kinds of polymer study. No new principles are involved, since the intramolecular forces that maintain structural integrity by holding the molecule together, as well as providing any other necessary internal interactions, are treated in the same way as intermolecular forces. In Chapter 10 we will consider a more complex model in which a molecule exhibits a certain amount of flexibility but is also subject to various structural constraints that restrict the internal motions.

9.2 Description of molecule

9.2.1 Polymer chains

Owing to the central role played by polymers in a variety of fields, biochemistry and materials engineering are just two examples, model polymer systems have been the subject of extensive study, both by MD and by other methods such as Monte Carlo [bin95]. Of the many kinds of polymer topology that occur, chains have received the most attention, but other types, including stars [gre94] and membranes [abr89], have not been neglected. Chain properties can be divided into two categories, equilibrium and dynamical; much of the equilibrium behavior – especially in the case of long chains – actually falls under the heading of critical phenomena, and here MD is unable to compete with lattice-based

222

methods because of their far less demanding nature, but when it comes to transport phenomena MD is, once again, the only viable method.

Polymer chain models can be studied for different reasons. At one extreme is the attempt to reproduce the behavior of a real polymer, an example being the alkane model we will meet in Chapter 10, or complex biopolymers such as proteins [ber86a, bro88, bro90a]. Here we concentrate on a much simpler model that aims at capturing some of the more general aspects of chain behavior, rather than all the myriad quantitative details. One can regard this model as the analogy of a soft-sphere fluid, but while for simple fluids there is just one basic model, for polymers there are a number of different systems that can be regarded as basic. The simplest is a single chain in the vacuum, used in examining the configurational properties of an isolated polymer. This is followed by a chain in an inert soft-sphere solvent, the purpose of the solvent being to introduce a certain amount of hydrodynamic coupling into the motion of the chain [pie92, smi92, dun93]. Then there are multiple-chain fluids [kre92]; here the chain density is an important parameter, because it determines how much of the dynamics is due to the chain interacting with itself and how much is due to interactions between chains. In each instance the details of the interactions between chain atoms as well as the nature of the solvent, if present, must be addressed.

A problem that must be faced when studying polymers is the range of timescales over which configurational change occurs. At one extreme are the localized changes in internal arrangement that involve only short segments of the chain; at the other are large-scale conformational changes and chain diffusion, processes that are seriously impeded by effects such as mutual obstruction and entanglement. This means that some of the more interesting rheological properties of polymer liquids and the challenging problems of protein folding appear to be beyond the limits of what can be simulated by MD. But a great deal can still be done within the timescales that are currently accessible.

9.2.2 Chain structure

The simplest model attempts to represent the excluded volume of the individual monomers out of which the polymer is constructed and the bonds that link them into chains. The monomers can be simple atoms modeled using a soft-sphere potential, while bonds with limited length variation can be produced by means of an attractive interaction between chain neighbors. Single or multiple chains can be included, and a soft-sphere solvent is readily added. Chains constructed in this way are totally flexible, within the limits set by the repulsive potential; a controlled degree of stiffness can be introduced by means of an interaction regulating the separation of next-nearest neighbors, although we will not do this here. More specific structural requirements are best addressed using the methods described in Chapter 10.

In the model treated here all pairs of atoms interact via the familiar soft-sphere repulsive force f_{ss}; in addition, there is an attractive interaction between each pair of adjacent bonded atoms of the form

$$f(r) = \begin{cases} f_{ss}\big((1 - r_m/r)r\big) & r_m - r_c < r < r_m \\ 0 & \text{otherwise} \end{cases} \tag{9.1}$$

In (9.1) the direction of the soft-sphere force has been reversed and its origin shifted to produce a force that limits the separation of bonded atoms; in practice the bond length variation can be restricted to a (not too) narrow range by a suitable choice of r_m ($> r_c$). The energy and length scales (ϵ and σ) characterizing the potential are kept unchanged.

9.3 Implementation details

9.3.1 Interactions

The evaluation of the forces between nonbonded atoms belonging to the same chain, as well as between atoms in different chains and between solvent–chain and solvent–solvent atom pairs, are all handled by the soft-sphere functions of Chapter 3, with just one minor alteration. If we assume that neighbor lists are used, the change affects the condition for selecting atom pairs in BuildNebrList; the modified form is

```
if ((m1 != m2 || j2 < j1) && (inPoly[j1] == 0 ||
    inPoly[j1] != inPoly[j2] || abs (j1 - j2) > 1))
```

so that bonded atom pairs are excluded – they will be treated separately. Each element of the array inPoly indicates whether the particular atom belongs to a polymer chain and, if so, which one.

We then require an additional function to evaluate the forces between bonded atoms. The total number of chains is given by nChain, the number of atoms per chain – assuming all chains to have the same length – by chainLen, and r_m is represented by the variable bondLim:

```
ComputeChainBondForces () {
  real dr[NDIM + 1], f, fcVal, rr, rrCut, rri, rri3,
      uVal, w;
  int i, j1, j2, k, n;
```

```
5    rrCut = Sqr (rCut);
     for (n = 1; n <= nChain; n ++) {
      for (i = 1; i <= chainLen - 1; i ++) {
        j1 = (n - 1) * chainLen + i;    j2 = j1 + 1;
        for (k = 1; k <= NDIM; k ++) {
10        dr[k] = r[k][j1] - r[k][j2];
          if (fabs (dr[k]) > regionH[k])
             dr[k] = dr[k] - SignR (region[k], dr[k]);
        }
        rr = Sqr (dr[1]) + Sqr (dr[2]) + Sqr (dr[3]);
15      if (rr < rrCut) {
          ... (same as ComputeForces) ...
        }
        w = 1. - bondLim / sqrt (rr);
        if (w > 0.) ErrExit ("bond snapped");
20      rr = rr * Sqr (w);
        if (rr < rrCut) {
          rri = 1. / rr;    rri3 = rri * rri * rri;
          fcVal = 48. * w * rri3 * (rri3 - 0.5) * rri;
          ... (same as ComputeForces) ...
25   } } }
     }
```

In computing the attractive part of the bond interaction a safety check is included to ensure that the bond has not 'snapped' because of numerical error or incorrectly formulated initial conditions (see below).

9.3.2 Initial state

When preparing the initial state it is essential that the atoms of each chain be positioned so that the bond lengths are all within their permitted ranges, and that no significant overlap occurs between atoms belonging to either the same or different chains. In this particular case study, neither of these issues presents any difficulty, especially if the density is not too high, but questions of how to pack molecules correctly into a reasonably low-energy state while avoiding overlap between molecules can arise in other situations [mck92]. Solvent atoms present less of a problem because they can be added after the chains are in place.

Possible initial chain states include fully stretched and planar zigzag configurations; another option is the linear helix that is even more compact than the zigzag form, a useful feature when chain packing becomes problematic at higher densities. The following function arranges the atoms of each chain in a zigzag state, with the major axis of the chain aligned in the x-direction. The chains themselves are organized as a body-

centered cubic (BCC) lattice, and after the chains have been positioned
the coordinates are corrected to allow for any periodic wraparound. We
also show how the solvent is added; the simple but inefficient approach
demonstrated here attempts to place solvent atoms at the sites of a simple
cubic lattice by checking whether the proposed location overlaps any of
the chain atoms already in position, and if overlap is found to occur the
tentative solvent atom is discarded (for large systems a method based on
the use of cells would be preferable):

```
   InitCoords () {
     real c[4], gap[4], bY, bZ;
     int i, j, k, m, n, nX, nY, nZ, overlap;
     bY = rCut * cos (pi / 4.);    bZ = rCut * sin (pi / 4.);
5    n = 0;
     for (k = 1; k <= NDIM; k ++)
        gap[k] = region[k] / initUchain[k];
     for (nZ = 1; nZ <= initUchain[3]; nZ ++) {
       c[3] = (nZ - 0.75) * gap[3] - regionH[3];
10     for (nY = 1; nY <= initUchain[2]; nY ++) {
         c[2] = (nY - 0.75) * gap[2] - regionH[2];
         for (nX = 1; nX <= initUchain[1]; nX ++) {
           c[1] = (nX - 0.75) * gap[1] - regionH[1];
           for (j = 1; j <= 2; j ++) {
15           for (m = 1; m <= chainLen; m ++) {
               n = n + 1;
               r[1][n] = c[1] + (j - 1) * gap[1] * 0.5;
               r[2][n] = c[2] + (j - 1) * gap[2] * 0.5 +
                 (1 - m % 2) * bY;
20             r[3][n] = c[3] + (j - 1) * gap[3] * 0.5 +
                 (m - 1) * bZ;
     } } } } }
     nAtom = n;
     ApplyBoundaryCond ();
25   for (k = 1; k <= NDIM; k ++)
        gap[k] = region[k] / initUcell[k];
     for (nZ = 1; nZ <= initUcell[3]; nZ ++) {
       ... (as for simple cubic lattice) ...
         c[1] = (nX - 0.5) * gap[1] - regionH[1];
30       overlap = 0;
         for (i = 1; i <= nChain * chainLen; i ++) {
           overlap = 1;
           for (k = 1; k <= NDIM; k ++) {
             if (fabs (r[k][i] - c[k]) > rCut) {
35             overlap = 0;
               break;
           } }
```

```
                  if (overlap) break;
               }
40             if (overlap == 0) {
                  n = n + 1;
                  for (k = 1; k <= NDIM; k ++) r[k][n] = c[k];
      } } } }
      nAtom = n;
45 }
```

The variables introduced here are

```
real bondLim;
int *inPoly, initUchain[NDIM + 1], chainLen, nChain;
```

and the additional input data consist of

```
RNAME (bondLim),
INAME (chainLen),
INAME (initUchain),
```

The number of chains, assuming that a BCC arrangement is used, is computed in SetParams:

```
nChain = 2 * initUchain[1] * initUchain[2] * initUchain[3];
if (nChain == 2) nChain = 1;
```

where the values in initUchain specify the number of unit cells that contain the chains. To enable the study of just a single chain we will assume that if a single unit cell is specified the intention is to have just one chain; to accommodate this case a change is needed in InitCoords:

```
   if (nChain == 1) {
     for (m = 1; m <= chainLen; m ++) {
       n = n + 1;
       r[1][n] = - 0.25 * region[1];
5      r[2][n] = - 0.25 * region[2] + (1 - m % 2) * bY;
       r[3][n] = - 0.25 * region[3] + (m - 1) * bZ;
     }
   } else {
   ... (as before) ...
10 }
```

The maximum possible number of atoms, subject to later reduction because of overlap between solvent and chain atoms, is set in SetParams:

```
nAtom = initUcell[1] * initUcell[2] * initUcell[3] +
    nChain * chainLen;
```

where initUcell now specifies the number of unit cells containing solvent atoms. The only additional array allocation is

```
inPoly = AllocVecI (nAtom);
```

The final stage of the initialization process involves explicit assignment of atoms to chains for use in the interaction calculations; since the chains are constructed consecutively this is a trivial task:

```
AssignToChain () {
  int i, j, n;
  n = 0;
  for (i = 1; i <= nChain; i ++) {
5   for (j = 1; j <= chainLen; j ++) {
      n = n + 1;    inPoly[n] = i;
  } }
  for (n = nChain * chainLen + 1; n <= nAtom; n ++)
      inPoly[n] = 0;
10 }
```

9.4 Properties

9.4.1 Chain conformation

Three spatial properties of polymer chains are frequently studied because of their experimental relevance. The first is the mean-square end-to-end distance $\langle R^2 \rangle$ from which it is possible to learn whether, on average, the chain is in an open or compact configuration; the distribution of R^2 values (or at least the moments of the distribution) can be used to determine the importance of effects such as excluded volume. Then there is the mean-square radius of gyration $\langle S^2 \rangle$ that provides information on the entire mass distribution of the chain, and plays a central role in interpreting light scattering and viscosity measurements. Lastly, since the actual mean spatial distribution of the chain mass – essentially its 'shape' – need not be spherical, details of the moments of the mass distribution are important.

For a chain of n_s monomers,

$$\langle R^2 \rangle = \langle |r_{n_s} - r_1|^2 \rangle \tag{9.2}$$

and, if all monomers have the same mass,

$$\langle S^2 \rangle = \frac{1}{n_s} \left\langle \sum_{i=1}^{n_s} |r_i - \bar{r}|^2 \right\rangle \tag{9.3}$$

where $\bar{r}$ is the center of mass. Elements of the tensor describing the mass distribution have the form

$$G_{xy} = \frac{1}{n_s} \sum_{i=1}^{n_s} (r_{xi} - \bar{r}_x)(r_{yi} - \bar{r}_y) \tag{9.4}$$

The three eigenvalues of **G** are denoted by g_1, g_2, g_3; their sum is just $\langle S^2 \rangle$, but it is their ratios that are of interest because if they are not equal to unity it means that the distribution is nonspherical. (Note that the inertia tensor [gol80] has components $S^2 \delta_{xy} - G_{xy}$.) Rearrangement of (9.4) leads to an alternative expression that is used in the computations, namely,

$$G_{xy} = \frac{1}{n_s} \sum_{i=1}^{n_s} r_{xi} r_{yi} - \frac{1}{n_s^2} \Big[\sum_{i=1}^{n_s} r_{xi} \Big] \Big[\sum_{i=1}^{n_s} r_{yi} \Big] \tag{9.5}$$

The function shown below accumulates these chain properties over a sequence of configurations; it is called from SingleStep by

```
if (stepCount >= stepEquil && (stepCount - stepEquil)
    stepChainProps == 0) EvalChainProps ();
```

The new variables needed are

```
real aaDistSq, eeDistSq, gMomRatio1, gMomRatio2, radGyrSq;
int countChainProps, limitChainProps, stepChainProps;
```

additional input data items are

```
INAME (limitChainProps),
INAME (stepChainProps),
```

and initialization:

```
countChainProps = 0;
```

The following function measures and averages the end-to-end distance, the radius of gyration, the eigenvalue ratios, and the actual bond lengths. Evaluating the eigenvalues $\{g_i\}$ requires diagonalizing a 3×3 matrix; to do this simply expand the appropriate determinant $(\det |\mathbf{G} - g\mathbf{I}|)$ to obtain the cubic characteristic equation, and the $\{g_i\}$ are just the solutions of this equation obtained by a call to SolveCubic (see Appendix). The organization of this function adheres to a pattern that should be familiar by now; the output function is trivial:

```
  EvalChainProps () {
    real c[4], g[7], gVal[4], shift[4], sumR[4], sumRR[4],
      a1, a2, a3, dr, ee, sumXY, sumYZ, sumZX, t;
    int i, j, k, n, n1;
5   countChainProps = countChainProps + 1;
    if (countChainProps == 1) aaDistSq = eeDistSq =
```

```
          radGyrSq = gMomRatio1 = gMomRatio2 = 0.;
      n = 0;
      for (i = 1; i <= nChain; i ++) {
10      ee = sumXY = sumYZ = sumZX = 0.;
        for (k = 1; k <= 3; k ++)
          shift[k] = sumR[k] = sumRR[k] = 0.;
        for (j = 1; j <= chainLen; j ++) {
          n = n + 1;
15        if (j > 1) {
            for (k = 1; k <= 3; k ++) {
              dr = r[k][n] - r[k][n - 1];
              if (fabs (dr) > regionH[k]) {
                t = SignR (region[k], dr);
20              shift[k] = shift[k] - t;    dr = dr - t;
              }
              aaDistSq = aaDistSq + Sqr (dr);
            }
          } else n1 = n;
25        for (k = 1; k <= 3; k ++) {
            c[k] = r[k][n] + shift[k];
            sumR[k] = sumR[k] + c[k];
            sumRR[k] = sumRR[k] + Sqr (c[k]);
          }
30        sumXY = sumXY + c[1] * c[2];
          sumYZ = sumYZ + c[2] * c[3];
          sumZX = sumZX + c[3] * c[1];
        }
        for (k = 1; k <= 3; k ++) {
35        ee = ee + Sqr (c[k] - r[k][n1]);
          sumR[k] = sumR[k] / chainLen;
          g[k] = sumRR[k] / chainLen - Sqr (sumR[k]);
        }
        eeDistSq = eeDistSq + ee;
40      g[4] = sumXY / chainLen - sumR[1] * sumR[2];
        g[5] = sumZX / chainLen - sumR[3] * sumR[1];
        g[6] = sumYZ / chainLen - sumR[2] * sumR[3];
        a1 = - g[1] - g[2] - g[3];
        a2 = g[1] * g[2] + g[2] * g[3] + g[3] * g[1] -
45        Sqr (g[4]) - Sqr (g[5]) - Sqr (g[6]);
        a3 = g[1] * Sqr (g[6]) + g[2] * Sqr (g[5]) +
          g[3] * Sqr (g[4]) - 2. * g[4] * g[5] * g[6] -
          g[1] * g[2] * g[3];
        SolveCubic (a1, a2, a3, g);
50      gVal[1] = Max3R (g[1], g[2], g[3]);
        gVal[3] = Min3R (g[1], g[2], g[3]);
        gVal[2] = g[1] + g[2] + g[3] - gVal[1] - gVal[3];
        radGyrSq = radGyrSq + gVal[1] + gVal[2] + gVal[3];
```

```
        gMomRatio1 = gMomRatio1 + gVal[2] / gVal[1];
55      gMomRatio2 = gMomRatio2 + gVal[3] / gVal[1];
     }
     if (countChainProps == limitChainProps) {
        aaDistSq = aaDistSq / (nChain * (chainLen - 1) *
           limitChainProps);
60      eeDistSq = eeDistSq / (nChain * limitChainProps);
        ... (ditto for radGyrSq, gMomRatio1, gMomRatio2) ...
        PrintChainProps (stdout);
        countChainProps = 0;
     }
65 }
```

9.4.2 Measurements

The results shown here are for a single chain in a soft-sphere solvent. We consider chains consisting of $n_s = 8$, 16, and 24 monomers. The input data for a chain with $n_s = 8$ includes

```
initUcell         10 10 10
initUchain        1 1 1
bondLim           2.1
chainLen          8
limitChainProps   100
stepAdjustTemp    1000
stepChainProps    20
```

as well as $\rho = 0.5$, $T = 2$, and $\Delta t = 0.005$. A simple cubic lattice is used for the initial positions of the solvent atoms; the maximum number of solvent atoms is therefore 1000 (the values in initUcell determine the region size), although overlap with chain monomers may reduce this number very slightly. For $n_s = 16$ the values in initUcell are increased to 12, and for $n_s = 24$ to 16 – the region must be large enough both to hold the chain in its initial state and to prevent unwanted wraparound effects. Constant temperature dynamics and PC integration are employed.

Results obtained from runs of 5×10^5 timesteps are listed in Table 9.1 for the three chain lengths studied. The mean bond lengths ($\langle l \rangle$) are practically the same in each case; the value of $\sigma(l)$ is typically 0.0002, so that bond length is seen to be tightly controlled. The values of $\langle R^2 \rangle$ and $\langle S^2 \rangle$ are comparable to published results, although the values do depend on solvent density [smi92]. The eigenvalue ratios $\langle g_2/g_1 \rangle$ and $\langle g_3/g_1 \rangle$ provide clear evidence that the mean shape of the chain is far from spherical, more closely resembling a flattened cigar.

Table 9.1. *Measurements of chain properties*

n_s	$\langle l \rangle$	$\langle R^2 \rangle$	$\sigma(R^2)$	$\langle S^2 \rangle$	$\sigma(S^2)$	$\langle g_2/g_1 \rangle$	$\langle g_3/g_1 \rangle$
8	1.0531	12.39	0.56	2.064	0.057	0.2623	0.0823
16	1.0535	32.53	3.92	5.118	0.309	0.2528	0.0832
24	1.0534	59.84	12.63	8.981	1.112	0.2292	0.0732

Table 9.2. *Block averaged estimates of standard deviation*

| | $n_s = 8$ | | $n_s = 16$ | | $n_s = 24$ | |
b	$\sigma(\langle R^2 \rangle)$	$\sigma(\langle S^2 \rangle)$	$\sigma(\langle R^2 \rangle)$	$\sigma(\langle S^2 \rangle)$	$\sigma(\langle R^2 \rangle)$	$\sigma(\langle S^2 \rangle)$
1	0.220	0.020	0.901	0.083	2.042	0.181
2	0.240	0.021	1.000	0.093	2.592	0.229
4	0.216	0.019	1.079	0.101	3.245	0.281
8	0.200	0.019	1.186	0.105	3.863	0.336
16	0.259	0.025	1.334	0.126	4.779	0.413
32	0.269	0.026	1.511	0.136	4.412	0.422
64	0.085	0.008	0.993	0.064	7.351	0.718

In order to obtain error estimates for $\langle R^2 \rangle$ and $\langle S^2 \rangle$ we resort to the block averaging described in Chapter 4. The results of this analysis over a series of block sizes (b) are shown in Table 9.2. The quality of the estimates is seen to decrease as the chains become longer, suggesting the need for even longer runs.

9.5 Further work

9.1 How long must these simulations be to ensure that block averaging converges?

9.2 The length dependence of $\langle R^2 \rangle$ and $\langle S^2 \rangle$ has been studied extensively for chains on lattices [kre88], and, while MD cannot reach the extremely long chains that lattice-based Monte Carlo methods can handle, the results for shorter chains are still of interest; investigate.

9.3 Study the rate at which chain structure relaxes by examining the time-dependent autocorrelation function of a quantity such as $\langle R^2 \rangle$; relaxation rates are very sensitive to chain length and solvent density [smi92].

9.4 How does the presence of a solvent affect the chain dynamics [dun93]?

9.5 Model a pure polymer liquid; here reptation is considered to be an important mechanism for molecular motion [kre92]. A suitable initial state must be constructed for this problem.

10
Geometrically constrained molecules

10.1 Introduction

Some internal degrees of freedom are important to molecular motion, while others can be regarded as frozen. Classical mechanics allows geometrical relations between coordinates to be included as holonomic constraints. We have already encountered constraints in connection with non-Newtonian modifications of the dynamical equations (Chapter 6); here the constraints occur in a Newtonian context, so that there is little doubt as to the physical nature of the trajectories.

In this chapter we focus on a class of model where constraints play an important role, namely, the polymer models used in studying alkane chains and more complex molecules, in which a combination of geometrical constraints and internal motion is required. The treatment of constraints is not the only new feature of such models; the interactions responsible for bond bending and torsion are essentially three- and four-body potentials, and some rather intricate vector algebra is required to determine the forces. The particular alkane model described here incorporates one further simplification, namely, the use of the often encountered 'united-atom' approximation – the hydrogen atoms attached to each carbon atom in the backbone are absorbed into the carbon and are thereby eliminated from the problem

10.2 Geometric constraints

10.2.1 Role of constraints

The notion of a constraint acting at the molecular level is merely an attempt at simplification; the justification for assuming that certain bond lengths and angles are constant is that, at the prevailing temperature, there is insufficient energy to excite the associated vibrational degrees of freedom, or modes, out of their quantum ground states. Or, adopting a classical perspective, the potential function responsible for limiting the variation of the bond length or angle must involve a very deep and narrow well; the natural frequency associated with such a potential will be much higher than those of other kinds of internal motion and is therefore likely to demand an intolerably small integration timestep. To avoid this situation it is customary to eliminate such degrees of freedom entirely by the simple expedient of replacing them with constraints.

The only unanswered question is whether a completely frozen mode is an accurate way of representing a mode that is really only 'stiff', in the sense that its vibration frequency is much greater than that of other modes and coupling with the rest of the system is weak; there is no completely satisfactory answer since constraints and stiff potentials are both attempts to describe what is fundamentally a quantum problem. The distinction between stiff and frozen modes is important in statistical mechanics, and configurational averages depend on the choice [hel79]; the same is true for dynamical properties [van82].

10.2.2 Problem formulation

Consider a molecule whose structure is subject to one or more geometrical constraints; fixing the distance between any two atoms introduces a constraint of the form

$$|r_i - r_j|^2 = d_{ij}^2 \tag{10.1}$$

thereby eliminating one degree of freedom. If i and j are bonded neighbors within a molecule, then this constraint amounts to fixing the bond length; if they are next-nearest neighbors, and the two intervening bonds also have constant length, then it is the bond angle that is fixed. While these are examples of replacing stiff interactions between pairs and triplets of atoms, there are other types of structural constraint, such as those used for maintaining the planarity of a molecule; constraints must be formulated with care to ensure the correct selection is made [cic82]. Assuming there are a total of n_c distance constraints imposed on a particular molecule, then if the kth constraint acts between atoms $i(k)$ and $j(k)$, the constraints can be summarized by the set of equations

$$\sigma_k \equiv r_{i(k)j(k)}^2 - d_{i(k)j(k)}^2 = 0, \quad k = 1, \ldots, n_c \tag{10.2}$$

For simplicity, the indexing used here considers just a single molecule, but this is readily extended. Note that, because constraints remove degrees of freedom that would otherwise contribute to the temperature, allowance must be made when relating temperature to kinetic energy.

The equations of motion follow directly from the Lagrangian formulation described in Chapter 3. The result (now allowing for different masses) is

$$m_i \ddot{r}_i = F_i + G_i \tag{10.3}$$

where F_i is the usual force term, m_i the mass of the ith atom, or group of atoms combined into a single monomer, and the additional force-like term G_i that expresses the effect of the constraints on atom i can be written

$$G_i = - \sum_{k \in C(i)} \lambda_k \nabla_i \sigma_k \tag{10.4}$$

Here $C(i)$ denotes the set of constraints that directly involve r_i and the $\{\lambda_k\}$ are the Lagrange multipliers introduced into the problem (the reversed sign in (10.4) follows custom [ryc77]). The force F_i includes all (non-constraint) interactions within the molecule, as well as the intermolecular forces acting on individual atoms (or monomers). There are three scalar equations of motion for each atom, as well as n_c constraint equations for the molecule as a whole, exactly the number needed to evaluate the Lagrange multipliers and integrate the equations of motion.

Solving the problem can be carried out in various ways. A particularly simple method is to advance the system over a single timestep by integrating the unconstrained equations of motion (ignoring G_i) and then adjusting all the coordinates (in practice by only a small amount) so that the constraints are satisfied in the new state [ryc77]. This adjustment is carried out by means of an iterative relaxation procedure that modifies each pair of constrained coordinates in turn until all constraints are satisfied to the required accuracy. The alternative is to solve the full problem, by first computing the Lagrange multipliers from the time-differentiated constraint equations and then using these values in solving the equations of motion [edb86]. But, unlike the relaxation approach, which restores the constraints to their correct values, here the constraints are subject to numerical integration error. In practice the error is small and can be corrected by, for example, including an occasional series of relaxation cycles. Both methods will be described below, but first the subject of how to label the atoms and constraints systematically must be addressed.

10.2.3 Atom and constraint indexing

For the linear chain molecules discussed in this chapter, the indexing problem has a simple solution. For more complex molecular structures, that can involve both tree- and ring-like forms, the problem is a little more difficult [mor91]. We concentrate on the case of a simple chain subjected to bond-length constraints, and optionally, to bond-angle constraints as well. Once the constraints have been identified the remainder of the processing need not be concerned with the topology of the molecule.

Consider a polymer chain consisting of n_s monomers – atoms for short. If only the bond lengths are constrained there will be a total of $n_c = n_s - 1$ constraints, with constraint k relating the coordinates of atoms k and $k + 1$. If, on the other hand, the chain is subject to both length and angle constraints, there will be $n_s - 1$ of the former and $n_s - 2$ of the latter, so $n_c = 2n_s - 3$. Each of the constraints acting on atom i then involves one of the four atoms $j = i \pm 1,\ i \pm 2$; length and angle constraints can be indexed in alternating fashion, leading to the simple result that the kth constraint acts between atoms $\lfloor (k + 1)/2 \rfloor$ and $\lfloor (k + 4)/2 \rfloor$.

10.3 Solving the constraint problem
10.3.1 Matrix method

Of the two methods, solving the equations of motion together with the constraints seems to be the more appealing approach from a strictly aesthetic point of view. This entails expressing the constraint equations in matrix form and then solving the resulting linear algebra problem using standard numerical techniques. The constraints will of course be subject to numerical error, but if this turns out to be sufficiently small the results can be corrected from time to time using the relaxation method discussed later in this section; such corrections can also be carried out by, for example, using standard optimization methods to minimize a penalty function that measures constraint deviations [edb86].

The constraint forces can be rewritten in the form

$$G_i = -2 \sum_{k \in C(i)} \lambda_k r_{i(k)j(k)} = \sum_{k=1}^{n_c} M_{ik} \lambda_k s_k \qquad (10.5)$$

where

$$s_k = r_{\min(i,j)} - r_{\max(i,j)} \tag{10.6}$$

and the elements of the matrix $\mathbf{M}$, which has n_s rows and n_c columns, are

$$M_{ik} = \begin{cases} +2 & k \in C\big(i(k)\big), \ j(k) < i(k) \\ -2 & k \in C\big(i(k)\big), \ j(k) > i(k) \\ 0 & k \notin C\big(i(k)\big) \end{cases} \tag{10.7}$$

Since s_k^2 is constant, it follows that

$$\ddot{s}_k \cdot s_k + \dot{s}_k^2 = 0 \tag{10.8}$$

The acceleration $\ddot{s}_k$ appearing in (10.8) can be replaced by the actual equation of motion obtained from (10.3). If the indices of the atoms associated with the kth constraint are arranged so that $i(k) < j(k)$ and we define a new matrix

$$L_{kk'} = \big(M_{i(k)k'}/m_{i(k)} - M_{j(k)k'}/m_{j(k)}\big)s_k \cdot s_{k'} \tag{10.9}$$

then the result of this replacement is

$$\sum_{k'=1}^{n_c} L_{kk'}\lambda_{k'} = -\big(F_{i(k)}/m_{i(k)} - F_{j(k)}/m_{j(k)}\big) \cdot s_k - \dot{s}_k^2 \tag{10.10}$$

for $k = 1, \ldots, n_c$. The matrix $\mathbf{L}$ is of size $n_c \times n_c$, and the only unknowns in (10.10) are the values of $\lambda_{k'}$.

If bond lengths only are constrained, there will be exactly two nonzero elements in the ith row of $\mathbf{M}$, corresponding to the two constraints that involve atom i:

$$M_{i,i-1} = +2, \ M_{ii} = -2 \tag{10.11}$$

If both lengths and angles are constrained there are four nonzero elements per row, namely,

$$M_{i,2i-4} = M_{i,2i-3} = +2, \ M_{i,2i-1} = M_{i,2i} = -2 \tag{10.12}$$

As an example, the equation of motion for an atom that is subject to both types of constraints and that is not located at the chain ends is

$$\ddot{r}_i = F_i + 2\lambda_{2i-4}r_{i-2,i} + 2\lambda_{2i-3}r_{i-1,i} - 2\lambda_{2i-1}r_{i,i+1}$$
$$- 2\lambda_{2i}r_{i,i+2} \tag{10.13}$$

where we assume that all masses are the same and use MD units. The corresponding equations for the two atoms at either end of the chain omit those terms that refer to nonexistent atoms.

The function that constructs **M** for the case of bond-length constraints follows. The matrix is stored columnwise as a linear array cMat, so that the array element with index $(k - 1)n_s + i$ corresponds to M_{ik}:

```
   BuildConstraintMatrix () {
     int i, m;
     for (i = 1; i <= chainLen * nConstraint; i ++)
       cMat[i] = 0.;
5    for (i = 1; i <= chainLen; i ++) {
       m = i - 1;
       if (m > 0) cMat[(m - 1) * chainLen + i] = 2;
       m = m + 1;
       if (m <= nConstraint)
10        cMat[(m - 1) * chainLen + i] = -2;
     }
     for (m = 1; m <= nConstraint; m ++) {
       cDistSq[m] = Sqr (bondLen);
       cAtom1[m] = m;    cAtom2[m] = m + 1;
15   }
   }
```

Here cDistSq contains the squares of the constrained distances, chainLen corresponds to n_s, and nConstraint to n_c. The version that includes angle constraints as well is

```
   BuildConstraintMatrix () {
     int i, m;
     for (i = 1; i <= chainLen * nConstraint; i ++)
       cMat[i] = 0.;
5    for (i = 1; i <= chainLen; i ++) {
       m = 2 * i - 4;
       if (m > 0) cMat[(m - 1) * chainLen + i] = 2;
       m = m + 1;
       if (m > 0) cMat[(m - 1) * chainLen + i] = 2;
10      m = m + 2;
       if (m <= nConstraint)
         cMat[(m - 1) * chainLen + i] = -2;
       m = m + 1;
       if (m <= nConstraint)
15        cMat[(m - 1) * chainLen + i] = -2;
     }
     for (m = 1; m <= nConstraint; m ++) {
       cDistSq[m] = Sqr (bondLen);
       if (m % 2 == 0) cDistSq[m] = cDistSq[m] * 2. *
```

```
20           (1. - cos (bondAngle));
         cAtom1[m] = (m + 1) / 2;    cAtom2[m] = (m + 4) / 2;
     }
   }
```

Evaluating the Lagrange multipliers and including the constraint forces
in the equations of motion is the task of the function given below. In
the course of the processing the matrix **L**, represented by cvMat, is
constructed and the linear equations (10.10) solved using a standard
method such as LU decomposition [pre92]; here the solution function is
simply referred to as SolveLineq (see Appendix). Both the right-hand
side of (10.10) and subsequently the solution (the set of $\lambda_{k'}$) are stored in
vVec; cVec holds the constraint vectors s_k:

```
   ComputeConstraints () {
     real dv, w;
     int i, k, m, mDif, m1, m2, n, nn;
     for (n = 1; n <= nChain; n ++) {
5      nn = (n - 1) * chainLen;
       for (m = 1; m <= nConstraint; m ++) {
         for (k = 1; k <= NDIM; k ++) {
           cVec[k][m] = r[k][nn + cAtom1[m]] -
             r[k][nn + cAtom2[m]];
10         if (fabs (cVec[k][m]) > regionH[k]) cVec[k][m] =
             cVec[k][m] - SignR (region[k], cVec[k][m]);
       } }
       m = 0;
       for (m1 = 1; m1 <= nConstraint; m1 ++) {
15       for (m2 = 1; m2 <= nConstraint; m2 ++) {
           m = m + 1;
           mDif = cMat[(m1 - 1) * chainLen + cAtom1[m2]] -
             cMat[(m1 - 1) * chainLen + cAtom2[m2]];
           cvMat[m] = 0.;
20         if (mDif != 0) cvMat[m] = mDif * (cVec[1][m1] *
             cVec[1][m2] + cVec[2][m1] * cVec[2][m2] +
             cVec[3][m1] * cVec[3][m2]);
       } }
       for (m = 1; m <= nConstraint; m ++) {
25       vVec[m] = 0.;
         for (k = 1; k <= NDIM; k ++) {
           dv = rv[k][nn + cAtom1[m]] - rv[k][nn + cAtom2[m]];
           vVec[m] = vVec[m] - (ra[k][nn + cAtom1[m]] -
             ra[k][nn + cAtom2[m]]) * cVec[k][m] - Sqr (dv);
30     } }
       SolveLineq (cvMat, vVec, nConstraint);
       for (m = 1; m <= nConstraint; m ++) {
```

```
           for (i = 1; i <= chainLen; i ++) {
             w = cMat[(m - 1) * chainLen + i];
35           if (w != 0.) {
               for (k = 1; k <= NDIM; k ++)
                 ra[k][nn + i] = ra[k][nn + i] + w * vVec[m] *
                    cVec[k][m];
     } } } }
40 }
```

Any residual drift in the constraints can be removed when necessary (see later), but the drift should be sufficiently small that quite a few timesteps can intervene between such adjustments.

A list of the new quantities appearing in the calculations (including some used later on) is as follows:

```
real **cVec, *cDistSq, *curBondLenSq, *cvMat, *vVec,
   bondAngle, bondLen, constraintDevA, constraintDevL,
   constraintPrec;
int *cAtom1, *cAtom2, *cMat, chainLen, cycleR, cycleV,
5   nChain, nConstraint, stepRestore;
```

Required array allocations (in AllocArrays) are

```
cVec = AllocMatR (NDIM, nConstraint);
cDistSq = AllocVecR (nConstraint);
curBondLenSq = AllocVecR (nConstraint);
cvMat = AllocVecR (nConstraint * nConstraint);
5 vVec = AllocVecR (nConstraint);
cAtom1 = AllocVecI (nConstraint);
cAtom2 = AllocVecI (nConstraint);
cMat = AllocVecI (chainLen * nConstraint);
```

The deviations of the supposedly constrained bond lengths from the correct values are easily monitored (here only averages are computed):

```
AnlzConstraintDevs () {
   real dr1[NDIM + 1], sumL;
   int i, k, n, ni;
   sumL = 0.;
5  for (n = 1; n <= nChain; n ++) {
     for (i = 1; i <= chainLen - 1; i ++) {
       ni = (n - 1) * chainLen + i;
       for (k = 1; k <= NDIM; k ++) {
         dr1[k] = r[k][ni + 1] - r[k][ni];
10       if (fabs (dr1[k]) > regionH[k])
             dr1[k] = dr1[k] - SignR (region[k], dr1[k]);
       }
       curBondLenSq[i] = Sqr (dr1[1]) + Sqr (dr1[2]) +
         Sqr (dr1[3]);
```

```
15          sumL = sumL + curBondLenSq[i];
      } }
      constraintDevL = sqrt (sumL / (nChain *
         (chainLen - 1))) - bondLen;
   }
```

If the bond angles are also constrained, add

```
      real ... dr2[NDIM + 1], sumA;
      ...
      sumA = 0.;
      ...
5     for (i = 2; i <= chainLen - 1; i ++) {
         ni = (n - 1) * chainLen + i;
         for (k = 1; k <= NDIM; k ++) {
            dr1[k] = r[k][ni + 1] - r[k][ni];
            if (fabs (dr1[k]) > regionH[k]) ...
10          dr2[k] = r[k][ni - 1] - r[k][ni];
            if (fabs (dr2[k]) > regionH[k]) ...
         }
         sumA = sumA + Sqr (dr1[1] * dr2[1] +
            dr1[2] * dr2[2] + dr1[3] * dr2[3]) /
15          (curBondLenSq[i - 1] * curBondLenSq[i]);
      }
      ...
      constraintDevA = sqrt (sumA / (nChain *
         (chainLen - 2))) - cos (pi - bondAngle);
```

10.3.2 Relaxation method

This approach to dealing with constraints – the so-called 'shake' method – [ryc77] begins by advancing the system over a single timestep while ignoring the constraints. If the simple Verlet integration method (Chapter 3) is used we obtain a set of uncorrected coordinates:

$$r_i'(t + h) = 2r_i(t) - r_i(t - h) + (h^2/m_i)F_i(t) \tag{10.14}$$

We now want to adjust all the r_i' to obtain corrected coordinates r_i that satisfy the constraints. This can be done by adding in the missing constraint force term (10.4); since $\nabla_i \sigma_k = 2r_{ij}(t)$ this leads to

$$r_i(t + h) = r_i'(t + h) - 2(h^2/m_i) \sum_{k \in C(i)} \lambda_k r_{ij}(t) \tag{10.15}$$

At this point we change the meaning of λ_k; it will no longer be regarded as a Lagrange multiplier, but rather as an additional variable whose value is determined by having the constraint satisfied to full numerical accuracy and not subject to the truncation error of the numerical integration. This will ensure that, despite the numerical error experienced by the atomic trajectories, the constrained bond lengths and angles always maintain their correct values.

Implementation of the iterative method begins by setting $r_i' = r_i'(t+h)$ and then, for each constraint, applying corrections along the direction of $r_{ij}(t)$

$$r_i'' = r_i' - 2(h^2/m_i)\gamma r_{ij}(t), \quad r_j'' = r_j' + 2(h^2/m_i)\gamma r_{ij}(t) \tag{10.16}$$

The correction γ is determined from the solution of $r_{ij}''^2 = d_{ij}^2$, namely,

$$\left[r_{ij}' - 2h^2(1/m_i + 1/m_j)\gamma r_{ij}\right]^2 = d_{ij}^2 \tag{10.17}$$

which, to lowest order in h^2, is

$$\gamma = \frac{r_{ij}'^2 - d_{ij}^2}{4h^2(1/m_i + 1/m_j)r_{ij}' \cdot r_{ij}} \tag{10.18}$$

The estimated coordinates r_i' and r_j' are then updated using this value of γ in (10.16). The process is repeated, cycling through each of the constraints in turn, until all the constraints satisfy $|r_{ij}'^2 - d_{ij}^2| < \epsilon_r d_{ij}^2$, where ϵ_r is the specified tolerance. (The value of λ_k is just the sum of all the γ-corrections for that constraint, but the value itself is not needed in the calculation.)

As shown here the method is tied to a specific integration method, but a very similar result can be used for restoring constraints in general. Simply write

$$r_i' = r_i - \gamma r_{ij}, \quad r_j' = r_j + \gamma r_{ij} \tag{10.19}$$

where γ is now just a small number, and then solve the equations $r_{ij}'^2 = d_{ij}^2$ iteratively as before; here we assume that all atoms have the same mass, otherwise the inverse mass terms must be included to avoid moving the center of mass. If terms quadratic in γ are neglected, the solution is reminiscent of the one above:

$$\gamma = \frac{r_{ij}^2 - d_{ij}^2}{4r_{ij}^2} \tag{10.20}$$

(d_{ij}^2 can replace r_{ij}^2 in the denominator).

The velocities can be corrected in a similar manner, thereby ensuring that the atoms have zero relative velocity along the direction of their mutual constraint; such corrections can also be incorporated in the original 'shake' method [and83]. Each such restriction is equivalent to

$$\dot{\sigma}_k = 2\dot{r}_{ij} \cdot r_{ij} = 0 \qquad (10.21)$$

Following the same approach as before, the velocities are adjusted by iterating the equations

$$\dot{r}'_i = \dot{r}_i - \gamma r_{ij}, \quad \dot{r}'_j = \dot{r}_j + \gamma r_{ij} \qquad (10.22)$$

where the value of γ is now chosen to ensure that $\dot{r}'_{ij} \cdot r_{ij} = 0$:

$$\gamma = \frac{\dot{r}_{ij} \cdot r_{ij}}{2r_{ij}^2} \qquad (10.23)$$

(since r_{ij} already satisfies the constraint, d_{ij}^2 can be used in the denominator). The process is repeated until all corrections fall below a specified tolerance.

The function for restoring the coordinates and velocities to their constrained values in this more general case is

```
RestoreConstraints () {
   real dr[NDIM + 1], cDev, cDevR, cDevV, g, ga;
   int changed, k, m, maxCycle, m1, m2, n, nn;
   maxCycle = 200;      cDevR = cDevV = 0.;
5  for (n = 1; n <= nChain; n ++) {
      nn = (n - 1) * chainLen;
      cycleR = 0;      changed = 1;
      while (cycleR < maxCycle && changed) {
      cycleR = cycleR + 1;      changed = 0;
10    cDev = 0.;
      for (m = 1; m <= nConstraint; m ++) {
        m1 = nn + cAtom1[m];      m2 = nn + cAtom2[m];
        for (k = 1; k <= NDIM; k ++) {
          dr[k] = r[k][m1] - r[k][m2];
15        if (fabs (dr[k]) > regionH[k])
              dr[k] = dr[k] - SignR (region[k], dr[k]);
        }
        g = (Sqr (dr[1]) + Sqr (dr[2]) + Sqr (dr[3]) -
            cDistSq[m]) / (4. * cDistSq[m]);
20      ga = fabs (g);
        if (ga > cDev) cDev = ga;
        if (ga > constraintPrec) {
          changed = 1;
          for (k = 1; k <= NDIM; k ++) {
25          r[k][m1] = r[k][m1] - g * dr[k];
            r[k][m2] = r[k][m2] + g * dr[k];
      } } } } }
```

```
      if (cDev > cDevR) cDevR = cDev;
      cycleV = 0;     changed = 1;
30    while (cycleV < maxCycle && changed) {
        cycleV = cycleV + 1;     changed = 0;
        cDev = 0.;
        for (m = 1; m <= nConstraint; m ++) {
          m1 = nn + cAtom1[m];     m2 = nn + cAtom2[m];
35        g = 0.;
          for (k = 1; k <= NDIM; k ++) {
            dr[k] = r[k][m1] - r[k][m2];
            if (fabs (dr[k]) > regionH[k])
              dr[k] = dr[k] - SignR (region[k], dr[k]);
40          g = g + (rv[k][m1] - rv[k][m2]) * dr[k];
          }
          g = g / (2. * cDistSq[m]);
          ga = fabs (g);
          if (ga > cDev) cDev = ga;
45        if (ga > constraintPrec) {
            changed = 1;
            for (k = 1; k <= NDIM; k ++) {
              rv[k][m1] = rv[k][m1] - g * dr[k];
              rv[k][m2] = rv[k][m2] + g * dr[k];
50  } } } }
      if (cDev > cDevV) cDevV = cDev;
    }
  }
```

Here `constraintPrec` is the tolerance used in establishing convergence. The limit `maxCycle` is introduced as a safety measure; the number of iterations should generally not be much greater than ten or so (for small molecules), otherwise one might be inclined to suspect the reliability of the whole approach.

10.4 Internal forces

10.4.1 Bond-torsion force

The torsional force associated with twisting around a bond is another example of an effective interaction. This particular motion provides the means for local changes in spatial arrangement of the polymer chain; simultaneous twisting around two bonds, for example, is all that is needed for a crankshaft type of motion. The force associated with the twist, or torsional degree of freedom, is defined in terms of the relative coordinates of four consecutive atoms, here for convenience labeled 1–4; if bond lengths and angles are fixed this force depends on the angle of

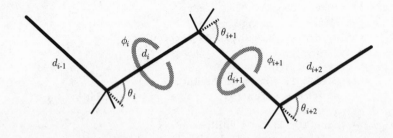

Fig. 10.1 The bonds, bond angles, and dihedral angles in a portion of an alkane chain.

rotation around the bond between atoms 2 and 3 – the so-called dihedral angle. The dihedral angle is defined as the angle between the planes formed by atoms 1,2,3 and 2,3,4 measured in the plane normal to the 2−3 bond; it is zero when all four atoms are coplanar and atoms 1 and 4 are on opposite sides of the bond. Only these four atoms are subject to this particular torsion force, and the purpose of the following analysis is to determine the force on each.

Labeling the atoms, bonds, and angles of a linear polymer in a systematic manner is trivial for linear chains; for molecules with other topologies the problem is more complex, requiring an algorithm (not addressed here) that systematically traverses the graph describing the connectivity of the molecule. Here bond i joins atoms $i-1$ and i, and is denoted by the vector

$$d_i = r_i - r_{i-1} \tag{10.24}$$

As shown in Figure 10.1 the angle between bonds $i-1$ and i is given by

$$\cos \theta_i = \frac{d_{i-1} \cdot d_i}{|d_{i-1}||d_i|} \tag{10.25}$$

where $\theta_i = 0$ when the bonds are parallel; by convention, the 'bond angle' refers to $\pi - \theta_i$. The dihedral angle associated with bond i is obtained from

$$\cos \phi_i = -\frac{(d_{i-1} \times d_i) \cdot (d_i \times d_{i+1})}{|d_{i-1} \times d_i||d_i \times d_{i+1}|} \tag{10.26}$$

There are two parts to the torsional force calculation: the functional dependence on the dihedral angle ϕ_i and the vector algebra used to derive expressions for the forces on each of the four affected atoms. We begin with the second part [pea79, dun92]. If we define

$$c_{ij} = d_i \cdot d_j \tag{10.27}$$

then we can express the bond and dihedral angles as

$$\cos \theta_i = c_{i-1,i}/(c_{i-1,i-1}c_{ii})^{1/2}, \quad \cos \phi_i = a_i/b_i^{1/2} \tag{10.28}$$

where for conciseness we have introduced the quantities

$$a_i = c_{i-1,i+1}c_{ii} - c_{i-1,i}c_{i,i+1} \tag{10.29}$$

$$b_i = \left(c_{i-1,i-1}c_{ii} - c_{i-1,i}^2\right)\left(c_{ii}c_{i+1,i+1} - c_{i,i+1}^2\right) \tag{10.30}$$

The torque caused by a rotation about bond i produces forces on the four atoms $j = i - 2, \dots, i + 1$ equal to

$$-\nabla_{r_j} u(\phi_i) = -\left.\frac{du(\phi)}{d(\cos \phi)}\right|_{\phi = \phi_i} f_j^{(i)} \tag{10.31}$$

where $u(\phi)$ is the torsion potential and

$$f_j^{(i)} = \nabla_{r_j} \cos \phi_i \tag{10.32}$$

It is clear that the sums of the forces and torques are zero, therefore

$$\sum_{j=i-2}^{i+1} f_j^{(i)} = 0, \qquad \sum_{j=i-2}^{i+1} r_j \times f_j^{(i)} = 0 \tag{10.33}$$

or, equivalently,

$$(d_{i-1} + d_i) \times f_{i-2}^{(i)} + d_i \times f_{i-1}^{(i)} - d_{i+i} \times f_{i+1}^{(i)} = 0 \tag{10.34}$$

Since (10.33) provides two relations between the four $f_j^{(i)}$ we can write

$$f_{i-1}^{(i)} = \alpha f_{i-2}^{(i)} + \beta f_{i+1}^{(i)} \tag{10.35}$$

so that (10.34) becomes

$$(d_{i-1} + d_i + \alpha d_i) \times f_{i-2}^{(i)} + (\beta d_i - d_{i+1}) \times f_{i+1}^{(i)} = 0 \tag{10.36}$$

and because both $f_{i-2}^{(i)}$ and $f_{i+1}^{(i)}$ are normal to d_i it follows that

$$\alpha = -1 - c_{i-1,i}/c_{ii}, \quad \beta = c_{i,i+1}/c_{ii} \tag{10.37}$$

Hence,

$$f_{i-1}^{(i)} = -(1 + c_{i-1,i}/c_{ii}) f_{i-2}^{(i)} + (c_{i,i+1}/c_{ii}) f_{i+1}^{(i)} \tag{10.38}$$

and, since the four $f_j^{(i)}$ sum to zero,

$$f_i^{(i)} = (c_{i-1,i}/c_{ii}) f_{i-2}^{(i)} - (1 + c_{i,i+1}/c_{ii}) f_{i+1}^{(i)} \tag{10.39}$$

We next evaluate $f_{i-2}^{(i)}$ and $f_{i+1}^{(i)}$ by expanding (10.32):

$$f_j^{(i)} = b_i^{-3/2} \left(b_i \nabla_{r_j} a_i - a_i \nabla_{r_j} b_i / 2 \right) \tag{10.40}$$

In order to complete the evaluation we need the derivatives of all the scalar products $d_p \cdot d_q$ with respect to r_j; the full list is

$$\nabla_{r_j} c_{pq} = \begin{cases} 2d_j & p = q = j \\ -2d_{j+1} & p = q = j+1 \\ d_{j+1} - d_j & p = j, \ q = j+1 \\ d_q & p = j, \ q \neq j, j+1 \\ -d_q & p = j+1, \ q \neq j, j+1 \\ 0 & \text{otherwise} \end{cases} \tag{10.41}$$

A certain amount of rather tedious algebra using the above results (made easier if aided by suitable symbolic software) leads to

$$f_{i-2}^{(i)} = \frac{c_{ii}}{b_i^{1/2}(c_{i-1,i-1}c_{ii} - c_{i-1,i}^2)} \left[t_1 d_{i-1} + t_2 d_i + t_3 d_{i+1} \right] \tag{10.42}$$

$$f_{i+1}^{(i)} = \frac{c_{ii}}{b_i^{1/2}(c_{ii}c_{i+1,i+1} - c_{i,i+1}^2)} \left[t_4 d_{i-1} + t_5 d_i + t_6 d_{i+1} \right] \tag{10.43}$$

where

$$\begin{aligned} t_1 &= c_{i-1,i+1}c_{ii} - c_{i-1,i}c_{i,i+1} \\ t_2 &= c_{i-1,i-1}c_{i,i+1} - c_{i-1,i}c_{i-1,i+1} \\ t_3 &= c_{i-1,i}^2 - c_{i-1,i-1}c_{ii} \\ t_4 &= c_{ii}c_{i+1,i+1} - c_{i,i+1}^2 \\ t_5 &= c_{i-1,i+1}c_{i,i+1} - c_{i-1,i}c_{i+1,i+1} \\ t_6 &= -t_1 \end{aligned} \tag{10.44}$$

Both force vectors are normal to d_i, although this may be less than obvious from an expression such as (10.42). In certain cases considerable simplification is possible; if, for example, all $|d_i| = d$ and $\theta_i = \theta$, then

$$\cos \phi_i = \frac{d_{i-1} \cdot d_{i+1}/d^2 - \cos^2 \theta}{1 - \cos^2 \theta} \tag{10.45}$$

The force expressions are simplified, but the reduction in computational effort is probably not large enough to justify separate functions for individual cases.

The torsional potential function is typically expressed in polynomial

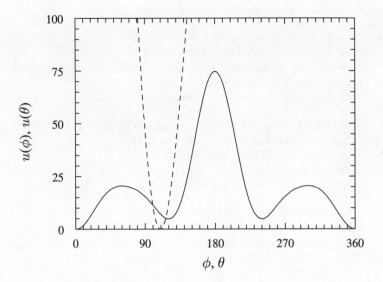

Fig. 10.2 Bond-torsion and bond-angle potential functions, $u(\phi)$ and $u(\theta)$.

form [ryc78]:

$$u(\phi) = \sum_{j \geq 0} w_j \cos^j \phi \tag{10.46}$$

so that the derivative used for the forces in (10.31) is

$$-\frac{du(\phi)}{d(\cos\phi)} = -\sum_{j \geq 1} j w_j \cos^{j-1} \phi \tag{10.47}$$

For the alkane model, the potential (whose coefficients, in energy units appropriate to the problem, are incorporated into the listing below) is shown in Figure 10.2. The deepest minimum is at the 'trans' angle $\phi = 0$, two secondary minima at the 'gauche' angles ($\pm 2\pi/3$), barriers at $\phi = \pm \pi/3$, and a maximum at π. Other similar functions are also in use for this model [cla90].

The torsional contribution to the interactions is computed by the following function (DotProd3 – see Appendix – evaluates the scalar product of two vectors):

```
ComputeChainTorsionForces () {
    real dr1[4], dr2[4], dr3[4], c, cA, cB1, cB2, cD,
        cR1, cR2, c11, c12, c13, c22, c23, c33, f, f1, f2,
        t1, t2, t3, t4, t5, t6,
 5      g0 = 1.000, g1 = 1.310, g2 = -1.414, g3 = -0.330,
```

```
              g4 = 2.828, g5 = -3.394, uCon = 15.50;
         int i, k, n, nn;
         for (n = 1; n <= nChain; n ++) {
          for (i = 1; i <= chainLen - 3; i ++) {
10          nn = (n - 1) * chainLen + i;
            for (k = 1; k <= 3; k ++) {
              dr1[k] = r[k][nn + 1] - r[k][nn];
              if (fabs (dr1[k]) > regionH[k])
                dr1[k] = dr1[k] - SignR (region[k], dr1[k]);
15            dr2[k] = r[k][nn + 2] - r[k][nn + 1];
              if (fabs (dr2[k]) > regionH[k]) ...
              dr3[k] = r[k][nn + 3] - r[k][nn + 2];
              if (fabs (dr3[k]) > regionH[k]) ...
            }
20          c11 = DotProd3 (dr1, dr1);
            c12 = DotProd3 (dr1, dr2);
            c13 = DotProd3 (dr1, dr3);
            c22 = DotProd3 (dr2, dr2);
            c23 = DotProd3 (dr2, dr3);
25          c33 = DotProd3 (dr3, dr3);
            cA = c13 * c22 - c12 * c23;
            cB1 = c11 * c22 - c12 * c12;
            cB2 = c22 * c33 - c23 * c23;
            cD = sqrt (cB1 * cB2);     c = cA / cD;
30          f = - uCon * (g1 + (2. * g2 + (3. * g3 + (4. * g4 +
              5. * g5 * c) * c) * c) * c);
            t1 = cA;     t2 = c11 * c23 - c12 * c13;
            t3 = - cB1;     t4 = cB2;
            t5 = c13 * c23 - c12 * c33;     t6 = - cA;
35          cR1 = c12 / c22;     cR2 = c23 / c22;
            for (k = 1; k <= 3; k ++) {
              f1 = f * c22 * (t1 * dr1[k] + t2 * dr2[k] +
                t3 * dr3[k]) / (cD * cB1);
              f2 = f * c22 * (t4 * dr1[k] + t5 * dr2[k] +
40              t6 * dr3[k]) / (cD * cB2);
              ra[k][nn] = ra[k][nn] + f1;
              ra[k][nn + 1] = ra[k][nn + 1] - (1. + cR1) * f1 +
                cR2 * f2;
              ra[k][nn + 2] = ra[k][nn + 2] + cR1 * f1 -
45              (1. + cR2) * f2;
              ra[k][nn + 3] = ra[k][nn + 3] + f2;
            }
            uSum = uSum + uCon * (g0 + (g1 + (g2 + (g3 + (g4 +
              g5 * c) * c) * c) * c) * c);
50        } }
         }
```

10.4.2 *Bond-angle force*

While bond lengths are generally held fixed by constraints, there is no clear preference for bond angles, and both constraints and potentials are in use [cla90, ryc90]. Here we treat the case where interactions ensure that bond angles have only limited variation; the angle and torsion forces are assumed fully independent. We use the same notation as before.

A change in the angle θ_i produces forces on the three atoms $j = i - 2, i - 1, i$ given by

$$-\nabla_{r_j} u(\theta_i) = -\left.\frac{du(\theta)}{d(\cos\theta)}\right|_{\theta=\theta_i} f_j^{(i)} \tag{10.48}$$

where $u(\theta)$ is the angle potential and $f_j^{(i)} = \nabla_{r_j} \cos\theta_i$. The sum of the three forces is zero. More of the above algebra leads to

$$f_{i-2}^{(i)} = (c_{i-1,i-1}c_{ii})^{-1/2}\left[(c_{i-1,i}/c_{i-1,i-1})d_{i-1} - d_i\right] \tag{10.49}$$

$$f_i^{(i)} = (c_{i-1,i-1}c_{ii})^{-1/2}\left[d_{i-1} - (c_{i-1,i}/c_{ii})d_i\right] \tag{10.50}$$

The potential associated with bond-angle changes for the alkane model is

$$u(\theta) = (w/2)(\cos\theta - \cos\theta_0)^2 \tag{10.51}$$

where w is a constant [cla90] and $\cos\theta_0 = 1/3$; a plot of the potential function is included in Figure 10.2.

The function that carries out the force and energy computations for this interaction follows:

```
     ComputeChainAngleForces () {
         real dr1[4], dr2[4], c, cCon, cD, c11, c12, c22, f,
             f1, f2, uCon = 868.6;
         int i, k, n, nn;
5        cCon = cos (pi - bondAngle);
         for (n = 1; n <= nChain; n ++) {
             for (i = 1; i <= chainLen - 2; i ++) {
                 nn = (n - 1) * chainLen + i;
                 for (k = 1; k <= 3; k ++) {
10                   dr1[k] = r[k][nn + 1] - r[k][nn];
                     if (fabs (dr1[k]) > regionH[k]) ...
                     dr2[k] = r[k][nn + 2] - r[k][nn + 1];
                     if (fabs (dr2[k]) > regionH[k]) ...
                 }
15               c11 = DotProd3 (dr1, dr1);
                 c12 = DotProd3 (dr1, dr2);
                 c22 = DotProd3 (dr2, dr2);
```

```
     cD = sqrt (c11 * c22);     c = c12 / cD;
     f = - uCon * (c - cCon);
20   for (k = 1; k <= 3; k ++) {
       f1 = f * ((c12 / c11) * dr1[k] - dr2[k]) / cD;
       f2 = f * (dr1[k] - (c12 / c22) * dr2[k]) / cD;
       ra[k][nn] = ra[k][nn] + f1;
       ra[k][nn + 1] = ra[k][nn + 1] - f1 - f2;
25     ra[k][nn + 2] = ra[k][nn + 2] + f2;
     }
     uSum = uSum + 0.5 * uCon * Sqr (c - cCon);
   } }
 }
```

10.4.3 Other interactions

So far we have discussed just two of the interactions in the model, namely, the bond-torsion and bond-angle forces. Pairs of atoms in each molecule that are neither directly linked by a constraint, nor jointly involved in these three- and four-body forces, interact with the usual LJ potential (the butane molecule studied later on is sufficiently small that there are no pairs in this category). Atoms in different molecules interact with the same force, and solvent atoms can also be included with similar or distinct interaction parameters, depending on what is being modeled; here, for faster computation, the LJ interactions are replaced by soft spheres [tox88].

If the neighbor-list method is used for computing the interactions between pairs of atoms not involved in constraints and bond forces, the only change required in BuildNebrList is the elimination of such pairs. This is done by modifying the condition used to select atom pairs for the list:

```
if ((m1 != m2 || j2 < j1) && (inPoly[j1] == 0 ||
    inPoly[j1] != inPoly[j2] || abs (j1 - j2) > 3))
```

The additional test checks whether both atoms belong to the same molecule (inPoly is used in the same way as for flexible chains, see Chapter 9) and if this is true then how far apart they are. The interaction functions called from SingleStep are (the first function only if relevant)

```
ComputeChainAngleForces ();
ComputeChainTorsionForces ();
ComputeConstraints ();
```

Adjustment of minor constraint deviations is carried out at suitable intervals by

```
  cycleR = cycleV = 0;
  if (stepCount % stepRestore == 0) {
    RestoreConstraints ();
    ApplyBoundaryCond ();
5 }
```

To obtain reports on how well the constraints are preserved add the following to SingleStep prior to any call to RestoreConstraints:

```
  if (stepCount % stepAvg == 0) AnlzConstraintDevs ();
```

and to PrintSummary add

```
  fprintf (fp, "constraint devs: %.3e %.3e cycles: %d %d\n",
    constraintDevL, constraintDevA, cycleR, cycleV);
```

The function RestoreConstraints should also be called at the beginning of the run to correct the randomly assigned initial velocities.

10.5 Implementation details

10.5.1 Initial state and parameters

The initial state uses the same BCC lattice arrangement and planar zigzag (or trans) conformation used previously for flexible chains (Chapter 9) but with distances and angles modified (InitCoords):

```
  bY = bondLen * cos (bondAngle / 2.);
  bZ = bondLen * sin (bondAngle / 2.);
```

New input data items are

```
  INAME (initUchain),
  RNAME (bondAngle),
  RNAME (bondLen),
  INAME (chainLen),
5 RNAME (constraintPrec),
  INAME (stepRestore),
```

and initialization (SetupJob) requires

```
  AssignToChain ();
  BuildConstraintMatrix ();
```

No solvent is used here, so the values of initUchain replace initUcell when determining the region size. (We do not include a simulation that has both multiple chains and solvent; in this case there will be two independent densities – for chains and for solvent atoms – that have to be specified.)

The effect of the constraints must be taken into account when setting the initial velocities to correspond to a given temperature. The total number of degrees of freedom per chain is reduced from $3n_s$ to $2n_s + 1$ in the case of bond-length constraints, and to $n_s + 3$ if bond angles are also constrained. This, and the loss of three more degrees of freedom because of momentum conservation, are included in the variable vMag evaluated in SetParams (the value of nConstraint shown here is for the case of length and angle constraints):

```
nConstraint = 2 * chainLen - 3;
vMag = sqrt ((NDIM * (1. - 1. / nAtom) -
   (real) nConstraint / chainLen) * temperature);
```

Temperature adjustment early in the run uses the function InitAdjust-Temp.

The reduced length and energy units [ryc78] are $\sigma = 3.92\,\text{Å}$ and $\epsilon/k_B = 72\,\text{K}$. All atoms (or monomers) are assumed to have mass $2.411 \times 10^{-23}\,\text{g}$, and this is defined as the unit of mass in MD units. The unit of time is then $1.93 \times 10^{-12}\,\text{s}$. The bond length of $1.53\,\text{Å}$ used in the model corresponds to 0.390; the bond angle is $109.47°$. At the density of liquid butane ($0.675\,\text{g/cm}^3$) there are 0.365 molecules per unit volume, in MD units.

10.5.2 Structural properties

Properties of the chain fluid as a whole can be studied using the atomic RDF, as in Chapter 4. An extra test is needed in EvalRdf to eliminate the very sharp peaks at the fixed nearest-neighbor separation and at either the fixed or the narrowly spread next-nearest-neighbor distance:

```
if (inPoly[j1] == inPoly[j2] && inPoly[j1] != 0 &&
   abs (j1 - j2) < 3) continue;
```

The first example of a measurement specific to this chain model constructs a normalized histogram showing the dihedral-angle distribution averaged over all chains and over all the angles in each chain. If there is some reason to believe that the distribution depends on where the bond is located in the chain (not for the example studied here), then the results for each bond would have to be maintained separately. Use of this function follows the familiar pattern established in earlier case studies:

```
AccumDihedAngDistn (int icode) {
   real dr1[4], dr2[4], dr3[4], cosAngSq, c11, c12, c13,
      c22, dihedAng, t;
   int i, j, k, n, nn;
```

```
5    if (icode == 0) {
       for (j = 1; j <= sizeHistDihedAng; j ++)
         histDihedAng[j] = 0.;
     } else if (icode == 1) {
       for (n = 1; n <= nChain; n ++) {
10       for (i = 1; i <= chainLen - 3; i ++) {
           nn = (n - 1) * chainLen + i;
           for (k = 1; k <= 3; k ++) {
             ... (same as ComputeChainTorsionForces) ...
           }
15         c11 = DotProd3 (dr1, dr1);
           c12 = DotProd3 (dr1, dr2);
           c13 = DotProd3 (dr1, dr3);
           c22 = DotProd3 (dr2, dr2);
           cosAngSq = Sqr (c12) / (c11 * c22);
20         t = (- cosAngSq + c13 / Sqr (bondLen)) /
             (1. - cosAngSq);
           if (fabs (t) > 1.) t = SignR (1., t);
           dihedAng = acos (t);
           if (dr1[1] * (dr2[2] * dr3[3] - dr2[3] * dr3[2]) +
25             dr1[2] * (dr2[3] * dr3[1] - dr2[1] * dr3[3]) +
               dr1[3] * (dr2[1] * dr3[2] - dr2[2] * dr3[1]) <
               0.) dihedAng = 2. * pi - dihedAng;
           j = (int) (dihedAng * sizeHistDihedAng /
             (2. * pi)) + 1;
30         histDihedAng[j] = histDihedAng[j] + 1.;
       } }
     } else if (icode == 2) {
       t = 0.;
       for (j = 1; j <= sizeHistDihedAng; j ++)
35       t = t + histDihedAng[j];
       for (j = 1; j <= sizeHistDihedAng; j ++)
         histDihedAng[j] = histDihedAng[j] / t;
     }
   }
```

The function is called from `SingleStep` by

```
if (stepCount >= stepEquil && (stepCount - stepEquil)
  stepChainProps == 0) AccumDihedAngDistn (1);
```

New variables used here are

```
real *histDihedAng;
int sizeHistDihedAng, stepChainProps;
```

additional input data items are

```
INAME (sizeHistDihedAng),
INAME (stepChainProps),
```

and an array must be allocated:

```
histDihedAng = AllocVecR (sizeHistDihedAng);
```

After normalizing by a call to AccumDihedAngDistn(2) the results are output as part of PrintSummary (Chapter 2):

```
     fprintf (fp, "dihedral ang\n");
     for (n = 1; n <= sizeHistDihedAng; n ++) {
       hVal = (n - 0.5) * 360. / sizeHistDihedAng;
       fprintf (fp, "%5.1f %.4f\n", hVal, histDihedAng[n]);
5    }
```

Since we have chosen to output the results of this analysis together with the energy (etc.) summary, the accumulated results must be zeroed by a call AccumDihedAngDistn(0) whenever AccumProps(0) is called.

The next example considers the bond-angle distribution, and is obviously only relevant where bond angle is controlled by a potential rather than by a constraint. The computation is very similar to the preceding one, so we will omit most of the details:

```
     AccumBondAngDistn (int icode) {
       real bondAng ...
       ...
       if (icode == 0) {
5        ...
       } else if (icode == 1) {
         for (n = 1; n <= nChain; n ++) {
           for (i = 1; i <= chainLen - 2; i ++) {
             nn = (n - 1) * chainLen + i;
10           for (k = 1; k <= 3; k ++) {
               ... (same as ComputeChainAngleForces) ...
             }
             c11 = ...
             ...
15           bondAng = pi - acos (c12 / sqrt (c11 * c22));
             j = (int) (bondAng * sizeHistBondAng / pi) + 1;
             histBondAng[j] = histBondAng[j] + 1.;
       } }
       } else if (icode == 2) {
20       ... (based on AccumDihedAngDistn) ...
       }
     }
```

The final example considers the time-dependence of the dihedral-angle autocorrelation function. The quantity measured is

$$C(t) = \langle \cos(\phi_i(t) - \phi_i(0)) \rangle \tag{10.52}$$

and again no distinction is made between different bonds, although (for longer chains) it is quite likely that the time-dependence will vary with the position in the chain. We also omit any mention of overlapped data buffers (Chapter 5) that could be used to improve the quality of the results:

```
   EvalDihedAngCorr () {
     real dr1[4], dr2[4], dr3[4], cosAngSq, c11, c12, c13,
        c22, dihedAng, t;
     int i, j, k, n, nn;
 5   countDihedAngCorr = countDihedAngCorr + 1;
     dihedAngCorr[countDihedAngCorr] = 0.;
     j = 0;
     for (n = 1; n <= nChain; n ++) {
       for (i = 1; i <= chainLen - 3; i ++) {
10         j = j + 1;
         nn = (n - 1) * chainLen + i;
         ... (same as AccumDihedAngDistn) ...
         if (dr1[1] * ...) dihedAng = 2. * pi - dihedAng;
         if (countDihedAngCorr == 1)
15           dihedAngOrg[j] = dihedAng;
         dihedAngCorr[countDihedAngCorr] =
            dihedAngCorr[countDihedAngCorr] +
            cos (dihedAng - dihedAngOrg[j]);
     } }
20   if (countDihedAngCorr == limitDihedAngCorr) {
       for (n = 1; n <= limitDihedAngCorr; n ++)
          dihedAngCorr[n] = dihedAngCorr[n] / nDihedAng;
       PrintDihedAngCorr (stdout);
       countDihedAngCorr = 0;
25   }
   }
```

New variables needed are

```
   real *dihedAngCorr, *dihedAngOrg;
   int countDihedAngCorr, limitDihedAngCorr, nDihedAng,
     stepDihedAngCorr;
```

input data are

```
   INAME (limitDihedAngCorr),
   INAME (stepDihedAngCorr),
```

and a quantity that is computed in SetParams:

```
   nDihedAng = nChain * (chainLen - 3);
```

The array allocations are

```
   dihedAngCorr = AllocVecR (limitDihedAngCorr);
```

```
dihedAngOrg = AllocVecR (nDihedAng);
```

initialization:

```
countDihedAngCorr = 0;
```

and, finally, the call for the actual processing is

```
if (stepCount >= stepEquil && (stepCount - stepEquil)
   stepDihedAngCorr == 0) EvalDihedAngCorr ();
```

10.6 Measurements

10.6.1 Constraint preservation

The runs used in this case study include the following input data:

```
initUchain      3 3 3
density         0.365
temperature     4.17
deltaT          0.002
bondAngle       1.91063
bondLen         0.39
chainLen        4
constraintPrec  1.0e-5
stepAvg         4000
stepChainProps  5
stepRestore     200
```

Since a BCC lattice is used for the initial state the total number of chains is 54.

The above data are for a model butane liquid in which both bond-length and bond-angle constraints are applied. Constant-energy MD is used together with PC integration. No energy adjustments are made after the correct temperature is reached, but the energy drift over 70 000 steps is just 4%.

If we examine the degree to which the constraints are maintained over the first few thousand steps we find that if constraints are restored using the relaxation method every 200 timesteps (stepRestore), then the deviations measured by AnlzConstraintDevs are typically 2×10^{-4} for constraintDevL and 10^{-3} for constraintDevA. If restoration occurs every 100 steps, then the deviations are both reduced by a factor of three. Typical numbers of restoration cycles needed each time (cycleR, cycleV) are in the approximate range $3-15$.

The alternative is to replace the bond-angle constraint by a potential. Because of the very stiff nature of this interaction the timestep must be reduced by a factor of four to 0.0005 in order to achieve the same degree of energy conservation. If constraints are restored every 1600 steps (equivalent to 400 of the larger steps in the preceding test), then the deviation measured by `constraintDevL` is 6×10^{-5}; more frequent restoration at intervals of 800 or 400 steps leads to deviations of size 2×10^{-5} and 6×10^{-6} (close to the tolerance level) respectively. Typically, $2-4$ restoration cycles are required in this case.

10.6.2 Properties

The RDF obtained from the butane simulation is shown in Figure 10.3 for the case of both bond-length and bond-angle constraints. Additional input data needed for this computation are

```
limitRdf        100
rangeRdf        2.2
sizeHistRdf     110
stepLimit       21000
stepRdf         50
```

Only the final set of averaged RDF values is considered. The RDF computation excludes the contributions from nearest- and next-nearest-neighbor pairs within each chain whose separations are fixed by the constraints.

We now turn to the distribution of dihedral angles, $f(\phi)$, for both kinds of constraint – bond length only and length/angle – and for the former, the bond-angle distribution, $f(\theta)$, as well. Total run lengths are (a relatively short) 70 000 timesteps for length/angle constraints and four times this value for the length constraints. In the former case we omit the first two sets of output and average over the remaining 15 sets; in the latter the first eight sets are skipped, leaving 64 sets for computing the average distributions. The extra input data are

```
sizeHistBondAng  36
sizeHistDihedAng 36
```

The results of this analysis appear in Figure 10.4. Length/angle constraints produce a slightly sharper distribution at zero dihedral angle, although it is not clear from the data shown whether the deviations are statistically significant (the omitted error bars could also account for this difference). The bond-angle distribution is relatively narrow, as might be expected from the stiff potential involved.

In Figure 10.5 we show the behavior of the dihedral-angle autocorre-

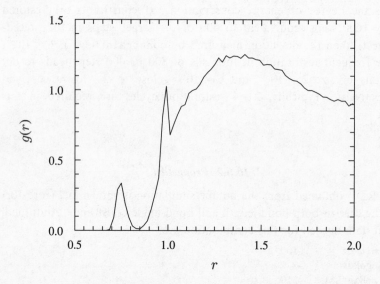

Fig. 10.3 RDF for liquid butane; intrachain pairs with fixed separations are excluded.

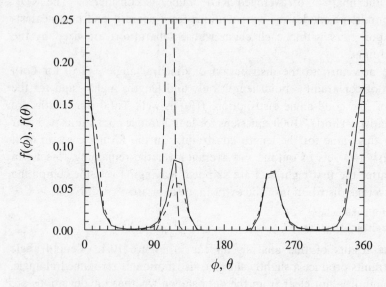

Fig. 10.4 Dihedral-angle distributions, $f(\phi)$, for butane chains subject to length/ angle constraints (solid curve), and to length constraints alone (short dashes); for the latter the narrow distribution of bond angles, $f(\theta)$, is also shown (long dashes).

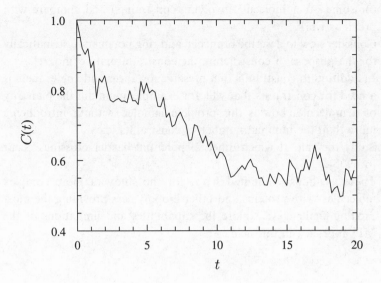

Fig. 10.5 Dihedral-angle autocorrelation function.

lation function for the case of length/angle constraints. Additional input data needed here are

```
limitDihedAngCorr 100
stepDihedAngCorr   100
```

Only a single series of measurements (covering a mere 10 000 steps) are represented here, and since there is no overlapped sampling the poor quality of the results is understandable. The large-t limit is determined by the average dihedral angle, but the results have not been adjusted to allow for this.

10.7 Further work

10.1 Compare the computational requirements of the matrix method for constraints with the 'shake' method using just relaxation; does the result depend on chain length?

10.2 Include a thermostat that acts on the centers of mass of the molecules [edb86]; how can one apply the thermostat to the intramolecular motion as well?

10.3 Pressure can be studied after establishing that the constraints do not contribute to the virial [cic86b]; measure the pressure for butane (or some other molecule involving constraints) and compare with previous work.

10.4 Consider how to describe branched and ring polymers systematically for the purpose of constructing the constraint matrix [mor91].

10.5 In addition to constraints that preserve distances and angles there is a need for constraints that will, for example, maintain the planarity of a molecule; how is this problem handled without introducing more than the minimal number of constraints [cic82]?

10.6 Study the rate at which dihedral potential barrier crossings occur [bro90b].

10.7 There is almost unlimited scope for the study of more complex molecules, with proteins and other biopolymers providing the most exciting challenges; explore the capabilities and limitations of the MD approach in this field.

11

Other interactions

11.1 Introduction

The range of problems amenable to study using MD knows few bounds, and as computers become more powerful the range will continue to expand. Because of the enormous breadth of the subject we have chosen to concentrate on the simplest of systems and avoid overly specialized models. The case studies so far have been based mainly on two-body interactions that are of comparatively short range; these permit a considerable variety of problems to be studied, but a few conspicuous gaps remain. Short-range forces, for example, exclude a very important group of problems involving electric charges and dipoles; long-range forces play an important role here, and there is no way their presence can be neglected. The two-body nature of the force introduces its own limitations, and while certain kinds of intermolecular interaction can be imitated by the appropriate combinations of pair potentials (as in Chapter 8), it is sometime essential to introduce many-body (in practice three-body) interactions to capture specific features of the 'real' intermolecular force [mai81].

In this chapter we introduce both long-range and three-body interactions. We cannot do justice to the range of applications to which these and other enhancements of the MD method contribute, but in the prevailing culinary atmosphere we hope the reader will gain at least a taste of what is involved.

11.2 Long-range forces

11.2.1 The problem

In the study of ionic crystals, where long-range electrostatic interactions dominate, the Ewald method [zim72] is able to take advantage of the periodic lattice structure to rearrange the expression for the total energy

263

into a form that can be readily evaluated. The same idea can be applied in MD simulations of charged and dipolar fluids with periodic boundaries, where, because the long-range force cannot be truncated without incurring serious error, it continues to act between the periodic replicas of the system as well. The Ewald technique thus eliminates the discontinuity arising from truncated long-range forces, although there are more subtle problems connected with how the properties are affected by the choice of boundary conditions [neu83, del86]. The computational effort depends to some extent on the degree of accuracy required but it grows considerably less rapidly than the all-pairs $O(N_a^2)$ rate.

While we will focus on an Ewald-type approach, other techniques do exist for dealing with long-range forces; space permits only the briefest mention. One is the reaction-field method, in which interactions are considered in full up to a certain range, and beyond this a mean-field approach is used [fri75]. This method is physically motivated, but there is another, more systematic, approach based on multipole expansions [gre89a, whi94]: by using a hierarchy of spatial subdivisions and a multipole expansion of the interaction contributions from the atoms in the cells generated at each level of subdivision, it is possible to evaluate interactions (including periodic boundaries) to any desired accuracy with little more than linear effort.

11.2.2 Ewald sums

Consider the system of N_a atoms, each of which now carries a charge (for more on Coulomb systems see [han86a]). A periodic array of replicated systems is created, in the spirit of the periodic boundary conditions used previously, but now, because of the long-range nature of the interactions, the energy of the replicated system includes contributions from all replicas since no truncation is imposed. The total interaction energy is

$$U_{qq} = \frac{1}{2}\sum_{n}' \sum_{i=1}^{N_a} \sum_{j=1}^{N_a} \frac{q_i q_j}{|r_{ij} + Ln|} \tag{11.1}$$

where q_i is the charge on atom i and L is the edge length (a cubic region is assumed). The sum is over all integer vectors n, and the prime indicates that terms with $i = j$ are omitted when $n = 0$ – self-interaction is prevented but atoms do interact with their replica images.

The Ewald formula is based on reorganizing this replica sum into sums over concentric spherical shells, assuming charge neutrality $\sum_j q_j = 0$:

$$U_{qq} = \sum_{i \le i < j \le N_a} q_i q_j \left[\sum_n \frac{\text{erfc}(\alpha|r_{ij} + Ln|)}{|r_{ij} + Ln|} \right.$$

$$\left. + \frac{1}{\pi L} \sum_{n \ne 0} \frac{1}{|n|^2} \exp\left(-\frac{\pi^2|n|^2}{\alpha^2 L^2} + \frac{2\pi i}{L} n \cdot r_{ij} \right) \right]$$

$$+ \frac{1}{2} \left[\sum_{n \ne 0} \left(\frac{\text{erfc}(\alpha L|n|)}{L|n|} + \frac{1}{\pi L|n|^2} \exp\left(-\frac{\pi^2|n|^2}{\alpha^2 L^2} \right) \right) \right.$$

$$\left. - \frac{2\alpha}{\pi^{1/2}} \right] \sum_{j=1}^{N_a} q_j^2 + \frac{2\pi}{3L^3} \left| \sum_{j=1}^{N_a} q_j r_j \right|^2 \qquad (11.2)$$

where $\text{erfc}(x) = (2/\pi^{1/2}) \int_x^\infty \exp(-t^2) dt$ is the complementary error function. Note that for $x \gg 1$, $\text{erfc}(x) \sim \exp(-x^2)$. There are various derivations of (11.2); one of these [del80] involves the introduction of a convergence factor into a series that is otherwise only conditionally convergent, followed by a Jacobi theta function transformation, and then the extraction of the leading-order asymptotic terms as the convergence factor tends to zero.

The rearranged sums include the free parameter α whose value must be determined (as shown later in this section) to maximize numerical accuracy; α can be chosen to ensure that terms of order $\exp(-\alpha^2 L^2)$ are negligible, and so (11.2) becomes

$$U_{qq} = \sum_{1 \le i < j \le N_a} \frac{q_i q_j \text{erfc}(\alpha|r_{ij}|)}{|r_{ij}|} - \frac{\alpha}{\pi^{1/2}} \sum_{j=1}^{N_a} q_j^2$$

$$+ \frac{1}{2\pi L} \sum_{n \ne 0} \frac{1}{|n|^2} \exp\left(-\frac{\pi^2|n|^2}{L^2 \alpha^2} \right) \left| \sum_{j=1}^{N_a} q_j \exp\left(\frac{2\pi i}{L} n \cdot r_j \right) \right|^2 \qquad (11.3)$$

The real-space terms in (11.3) are now short-ranged, so a spherical cutoff (with range $r_c < L/2$) can be used together with periodic boundaries. The Fourier-space sum (over n) will also prove amenable to truncation after only a limited number of terms. The squared dipole-moment sum in (11.2) has been dropped from the result; the physical implication of this is that the outermost replica shell is surrounded by a conducting medium, whereas including this term amounts to placing the system in a vacuum [del80, del86].

A typical value for the parameter α is $\alpha = 5/L$, though the result turns out to be relatively insensitive to the choice, provided there are sufficient terms in the n-sum. A spherical cutoff is imposed on the n-sum, namely, $|n|^2 \le n_c^2$; typically n_c is about 5. The accuracy of the Ewald result with the chosen parameters is readily checked numerically [kol92]. The invariance under the transformation $n \rightarrow -n$ can be used to halve the number of terms in the n-sum (by considering $n_z \geq 0$ only); the computational work can be reduced even further by restricting the sum over n to a single octant and calculating the sums for the four octants $(\pm n_x, \pm n_y, +n_z)$ at the same time.

Imposing a cutoff on the real-space sum at r_c leads to an error of order $\exp(-\alpha^2 r_c^2)$; truncating the Fourier-space sum at n_c produces an error of order $\exp(-\pi^2 n_c^2 / \alpha^2 L^2)$. Therefore, to obtain similarly sized errors in both real and Fourier contributions to U_{qq}, simply set $n_c = \alpha^2 r_c L / \pi$; then for the case $r_c = L/2$, if $\alpha = 5/L$ we obtain the very modest number $n_c = 25/2\pi \approx 4$ – the actual number of terms in the sum (before halving) is roughly $4\pi n_c^3 / 3$. For large systems it can be shown [per88] that if the optimal α is chosen for a specified numerical accuracy, then the computational effort grows as $N_a^{3/2}$; this represents a considerable saving over the original N_a^2 behavior, and of course there are no interaction cutoff errors.

Dipolar systems can be treated in a similar fashion. Even though the interaction energy of a pair of dipoles falls off with distance as $1/r^3$, this is still sufficiently slow in three dimensions for the sum over replica systems to remain conditionally convergent, so that the use of a cutoff is not possible; thus the same careful consideration given to the charge problem is required for dipoles as well.

The interaction of a pair of dipoles of strength μ is

$$-\mu^2(s_i \cdot \nabla)(s_j \cdot \nabla)(1/|r_{ij}|) \tag{11.4}$$

where s is a unit vector along the direction of the dipole. This differential operator can be applied to (11.2) to obtain the potential energy of a dipole system [ada76, del80]:

$$
\begin{aligned}
U_{dd} = \mu^2 \sum_{1 \leq i < j \leq N_a} & \left[\left(\frac{\operatorname{erfc}(\alpha|r_{ij}|)}{|r_{ij}|^3} + \frac{2\alpha \exp(-\alpha^2|r_{ij}|^2)}{\pi^{1/2}|r_{ij}|^2} \right)(s_i \cdot s_j) \right. \\
& - \left(\frac{3\operatorname{erfc}(\alpha|r_{ij}|)}{|r_{ij}|^5} + \left(2\alpha^2 + \frac{3}{|r_{ij}|^2} \right) \frac{2\alpha \exp(-\alpha^2|r_{ij}|^2)}{\pi^{1/2}|r_{ij}|^2} \right) \\
& \left. \times (s_i \cdot r_{ij})(s_j \cdot r_{ij}) \right] - \frac{2\alpha^3 \mu^2 N_a}{3\pi^{1/2}} + \frac{2\pi\mu^2}{3L^3} \left| \sum_{j=1}^{N_a} s_j \right|^2 \\
& + \frac{2\pi\mu^2}{L^3} \sum_{n \neq 0} \frac{1}{|n|^2} \exp\left(-\frac{\pi^2|n|^2}{L^2\alpha^2} \right) \left| \sum_{j=1}^{N_a} (n \cdot s_j) \exp\left(\frac{2\pi i}{L} n \cdot r_j \right) \right|^2
\end{aligned}
\tag{11.5}
$$

The derivation makes use of the results

$$d\,\text{erfc}(\alpha x)/dx = -2\alpha \exp(-\alpha^2 x^2)/\pi^{1/2} \tag{11.6}$$

and $\nabla(s \cdot r) = s$. Terms of order $\exp(-\alpha^2 L^2)$ will be dropped because α can be chosen to ensure their extreme smallness, and since the system is again assumed to be surrounded by a conducting medium the squared sum over s_j is also dropped. Several new functions are now introduced for conciseness:

$$a_1(r) = \text{erfc}(\alpha r)/r^3 + 2\alpha \exp(-\alpha^2 r^2)/\pi^{1/2} r^2 \tag{11.7}$$

$$a_n(r) = -r^{-1}da_{n-1}/dr, \quad n = 2,3 \tag{11.8}$$

$$e(n) = \exp(-\pi^2 n^2/L^2\alpha^2)/n^2 \tag{11.9}$$

$$\begin{Bmatrix} C \\ S \end{Bmatrix}(n) = \sum_{j=1}^{N_a}(n \cdot s_j)\begin{Bmatrix} \cos \\ \sin \end{Bmatrix}(2\pi n \cdot r_j/L) \tag{11.10}$$

The energy (11.5) can then be written as

$$U_{dd} = \mu^2 \sum_{1 \leq i < j \leq N_a} \left[a_1(r_{ij})(s_i \cdot s_j) - a_2(r_{ij})(s_i \cdot r_{ij})(s_j \cdot r_{ij}) \right]$$
$$+ \frac{2\pi\mu^2}{L^3} \sum_{n \neq 0} e(n)\left[C(n)^2 + S(n)^2 \right] - \frac{2\alpha^3\mu^2 N_a}{3\pi^{1/2}} \tag{11.11}$$

which, in the limit $\alpha \to 0$, reduces to the simple dipole result

$$U_{dd} = \mu^2 \sum_{i<j} \frac{1}{r_{ij}^3}\left[(s_i \cdot s_j) - \frac{3}{r_{ij}^2}(s_i \cdot r_{ij})(s_j \cdot r_{ij}) \right] \tag{11.12}$$

11.2.3 Dynamics

The MD model used in this case study attaches the dipoles to soft-sphere atoms; if the LJ potential is used for this problem instead [mai81], the interaction is known as the Stockmayer potential. The Lagrange equations of motion for translation involve the usual soft-sphere interaction together with a contribution from the force produced by the dipolar interactions. The latter is obtained by evaluating $-\nabla_{r_i} U_{dd}$, and consists of a sum over atoms truncated at the cutoff range r_c, together with a Fourier-space sum truncated at n_c. The dipole contribution to the force on a single atom is thus

$$F_i = \mu^2 \sum_{j(\neq i)} \left[(a_2(r_{ij})(s_i \cdot s_j) - a_3(r_{ij})(s_i \cdot r_{ij})(s_j \cdot r_{ij}))r_{ij} \right.$$

$$\left. + a_2(r_{ij})((s_j \cdot r_{ij})s_i + (s_i \cdot r_{ij})s_j) \right] + \frac{8\pi^2 \mu^2}{L^4} \sum_{n \neq 0} e(n)(n \cdot s_i)n$$

$$\times \left[C(n) \sin\left(\frac{2\pi}{L}n \cdot r_i\right) - S(n) \cos\left(\frac{2\pi}{L}n \cdot r_i\right) \right] \tag{11.13}$$

The equation of motion for the dipole vector s_i is just the rotation equation for a linear rigid molecule given in Chapter 8:

$$\ddot{s}_i = I^{-1}G_i - \left(I^{-1}(s_i \cdot G_i) + \dot{s}_i^2\right)s_i \tag{11.14}$$

where $G_i = -\nabla_{s_i} U_{dd}$ also consists of two sums:

$$G_i = \mu^2 \sum_{j(\neq i)} [-a_1(r_{ij})s_j + a_2(r_{ij})(s_j \cdot r_{ij})r_{ij}] - \frac{4\pi \mu^2}{L^3} \sum_{n \neq 0} e(n)n$$

$$\times \left[C(n) \cos\left(\frac{2\pi}{L}n \cdot r_i\right) + S(n) \sin\left(\frac{2\pi}{L}n \cdot r_i\right) \right] \tag{11.15}$$

In (11.14) we demonstrate the use of the second-order form of the rotational equation of motion; there is also an alternative method based on a pair of first-order equations for each dipole (see Chapter 8).

The real-space contributions to all the force terms F_i and G_i, and the potential energy, are evaluated by the following function; a cubic simulation region is assumed throughout, with the cutoff at $r_c = L/2$:

```
  ComputeForcesDipoleR () {
    real dr[4], alpha2, a1, a2, a3, d, irPi, rr, rrCut, rri,
      sr1, sr2, ss, t;
    int j1, j2, k, n;
5   rrCut = Sqr (regionH[1]);
    irPi = 1. / sqrt (pi);    alpha2 = Sqr (alpha);
    for (n = 1; n <= nAtom; n ++) {
      for (k = 1; k <= 3; k ++) sa[k][n] = 0.;
    }
10  for (j1 = 1; j1 <= nAtom - 1; j1 ++) {
      for (j2 = j1 + 1; j2 <= nAtom; j2 ++) {
        for (k = 1; k <= 3; k ++) {
          dr[k] = r[k][j1] - r[k][j2];
          if (fabs (dr[k]) > regionH[k])
15            dr[k] = dr[k] - SignR (region[k], dr[k]);
        }
        rr = Sqr (dr[1]) + Sqr (dr[2]) + Sqr (dr[3]);
        if (rr < rrCut) {
          d = sqrt (rr);    rri = 1. / rr;
```

```
20        t = 2. * dipoleInt * alpha * exp (- alpha2 * rr) *
             rri * irPi;
          a1 = dipoleInt * erfc (alpha * d) * rri / d + t;
          a2 = 3. * a1 * rri + 2. * alpha2 * t;
          a3 = 5. * a2 * rri + 4. * Sqr (alpha2) * t;
25        ss = s[1][j1] * s[1][j2] + s[2][j1] * s[2][j2] +
             s[3][j1] * s[3][j2];
          sr1 = s[1][j1] * dr[1] + s[2][j1] * dr[2] +
             s[3][j1] * dr[3];
          sr2 = s[1][j2] * dr[1] + s[2][j2] * dr[2] +
30           s[3][j2] * dr[3];
          for (k = 1; k <= 3; k ++) {
            t = (a2 * ss - a3 * sr1 * sr2) * dr[k] +
               a2 * (sr2 * s[k][j1] + sr1 * s[k][j2]);
            ra[k][j1] = ra[k][j1] + t;
35          ra[k][j2] = ra[k][j2] - t;
            sa[k][j1] = sa[k][j1] - a1 * s[k][j2] +
               a2 * sr2 * dr[k];
            sa[k][j2] = sa[k][j2] - a1 * s[k][j1] +
               a2 * sr1 * dr[k];
40        }
          uSum = uSum + a1 * ss - a2 * sr1 * sr2;
   } } }
   uSum = uSum - 2. * dipoleInt * pow (alpha, 3.) *
      nAtom * irPi / 3.;
45 }
```

The next function handles the Fourier-space part of the interactions.
Here there is a choice between versions that do and do not take advantage of the full symmetries in the *n*-sums. The latter is more concise, the former more efficient but even more tedious to read. We
will leave the faster version as an exercise for the concerned reader
and discuss the shorter form that only makes use of the shortcut
provided by the $\pm n$ symmetry. In the computation the sines and
cosines of the dot products are each expanded as triple-product sums
$\sin(x+y+z) = \sin x \cos y \cos z + ...$; the trigonometric functions are tabulated prior to the calculation using multiple-angle recurrence relations
as in Chapter 5.

```
   ComputeForcesDipoleF () {
     real cosX, cosY, cosZ, fMult, gR, gS, gU, pC, pS,
        sinX, sinY, sinZ, sumC, sumS, t, w;
     int n, nvv, nX, nY, nZ;
5    gU = 2. * pi * dipoleInt / pow (region[1], 3.);
     gR = 4. * pi * gU / region[1];    gS = 2. * gU;
     EvalSinCos ();
```

```
      w = Sqr (pi / (region[1] * alpha));
      for (nZ = 0; nZ <= fSpaceLimit; nZ ++) {
10      for (nY = - fSpaceLimit; nY <= fSpaceLimit; nY ++) {
          for (nX = - fSpaceLimit; nX <= fSpaceLimit; nX ++) {
            nvv = nX * nX + nY * nY + nZ * nZ;
            if (nvv == 0 || nvv > fSpaceLimit * fSpaceLimit)
              continue;
15          fMult = 2. * exp (- w * nvv) / nvv;
            if (nZ == 0) fMult = fMult * 0.5;
            sumC = sumS = 0.;
            cosX = cosY = cosZ = 1.;
            sinX = sinY = sinZ = 0.;
20          for (n = 1; n <= nAtom; n ++) {
              if (nX != 0) {
                cosX = cosV[abs (nX) * 3 - 2][n];
                sinX = sinV[abs (nX) * 3 - 2][n];
                if (nX < 0) sinX = - sinX;
25            }
              if (nY != 0) {
                cosY = cosV[abs (nY) * 3 - 1][n];
                sinY = sinV[abs (nY) * 3 - 1][n];
                if (nY < 0) sinY = - sinY;
30            }
              if (nZ != 0) {
                cosZ = cosV[nZ * 3][n];
                sinZ = sinV[nZ * 3][n];
              }
35            sumC = sumC + (nX * s[1][n] + nY * s[2][n] +
                nZ * s[3][n]) * (cosX * cosY * cosZ -
                cosX * sinY * sinZ - sinX * cosY * sinZ -
                sinX * sinY * cosZ);
              sumS = sumS + (nX * s[1][n] + nY * s[2][n] +
40              nZ * s[3][n]) * (sinX * cosY * cosZ +
                cosX * sinY * cosZ + cosX * cosY * sinZ -
                sinX * sinY * sinZ);
            }
            for (n = 1; n <= nAtom; n ++) {
45            if (nX != 0) {
                ... (as above) ...
              }
              if (nY != 0) ...
              if (nZ != 0) ...
50            pC = cosX * cosY * cosZ - cosX * sinY * sinZ -
                sinX * cosY * sinZ - sinX * sinY * cosZ;
              pS = sinX * cosY * cosZ + cosX * sinY * cosZ +
                cosX * cosY * sinZ - sinX * sinY * sinZ;
              t = gR * fMult * (nX * s[1][n] + nY * s[2][n] +
```

```
55              nZ * s[3][n]) * (sumC * pS - sumS * pC);
            ra[1][n] = ra[1][n] + nX * t;
            ra[2][n] = ra[2][n] + nY * t;
            ra[3][n] = ra[3][n] + nZ * t;
            t = gS * fMult * (sumC * pC + sumS * pS);
60          sa[1][n] = sa[1][n] - nX * t;
            sa[2][n] = sa[2][n] - nY * t;
            sa[3][n] = sa[3][n] - nZ * t;
          }
          uSum = uSum + gU * fMult * (Sqr (sumC) +
65            Sqr (sumS));
      } } }
  }
```

The function for filling the sine and cosine arrays is

```
EvalSinCos () {
    real t;
    int j, k, m, n;
    for (k = 1; k <= 3; k ++) {
5     t = 2. * pi / region[k];
      for (n = 1; n <= nAtom; n ++) {
        m = k;
        sinV[m][n] = sin (t * r[k][n]);
        cosV[m][n] = cos (t * r[k][n]);
10        m = m + 3;
        sinV[m][n] = 2. * cosV[k][n] * sinV[m - 3][n];
        cosV[m][n] = 2. * cosV[k][n] * cosV[m - 3][n] - 1.;
        for (j = 3; j <= fSpaceLimit; j ++) {
          m = m + 3;
15          sinV[m][n] = 2. * cosV[k][n] * sinV[m - 3][n] -
              sinV[m - 6][n];
          cosV[m][n] = 2. * cosV[k][n] * cosV[m - 3][n] -
              cosV[m - 6][n];
    } } }
20 }
```

The right-hand sides of the equations of motion (11.14) are computed by the following function; upon entry sa contains the values of G_i:

```
ComputeDipoleAccel () {
    real t;
    int k, n;
    for (n = 1; n <= nAtom; n ++) {
5     t = 0.;
      for (k = 1; k <= 3; k ++)
        t = t + sa[k][n] * s[k][n] + momInertia *
            Sqr (sv[k][n]);
      for (k = 1; k <= 3; k ++)
```

```
10      sa[k][n] = (sa[k][n] - t * s[k][n]) / momInertia;
     }
   }
```

New variables appearing in this program are

```
real **s, **sv, **sa, **sa1, **sa2, **so, **svo, **cosV,
   **sinV, alpha, dipoleInt, momInertia, vvsSum;
int fSpaceLimit, stepAdjustTemp;
```

where dipoleInt corresponds to μ^2. Additional input data items are

```
  RNAME (alpha),
  RNAME (dipoleInt),
  INAME (fSpaceLimit),
  RNAME (momInertia),
5 INAME (stepAdjustTemp),
```

and array allocations (apart from the usual coordinates, velocities, etc.) are

```
  cosV = AllocMatR (3 * fSpaceLimit, nAtom);
  sinV = AllocMatR (3 * fSpaceLimit, nAtom);
```

The initialization requires

```
  InitAngCoords ();
  InitAngVels ();
  InitAngAccels ();
```

where, assuming an aligned initial state, these functions are

```
  InitAngCoords () {
    int n;
    for (n = 1; n <= nAtom; n ++) {
      s[1][n] = s[2][n] = 0.;      s[3][n] = 1.;
5   }
  }

  InitAngVels () {
    real ang, angvFac;
    int n;
    angvFac = vMag / sqrt (1.5 * momInertia);
5   for (n = 1; n <= nAtom; n ++) {
      ang = 2. * pi * RandR (&randSeed);
      sv[1][n] = angvFac * cos (ang);
      sv[2][n] = angvFac * sin (ang);
      sv[3][n] = 0.;
10  }
  }

  InitAngAccels () {
```

```
    int k, n;
    for (n = 1; n <= nAtom; n ++) {
      for (k = 1; k <= 3; k ++)
5       sa[k][n] = sa1[k][n] = sa2[k][n] = 0.;
    }
  }
```

The factor of 1.5 in `InitAngVels` arises from the fact that each molecule has only two rotational degrees of freedom, and the initial values ensure that $\dot{s}_i = \omega_i \times s_i$. We use `nAtom` rather than `nMol` for the system size because of the absence of internal molecular structure (a mere semantic detail).

The additions to `SingleStep` – the locations should be obvious – are

```
PredictorStepS ();
  ...
ComputeForcesDipoleR ();
ComputeForcesDipoleF ();
5 ComputeDipoleAccel ();
  ...
CorrectorStepS ();
  ...
AdjustDipole ();
10 if (stepCount % stepAdjustTemp == 0) AdjustTemp ();
```

The integration functions `PredictorStepS` and `CorrectorStepS` for the rotational motion are the same as the corresponding (three-dimensional) translational functions – simply change the variable names. Calculations of kinetic energy, the Lagrange multiplier used for the thermostat, and the temperature adjustment must all allow for the rotational motion as described in Chapter 8. The function `AdjustDipole` is needed to restore the *s* vectors to unit length because of the small numerical drift:

```
AdjustDipole () {
  real sFac;
  int k, n;
  for (n = 1; n <= nAtom; n ++) {
5   sFac = 1. / sqrt (Sqr (s[1][n]) + Sqr (s[2][n]) +
      Sqr (s[3][n]));
    for (k = 1; k <= 3; k ++) s[k][n] = s[k][n] * sFac;
  }
}
```

11.2.4 Properties

In addition to the spatial correlations present in the fluid, the fact that each molecule has a dipole moment means that orientational order can be studied as well. The dipole directional order parameter is

$$M = \sum_{i=1}^{N_a} s_i \qquad (11.16)$$

and its magnitude $M = |M|$ is evaluated by an addition to `EvalProps`:

```
    real ... t;
    ...
    sdSum = 0.;
    for (k = 1; k <= 3; k ++) {
5     t = 0.;
      for (n = 1; n <= nAtom; n ++) t = t + s[k][n];
      sdSum = sdSum + Sqr (t);
    }
    sdSum = sqrt (sdSum) / nAtom;
```

The new variables used in computing $\langle M \rangle$ and its fluctuations (the latter are related to the dielectric constant [del86, han86b]) are

```
    real sdSum, sOrder, ssOrder;
```

and the additions to `AccumProps` for this calculation are

```
    sOrder = ssOrder = 0.;
    ...
    sOrder = sOrder + sdSum;
    ssOrder = ssOrder + Sqr (sdSum);
5   ...
    sOrder = sOrder / stepAvg;
    ssOrder = sqrt (ssOrder / stepAvg - Sqr (sOrder));
```

The order parameter M is a measure of long-range orientational order, but the question of local dipole alignment is a separate issue. The liquid state is characterized by short-range structural order, so that for a dipolar fluid it becomes possible to examine certain combinations of positional and orientational order to determine, for example, how the average relative orientation of the dipoles in the neighborhood of a given dipole depend on range. To be more specific, given the definition of some orientational quantity, compute its average values over the atoms in a series of concentric spherical shells centered at the position of the atom of interest; this amounts to an extension of the RDF computation in Chapter 4 which only considered shell occupancy.

The two quantities of interest here [del86, han86b] are the relative orientation of the dipole vectors $s_i \cdot s_j$ and the quantity responsible for the dipolar contribution to the energy: $3(s_i \cdot r_{ij})(s_j \cdot r_{ij})/r_{ij}^2 - s_i \cdot s_j$. The principal change to EvalRdf, aside from having to initialize and later normalize three sets of data rather than one, just as in the case of rigid molecules (Chapter 8), is the addition

```
  real ... , sr1, sr2, ss;
  ...
  if (rr < rrRange) {
    ss = s[1][j1] * s[1][j2] + s[2][j1] * s[2][j2] +
5      s[3][j1] * s[3][j2];
    sr1 = s[1][j1] * dr[1] + s[2][j1] * dr[2] +
      s[3][j1] * dr[3];
    sr2 = s[1][j2] * dr[1] + s[2][j2] * dr[2] +
      s[3][j2] * dr[3];
10  n = (int) (sqrt (rr) / deltaR) + 1;
    histRdf[1][n]= histRdf[1][n] + 1.;
    histRdf[2][n]= histRdf[2][n] + ss;
    histRdf[3][n]= histRdf[3][n] + 3. * sr1 * sr2 / rr - ss;
  }
```

11.2.5 Measurements

We begin with a test that provides some idea of the accuracy of the Ewald method for dipole systems. The computations use a set of 216 dipoles that are randomly positioned and oriented in a cube of edge $L = 10$. In order to prevent the results from being dominated by a few very close pairs, instead of using completely random coordinates the dipoles are placed at the sites of a simple cubic lattice and then each coordinate is randomly shifted in either direction by up to a quarter of the lattice spacing. The dipole energy and the sum of the absolute values of the force components (per dipole, and with $\mu = 1$) are then computed for various values of α and n_c. The results of the force calculations are shown in Figure 11.1. Even with $n_c = 5$ there is no problem in achieving forces accurate to about 0.01% and energies (not shown) accurate to 0.1%.

The MD run described here (compare [pol80, kus90], who use slightly different potentials) is carried out with $N_a = 108$, $\rho = 0.8$, $T = 1.35$, and additional data

```
  stepAdjustTemp 1000
  stepAvg        200
  stepEquil      1000
  stepLimit      86000
```

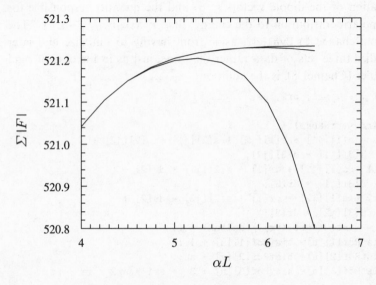

Fig. 11.1 Sum of force components for a random dipole system as a function of α for $n_c = 5, 6$, and 7.

alpha	5.5
dipoleInt	4.0
fSpaceLimit	5
momInertia	0.025
limitRdf	250
rangeRdf	2.5
sizeHistRdf	125
stepRdf	20

The initial state is an FCC lattice with all dipoles initially parallel. We use PC integration, $\Delta t = 0.0025$ (the value of the moment of inertia affects this choice), and constant-temperature dynamics; the temperature drift over 1000 timesteps is typically 0.5%.

The results for $g(r)$ and the two functions used to measure the short-range directional order are shown in Figure 11.2. The function $h_{110}(r)$ measures $\langle s_i \cdot s_j \rangle$ in terms of the interatomic distance, whereas $h_{112}(r)$ does the same for $\langle 3(s_i \cdot r_{ij})(s_j \cdot r_{ij})/r_{ij}^2 - s_i \cdot s_j \rangle$. Equilibration tends to be relatively slow for this system, so we ignore the RDF measurements made during the first 26 000 timesteps and average over all the remainder.

Properties associated with the presence of long-range order, in particular the dielectric constant, are more difficult to evaluate reliably [kus90]. One reason for this is the very long timescale over which fluctuations

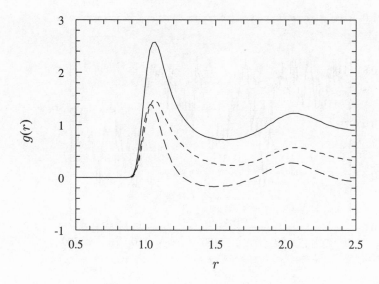

Fig. 11.2 RDF (solid curve) for the dipole fluid, together with the correlation functions $h_{112}(r)$ (short dashes) and $h_{110}(r)$ (long dashes).

occur. Figure 11.3 shows how M varies over the course of the entire run (see also [pay93]). Attempting to estimate a mean value from data of this kind is risky at best; while the results indicate $\langle M \rangle = 0.62$, a block variance analysis (Chapter 4) designed to obtain $\sigma(\langle M \rangle)$ shows no hint of convergence. For a reliable estimate a substantially longer run is required. The question of the nature of the surrounding medium alluded to earlier must also be addressed.

11.3 Three-body potentials

11.3.1 The problem

Even when regarded simply as effective potentials the capacity of the pair potential to reproduce known behavior has its limitations. We have already encountered situations where the potential extends beyond the basic two-body form: the interaction-site method used for rigid molecules in which forces between molecules (as opposed to the sites in the molecules) involves both distance and orientation (Chapter 8), and the intramolecular forces associated with the internal degrees of freedom of partially rigid molecules that depend on three or four atoms (Chapter 10). The situation described here is entirely different; the force

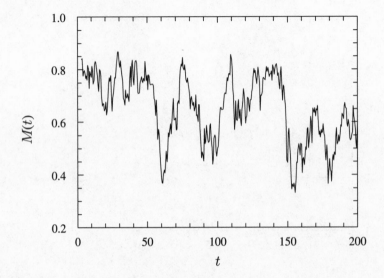

Fig. 11.3 Time-dependence of the order parameter.

between two atoms will now depend on the positions of all other atoms in the vicinity. Thus, in addition to the forces between pairs of atoms, a new type of force is introduced that acts on triplets of atoms and whose strength is a function of the three interatomic distances.

The crystalline state of silicon is a four-coordinated diamond lattice whose density increases upon melting; this reflects the fact that the liquid is more closely packed than the solid, exactly the opposite of what normally occurs (with water being another prominent exception). Since LJ-type potentials can only produce closely packed solids, the question is how to augment the potential function so that it incorporates a preferred set of bond directions, such as the tetrahedral arrangement needed for silicon. The simplest way to do this is to introduce three-body interactions chosen in a manner which stabilizes particular bond angles; this represents yet another attempt to imitate classically what is at heart a quantum effect.

11.3.2 Formulation

The model proposed for liquid silicon [sti85] has been kept as simple as possible – to the extent that any three-body interaction can be considered 'simple' from the computational perspective. The interactions include both two- and three-body contributions, expressed in suitable reduced units.

The two-body part of the potential has the form

$$u_2(r) = A\Big(\frac{B}{r^4} - 1\Big)\exp\Big(\frac{1}{r - r_c}\Big) \tag{11.17}$$

for $r < r_c$, and zero otherwise. The corresponding force is

$$-\nabla_r u_2 = \frac{A}{r}\Big[\frac{4B}{r^5} + \Big(\frac{B}{r^4} - 1\Big)\Big/(r - r_c)^2\Big]\exp\Big(\frac{1}{r - r_c}\Big)r \tag{11.18}$$

Observe the design of the function: both u_2 and its derivatives are zero at the cutoff range r_c; while this has the advantage of reducing spurious effects associated with cutoffs in general, care is required in the computation to avoid any risk of numerical overflow when $r \approx r_c$.

The three-body part is symmetric under permutations of atom indices, and is of course invariant under translation and rotation:

$$u_3(r_1, r_2, r_3) = h(r_{12}, r_{13}) + h(r_{21}, r_{23}) + h(r_{31}, r_{32}) \tag{11.19}$$

with each function h having the form

$$h(r_{12}, r_{13}) = \lambda \exp\Big(\frac{\gamma}{r_{12} - r_c} + \frac{\gamma}{r_{13} - r_c}\Big)(\cos\theta_{213} + 1/3)^2 \tag{11.20}$$

provided that both $r_{12} < r_c$ and $r_{13} < r_c$, and zero if either condition is violated; θ_{213} is the angle between r_{12} and r_{13}. The cutoff at r_c is smooth, as in the two-body part. An important feature built into the functional form of h is that it has a minimum when θ_{213} equals the tetrahedral angle. Not only does this introduce the desired angular correlations, it helps to ensure that each atom prefers a considerably smaller number of immediate neighbors (namely four) than is allowed by close packing.

We introduce unit vectors $d_{ij} = r_{ij}/r_{ij}$, so that $\cos\theta_{213} = d_{12} \cdot d_{13}$. The force contribution of each h function is evaluated separately; the forces due to a typical function $h(r_{12}, r_{13})$ act on three atoms, and, since $\nabla_{r_1} h = -\nabla_{r_2} h - \nabla_{r_3} h$, only two of the derivatives need be computed. For $m = 2, 3$ we have

$$-\nabla_{r_m} h(r_{12}, r_{13}) = -\lambda(c + 1/3)\exp\Big(\frac{\gamma}{r_{12} - r_c} + \frac{\gamma}{r_{13} - r_c}\Big)$$
$$\times \Big(\frac{\gamma(c + 1/3)}{(r_{1m} - r_c)^2}d_{1m} + 2\nabla_{r_m}c\Big) \tag{11.21}$$

where $c \equiv \cos\theta_{213}$. The derivatives $\nabla_{r_m}c$ are computed in exactly the same way as the bond-angle forces discussed in Chapter 10:

$$\nabla_{r_2}c = (cd_{12} - d_{13})/r_{12}, \quad \nabla_{r_3}c = (cd_{13} - d_{12})/r_{13} \tag{11.22}$$

Numerical values for the constants appearing in (11.17) and (11.20) are specified in the function ComputeForces below.

11.3.3 Implementation details

The neighbor-list method is once again the preferred choice, although with some modification. In order to identify all interacting atom triplets the list will have to be scanned twice inside a doubly nested loop, and the list itself must also include each atom pair twice, both as *ij* and *ji*. The neighbor list will be stored in an alternative form better suited for this computation: instead of simply listing the possibly interacting pairs, the information will now be stored in two parts, one a table of serial numbers of the neighboring atoms, the other a set of pointers to the first entry in the table corresponding to the neighbors of each atom. Such an approach could have been used for the original treatment in Chapter 3, although the extra data operations and the low repetition count of the resulting inner loop could reduce the efficiency.

Neighbor-list construction is carried out by the following function derived from the original version. Here, atoms are scanned first, then the set of cells surrounding the cell in which the particular atom resides, and finally the contents of each of these cells:

```
     BuildNebrList () {
        ...
        int...
           iofX[] = {0,-1,0,1,-1,0,1,-1,0,1,-1,0,1,-1,0,1,
5             -1,0,1,-1,0,1,-1,0,1,-1,0,1},
           iofY[] = {0,-1,-1,-1,0,0,0,0,1,1,1,-1,-1,-1,0,0,0,
              1,1,1,-1,-1,-1,0,0,0,1,1,1},
           iofZ[] = {0,-1,-1,-1,-1,-1,-1,-1,-1,-1,0,0,0,0,0,0,
              0,0,0,1,1,1,1,1,1,1,1,1};
10      ...
        nebrTabLen = 0;
        for (j1 = 1; j1 <= nAtom; j1 ++) {
          m1X = (int) ((r[1][j1] + regionH[1]) * invWid[1]) + 1;
          m1Y = (int) ((r[2][j1] + regionH[2]) * invWid[2]) + 1;
15        m1Z = (int) ((r[3][j1] + regionH[3]) * invWid[3]) + 1;
          nebrTabPtr[j1] = nebrTabLen + 1;
          for (offset = 1; offset <= 27; offset ++) {
            m2X = ...
            ...
20          m2 = ...
            j2 = cellList[m2];
            while (j2 > 0) {
```

```
          if (j2 != j1) {
             ...
25          if (rr < rrNebr) {
               nebrTabLen = nebrTabLen + 1;
               nebrTab[nebrTabLen] = j2;
          } }
          j2 = cellList[j2];
30      } }
      }
      nebrTabPtr[nAtom + 1] = nebrTabLen + 1;
    }
```

Additional arrays used (note that `nebrTab` is only one-dimensional here) are

```
int *nebrTab, *nebrTabPtr;
```

and these are allocated by

```
nebrTab = AllocVecI (nebrTabMax);
nebrTabPtr = AllocVecI (nAtom + 1);
```

The interaction calculation, including both two- and three-body contributions, is

```
ComputeForces () {
   real dr[4], dr12[4], dr13[4], aCon, bCon, cR, eR, f,
      fcVal, fcVal2, fcVal3, gCon, lCon, p12, p13, ri, ri3,
      rm, rm12, rm13, rr, rrCut, rr12, rr13, uVal;
5  int i, j1, j2, j3, k, m2, m3;
   rrCut = Sqr (rCut) - 0.001;
   aCon = 7.0496;    bCon = 0.60222;
   lCon = 21.;    gCon = 1.2;
   for (i = 1; i <= nAtom; i ++) {
10    for (k = 1; k <= 3; k ++) ra[k][i] = 0.;
   }
   uSum = 0.;
   for (j1 = 1; j1 <= nAtom; j1 ++) {
     for (m2 = nebrTabPtr[j1]; m2 <= nebrTabPtr[j1 + 1] - 1;
15       m2 ++) {
       j2 = nebrTab[m2];
       if (j1 < j2) {
         for (k = 1; k <= 3; k ++) {
           dr[k] = r[k][j1] - r[k][j2];
20         if (fabs (dr[k]) > regionH[k])
             dr[k] = dr[k] - SignR (region[k], dr[k]);
         }
         rr = DotProd3 (dr, dr);
         if (rr < rrCut) {
25         rm = sqrt (rr);    eR = exp (1. / (rm - rCut));
```

```
              ri = 1. / rm;     ri3 = ri * ri * ri;
              fcVal = aCon * (4. * bCon * ri3 * ri3 + (bCon *
                  ri3 * ri - 1.) * ri / Sqr (rm - rCut)) * eR;
              uVal = aCon * (bCon * ri3 * ri - 1.) * eR;
30            for (k = 1; k <= 3; k ++) {
                f = fcVal * dr[k];
                ra[k][j1] = ra[k][j1] + f;
                ra[k][j2] = ra[k][j2] - f;
              }
35            uSum = uSum + uVal;
      } } } }
      for (j1 = 1; j1 <= nAtom; j1 ++) {
        for (m2 = nebrTabPtr[j1]; m2 <= nebrTabPtr[j1 + 1] - 2;
          m2 ++) {
40        j2 = nebrTab[m2];
          for (k = 1; k <= 3; k ++) {
            dr12[k] = r[k][j1] - r[k][j2];
            if (fabs (dr12[k]) > regionH[k]) ...
          }
45        rr12 = DotProd3 (dr12, dr12);
          if (rr12 > rrCut) continue;
          rm12 = sqrt (rr12);
          for (k = 1; k <= 3; k ++) dr12[k] = dr12[k] / rm12;
          for (m3 = m2 + 1; m3 <= nebrTabPtr[j1 + 1] - 1;
50          m3 ++) {
            j3 = nebrTab[m3];
            for (k = 1; k <= 3; k ++) {
              dr13[k] = r[k][j1] - r[k][j3];
              if (fabs (dr13[k]) > regionH[k]) ...
55          }
          rr13 = DotProd3 (dr13, dr13);
          if (rr13 > rrCut) continue;
          rm13 = sqrt (rr13);
          for (k = 1; k <= 3; k ++) dr13[k] = dr13[k] / rm13;
60        cR = DotProd3 (dr12, dr13);
          eR = 1Con * (cR + 1./3.) * exp (gCon / (rm12 -
              rCut) + gCon / (rm13 - rCut));
          p12 = gCon * (cR + 1./3.) / Sqr (rm12 - rCut);
          p13 = gCon * (cR + 1./3.) / Sqr (rm13 - rCut);
65        for (k = 1; k <= 3; k ++) {
            fcVal2 = - eR * (p12 * dr12[k] +
                2. * (cR * dr12[k] - dr13[k]) / rm12);
            fcVal3 = - eR * (p13 * dr13[k] +
                2. * (cR * dr13[k] - dr12[k]) / rm13);
70          ra[k][j1] = ra[k][j1] - fcVal2 - fcVal3;
            ra[k][j2] = ra[k][j2] + fcVal2;
            ra[k][j3] = ra[k][j3] + fcVal3;
```

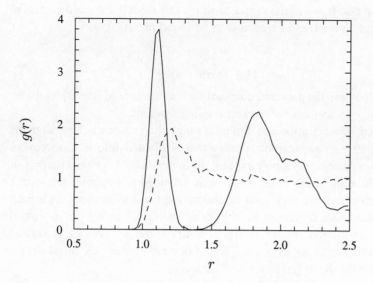

Fig. 11.4 RDFs for model silicon in the solid ($T = 0.08$) and liquid ($T = 0.12$) states.

```
        }
        uSum = uSum + eR * (cR + 1./3.);
75   } } }
    }
```

11.3.4 Measurements

The simulations shown here examine the RDFs in the crystalline and liquid phases cursorily. Runs are carried out at $T = 0.08$ and 0.12, with $\rho = 0.483$, and include the following additional input data:

```
initUcell        3 3 3
rCut             1.8
stepAdjustTemp   500
limitRdf         20
rangeRdf         2.5
sizeHistRdf      100
stepRdf          50
```

The initial state is a diamond lattice, so that $N_a = 216$. Constant-temperature MD is used, together with PC integration, and $\Delta t = 0.005$.

The RDF results shown in Figure 11.4 are based on a single average over 1000 steps after allowing 2000 steps for equilibration. A more detailed discussion of the degree to which this model captures the unusual structural properties of silicon is to be found in [sti85].

11.4 Further work

11.1 Compute the dielectric constant for the dipole fluid [kus90]; consider how to account for the surrounding medium.

11.2 An efficient approach for long-range interactions is based on the multipole expansion: The system is divided into a hierarchy of nested cells, multipole expansions are evaluated for the contents of the cells (using various algebraic relations to minimize the work), and these are then used in computing the interactions; while each atom sees its closest neighbors as individual atoms, more distant contributions to the interactions are based on cells of increasing size (this is an extremely brief summary – many technical details are involved) [gre89a, whi94]. Explore.

11.3 Another method for introducing orientation-dependent forces without resorting to multiple interaction sites is based on generalizing the LJ interaction to allow both σ and ϵ to depend on relative molecular alignment (the molecules are treated as rigid bodies) [gay81, sar93]. Ellipsoidal molecules used in the study of liquid crystals can be modeled in this way; examine the spatial and orientational order that occurs in such systems, and the expected transitions between the liquid, nematic (orientationally ordered), and crystalline states.

12

Step potentials

12.1 Introduction

Almost all the case studies in this book involve systems whose interactions are expressed in terms of continuous potentials. As a consequence the dynamical equations can be solved numerically with constant-timestep integration methods. If one is prepared to dispense with this continuity another route is available that offers several advantages, although it has its weak points as well. The alternative method is based on step potentials; hard spheres are the simplest example, but the method can be extended to include potential functions that have the shape of square wells or barriers, and even flexible 'molecules' can be built. Quantitative comparisons with specific real substances are obviously not the goal here, although comparisons with simple analytical models are possible. In fact, the earliest MD simulations [ald58, ald59, ald62] were of this kind, motivated by a desire to test basic theory.

A limitation of the constant-timestep methods used for continuous potentials is that they require the changes in interactions over each timestep (Δt) to be small, otherwise uncontrolled numerical errors can suddenly appear. While this does not usually affect equilibrium studies, because Δt can be made sufficiently (but not too) small that for a particular simulation (namely, a given potential function, temperature and density) the results are predictably stable, systems that are inhomogeneous because of, for example, a large imposed temperature gradient, may prove problematic unless Δt is made unacceptably small. As will become apparent in due course, the step-potential method is unaware of this problem because it advances by a series of discrete events. The need for explicit numerical integration is avoided by employing impulsive collisions whenever atoms interact, and between collisions each atom follows its own linear trajectory. There is no numerical integration error,

285

because trajectories are evaluated to the full precision of the computer hardware.

The use of step potentials introduces its own problems. First there is the increased complexity of the algorithm, since dealing with large numbers of collision events in an efficient manner requires careful attention to the question of data organization. The second problem is memory; storing the information describing events in a manner that is readily accessible and alterable tends to demand a good deal of extra memory.

The case study described here deals with the most basic kind of event-driven MD simulation, namely a hard-sphere fluid subject to periodic boundary conditions. At the end of the chapter we discuss generalizations of the basic method, some of which are used in a more complex case study in Chapter 13.

12.2 Computational approach

12.2.1 Dynamics

The physically interesting dynamics of the hard-sphere system is embodied in the collision rules; between collisions nothing of note happens and the atoms (as the spheres will subsequently be called) move in straight lines. Consider two identical atoms i and j currently separated by a distance $r = r_i - r_j$ and having a relative velocity $v = v_i - v_j$. These atoms will collide if and when their separation becomes equal to the atomic diameter σ; if this happens it will occur at some time τ in the future, where τ is the smaller positive solution of

$$|r + v\tau| = \sigma \tag{12.1}$$

The solution, if it exists, is

$$\tau = -\left(b + \sqrt{b^2 - v^2(r^2 - \sigma^2)}\right)/v^2 \tag{12.2}$$

where $b = r \cdot v$. In order for a solution to exist b must be negative and the argument of the square root positive. Solutions come in pairs; the larger positive solution reflects the fact that the separation σ occurs twice if the trajectories are extended beyond the collision point, although this is irrelevant for hard spheres. A negative solution corresponds to trajectories that (apparently) intersected in the past. The outcome of a collision between atoms is a simple change of velocities that preserves energy and momentum, namely,

$$\Delta v_i = -\Delta v_j = -br/\sigma^2 \tag{12.3}$$

If the analysis is extended to atoms that consist of a hard repulsive core surrounded by an attractive square well [ald59], then the generalized 'collision' corresponds to one of several kinds of event: (a) it can be a collision between the hard cores, as above; (b) it can correspond to two atoms either entering or leaving their mutual potential well; (c) it can correspond to two atoms, bound together in a well, bouncing as the well boundary is reached. We will discuss this extension briefly, although it is not part of the case study.

If σ_c and σ_w are the core and well diameters, then the most general expression for when a collision might occur is

$$\tau = \left(-b + s\sqrt{b^2 - v^2(r^2 - \sigma^2)}\right)/v^2 \tag{12.4}$$

For a core collision, only possible for atoms already inside their mutual well, $s = -1$ and $\sigma = \sigma_c$, as in (12.2). If the future event is to be either a bounce within a well or an escape from it (this depends on whether there is sufficient kinetic energy), $s = 1$ and $\sigma = \sigma_w$. Finally, when a well is about to be entered (a capture), $s = -1$ and $\sigma = \sigma_w$.

The velocity changes when the generalized collision occurs are

$$\Delta v_i = -\Delta v_j = \phi r \tag{12.5}$$

where

$$\phi = \left(-b + s\sqrt{b^2 - 4r^2 \Delta u/m}\right)/2r^2 \tag{12.6}$$

In (12.6), Δu denotes the potential energy change and m is the mass. For a core collision $\Delta u = 0$, so that $\phi = -b/\sigma_c^2$ as in (12.3). For well capture $\Delta u = -w$ – where w is the well depth – and $s = -1$. For well bounce and escape events, if $b^2 < 4\sigma_w^2 w/m$ the event is a bounce, so $\phi = -b/\sigma_w^2$, otherwise it is an escape, with $\Delta u = +w$ and $s = 1$. In order to obtain a true bound state a third atom is required to remove the excess kinetic energy from a pair of atoms whose separation has dropped below σ_w, otherwise, because their combined energy is positive, they will promptly escape from one another.

12.2.2 Cell subdivision

The simulation progresses by means of a time-ordered sequence of collision events [erp77]. Assuming that all possible future collisions have been examined it is a simple matter to determine which will occur first and advance all the atoms to that point in time. Such a scheme is correct in principle; in practice, the fact that determining the next collision of a given atom requires $O(N_a)$ work, because all atoms must be considered potential collision partners, rules out this simple approach. As in the case of soft-sphere MD, the use of cells (as well as a selective record of possible future collisions that have already been examined – see further) provides the means for reducing the number of atoms examined following a collision to a small value independent of N_a.

Assuming that the simulation region has been divided into cells whose edge length exceeds the sphere diameter σ, it is clear that collisions can only occur between atoms in the same and adjacent cells. By using relatively small cells the average occupancy can be reduced to just one atom (or even less) per cell (the lower bound depends on density), so the gain in computational efficiency is apparent. All that is needed is to keep track of which atoms belong to which cells, and, since it is hardly appropriate to recompute this information after each collision, the way this is done is to introduce a new kind of event that occurs whenever an atom moves from one cell to another. Determining which cell face an atom will cross next and when this is due to occur are simple problems that are solved in the program listing later on. It is true the cell-crossing events introduce additional work, but the overall reduction in effort more than compensates for this. Periodic boundaries are readily incorporated into the cell-based computations.

Another labor-saving device is the use of a local time variable associated with each atom (or, alternatively, with each cell). When a collision occurs, only atoms in the immediate neighborhood are of concern, and there is no point in updating the coordinates of atoms much further away. The use of a 'personal' time for each atom provides a record of when its coordinates were last updated, so that one of the few occasions when an update of the entire system is really necessary is prior to recording a snapshot of the system configuration.

12.2.3 Event calendar

We have tacitly assumed that the system always knows when the next event is due, whether it is a collision or a cell crossing, and the atom(s) involved. This implies the existence of an event calendar. Such a calendar must not only produce the next event but must also be easily modifiable: the calendar will include many future events, and as collisions occur changes must be made to its contents, both to incorporate newly predicted collisions and to remove previously predicted collisions that are no longer relevant because a participant has in the meantime undergone a different collision. Once the effort has been made to find a possible future collision this information should be retained for as long as it is potentially useful, but it must be recognized that if the calendar includes a few collision events per atom it is likely that most of this information will become obsolete before it has a chance to be used. Thus the calendar organization is central to the viability of this method; we will discuss its implementation in the next section.

When two atoms collide their velocities are changed, so that any information stored in the calendar regarding future events involving these atoms ceases to be valid. Such events will have to be erased from the calendar and replaced by whatever new events are predicted. The possible collisions that must be examined to determine these new events are between each of the colliding atoms and all other atoms in the neighboring cells, including the cell that the atom presently occupies. Similarly, when an atom crosses a cell boundary, the newly adjacent cells also contain potential collision partners that must be considered; in this case, however, existing calendar entries are still relevant and are retained. Following both kinds of event it is necessary to determine the next cell-crossing event for the atom(s) involved. If these details are taken care of correctly there is no way in which a collision can be missed.

12.2.4 Program details

At this point we describe those parts of the program that deal with the collisions and cell crossings. The handling of the event calendar will be discussed separately. Unlike the previous case studies that were built upon one another's programs, the hard-sphere simulation is organized completely differently, and so, with the exception of certain common utility functions, a separate program will be constructed. The reduced MD units used in this simulation are defined so that the atoms have unit mass and diameter.

The description begins with the main program, job initialization, and the function responsible for processing a single event. Most of the program is equally suitable for spheres in three dimensions and disks in two; elsewhere the three-dimensional version is shown, but the conversion task to two dimensions is principally one of erasure:

```
main (int argc, char **argv) {
  GetNameList (argc, argv);
```

```
    PrintNameList (stdout);
    SetParams ();
5   SetupJob ();
    moreCycles = 1;      eventCount = 0;
    while (moreCycles) {
      SingleEvent ();
      eventCount = eventCount + 1;
10    if (eventCount >= eventCountLimit) moreCycles = 0;
    }
  }


  SetupJob () {
    AllocArrays ();
    InitCoords ();
    InitVels ();
5   timeNow = nextSumTime = 0.;
    collCount = crossCount = 0.;
    StartRun ();
    ScheduleEvent (0, ATOM_LIMIT + 7, nextSumTime);
  }


  SingleEvent () {
    real vvSum;
    int k, n;
    NextEvent ();
5   if (evIdB <= ATOM_LIMIT) {
      ProcessCollision ();
      collCount = collCount + 1.;
    } else if (evIdB > ATOM_LIMIT + 100) {
      ProcessCellCrossing ();
10    crossCount = crossCount + 1.;
    } else {
      UpdateSystem ();
      nextSumTime = nextSumTime + intervalSum;
      ScheduleEvent (0, ATOM_LIMIT + 7, nextSumTime);
15    vSum = 0.;       vvSum = 0.;
      for (n = 1; n <= nAtom; n ++) {
        for (k = 1; k <= NDIM; k ++) {
          vSum = vSum + rv[k][n];
          vvSum = vvSum + Sqr (rv[k][n]);
20    } }
      vSum = fabs (vSum) / nAtom;
      sKinEnergy = 0.5 * vvSum / nAtom;
      PrintSummary (stdout);
    }
25 }
```

The call to NextEvent obtains the details of the next event; these include the two values evIdA and evIdB that are examined to determine the event type and atom(s) involved. If the event is a collision, so recognized because evIdB is less than some large constant value ATOM_LIMIT (that exceeds the maximum possible number of atoms), the two values identify the colliding atoms. For a cell crossing, signaled by a value of evIdB exceeding ATOM_LIMIT+100, evIdA is the atom and evIdB describes the cell face crossed. The only other kind of event expected here, corresponding to the value ATOM_LIMIT+7, is one which outputs a summary of the properties of the system. Other event classes can be accommodated (see Section 12.5).

The following variables are used in the program; those requiring explanation will receive it in due course:

```
   real **r, **rv, *atomTime, *treeTime, region[NDIM + 1],
      regionH[NDIM + 1], collCount, crossCount, density,
      intervalSum, nextSumTime, pi, sKinEnergy, temperature,
      timeNow, vMag, vSum;
5  int **inCell, **tree, *cellList, cellRange[2 * NDIM + 1],
      cells[NDIM + 1], initUcell[NDIM + 1], eventCount,
      eventCountLimit, eventMult, evIdA, evIdB, moreCycles,
      nAtom, poolSize, randSeed, runId;
   char *progId = "hs";
```

The list of data input by the program is

```
   NameList nameList[] = {
      INAME (runId),
      INAME (initUcell),
      RNAME (density),
5     RNAME (temperature),
      INAME (eventCountLimit),
      INAME (eventMult),
      RNAME (intervalSum),
      INAME (randSeed),
10 };
```

Processing of a single collision event is carried out by the function given below. Collisions between atoms on opposite sides of the periodic boundaries are treated correctly by pretending that the collision is with one of the periodic replica atoms. The hard-sphere collision dynamics are as described in (12.3). An array cellRange appears here; its values are used to determine which of the neighboring cells, out of the total of 27, should be examined for future collision events:

```
ProcessCollision () {
   real dr[NDIM + 1], fac, s1, s2;
```

```
     int k;
     UpdateAtom (evIdA);
5    UpdateAtom (evIdB);
     for (k = 1; k <= NDIM; k ++) {
       cellRange[2 * k - 1] = -1;    cellRange[2 * k] = 1;
     }
     for (k = 1; k <= NDIM; k ++) {
10     dr[k] = r[k][evIdA] - r[k][evIdB];
       if (fabs (dr[k]) > regionH[k])
         dr[k] = dr[k] - SignR (region[k], dr[k]);
     }
     s1 = s2 = 0.;
15   for (k = 1; k <= NDIM; k ++) {
       s1 = s1 + dr[k] * (rv[k][evIdA] - rv[k][evIdB]);
       s2 = s2 + Sqr (dr[k]);
     }
     fac = - s1 / s2;
20   for (k = 1; k <= NDIM; k ++) {
       rv[k][evIdA] = rv[k][evIdA] + dr[k] * fac;
       rv[k][evIdB] = rv[k][evIdB] - dr[k] * fac;
     }
     PredictEvent (evIdA, 0);
25   PredictEvent (evIdB, evIdA);
   }
```

Dealing with a cell-boundary crossing follows. Linked lists are used to record the atoms belonging to each cell; the atom concerned is removed from the list of the cell just exited and added to that of the newly entered cell. Periodic wraparound is applied where necessary, and the values in cellRange are used here to limit the cells examined for possible future collisions to the newly adjacent cells only:

```
   ProcessCellCrossing () {
     int k, n;
     UpdateAtom (evIdA);
     n = ((inCell[3][evIdA] - 1) * cells[2] +
5      inCell[2][evIdA] - 1) * cells[1] + inCell[1][evIdA] +
       nAtom;
     while (cellList[n] != evIdA) n = cellList[n];
     cellList[n] = cellList[evIdA];
     for (k = 1; k <= NDIM; k ++) {
10     cellRange[2 * k - 1] = -1;    cellRange[2 * k] = 1;
     }
     k = evIdB - ATOM_LIMIT - 100;
     if (rv[k][evIdA] > 0.) {
       cellRange[2 * k - 1] = 1;
15     inCell[k][evIdA] = inCell[k][evIdA] + 1;
```

```
       if (inCell[k][evIdA] == cells[k] + 1) {
         inCell[k][evIdA] = 1;
         r[k][evIdA] = - regionH[k];
       }
20    } else {
       cellRange[2 * k] = -1;
       inCell[k][evIdA] = inCell[k][evIdA] - 1;
       if (inCell[k][evIdA] == 0) {
         inCell[k][evIdA] = cells[k];
25       r[k][evIdA] = regionH[k];
       }
     }
     PredictEvent (evIdA, evIdB);
     n = ((inCell[3][evIdA] - 1) * cells[2] +
30      inCell[2][evIdA] - 1) * cells[1] + inCell[1][evIdA] +
       nAtom;
     cellList[evIdA] = cellList[n];    cellList[n] = evIdA;
   }
```

Predicting future events after a collision or cell crossing is carried out by the function `PredictEvent`. The first part of this function looks at possible cell-boundary crossings in all directions and schedules the earliest one. The second part examines every atom in the cells that must be scanned for possible collisions and determines whether a collision is possible using (12.2). Much of the code handles the special requirements of periodic boundaries. The reason why two arguments are needed by this function (the first is an atom number, the second an atom number, zero, or -1) should be apparent from the listing:

```
   PredictEvent (int na, int nb) {
     real dr[NDIM + 1], dv[NDIM + 1], shift[NDIM + 1],
       tm[NDIM + 1], b, d, t, tInt, vv;
     int signDir[NDIM + 1], evCode, iX, iY, iZ, jX, jY, jZ,
5      k, n;
     for (k = 1; k <= NDIM; k ++) {
       if (rv[k][na] != 0.) {
         if (rv[k][na] > 0.) signDir[k] = 0;
         else signDir[k] = 1;
10       tm[k] = ((inCell[k][na] - signDir[k]) * region[k] /
           cells[k] - r[k][na] - regionH[k]) / rv[k][na];
       } else tm[k] = 1e12;
     }
     if (tm[2] <= tm[3]) {
15      if (tm[1] <= tm[2]) k = 1;
       else k = 2;
     } else {
```

```
           if (tm[1] <= tm[3]) k = 1;
           else k = 3;
20       }
         evCode = 100 + k;
         ScheduleEvent (na, ATOM_LIMIT + evCode, timeNow + tm[k]);
         for (iZ = cellRange[5]; iZ <= cellRange[6]; iZ ++) {
           jZ = inCell[3][na] + iZ;      shift[3] = 0.;
25         if (jZ == 0) {
             jZ = cells[3];     shift[3] = - region[3];
           } else if (jZ > cells[3]) {
             jZ = 1;      shift[3] = region[3];
           }
30         for (iY = cellRange[3]; iY <= cellRange[4]; iY ++) {
             jY = ...
             ... (similar to above) ...
             for (iX = cellRange[1]; iX <= cellRange[2]; iX ++) {
               ... (similar to above) ...
35             n = ((jZ - 1) * cells[2] + jY - 1) * cells[1] +
                 jX + nAtom;
               for (n = cellList[n]; n > 0; n = cellList[n]) {
                 if (n != na && n != nb && (nb >= 0 || n < na)) {
                   tInt = timeNow - atomTime[n];
40                 for (k = 1; k <= NDIM; k ++) {
                     dr[k] = r[k][na] - (r[k][n] + rv[k][n] *
                       tInt) - shift[k];
                     dv[k] = rv[k][na] - rv[k][n];
                   }
45                 b = dr[1] * dv[1] + dr[2] * dv[2] +
                     dr[3] * dv[3];
                   if (b < 0.) {
                     vv = Sqr (dv[1]) + Sqr (dv[2]) + Sqr (dv[3]);
                     d = Sqr (b) - vv * (Sqr (dr[1]) +
50                     Sqr (dr[2]) + Sqr (dr[3]) - 1.);
                     if (d >= 0.) {
                       t = - (sqrt (d) + b) / vv;
                       ScheduleEvent (na, n, timeNow + t);
         } } } } } }
55 }
```

The following function is called at the start of the computation to create the cell lists and produce the initial event calendar:

```
StartRun () {
  int j, k, n;
  for (j = 1; j <= cells[1] * cells[2] * cells[3] +
    nAtom; j ++) cellList[j] = 0;
5  for (n = 1; n <= nAtom; n ++) {
    atomTime[n] = timeNow;
```

```
         for (k = 1; k <= NDIM; k ++)
            inCell[k][n] = (r[k][n] + regionH[k]) * cells[k] /
               region[k] + 1;
10       j = ((inCell[3][n] - 1) * cells[2] +
            inCell[2][n] - 1) * cells[1] + inCell[1][n] + nAtom;
         cellList[n] = cellList[j];    cellList[j] = n;
       }
       InitEventList ();
15     for (k = 1; k <= NDIM; k ++) {
         cellRange[2 * k - 1] = -1;    cellRange[2 * k] = 1;
       }
       for (n = 1; n <= nAtom; n ++) PredictEvent (n, -1);
     }
```

The function that updates an atom's coordinates and time variable is

```
UpdateAtom (int id) {
  int k;
  for (k = 1; k <= NDIM; k ++)
    r[k][id] = r[k][id] + rv[k][id] * (timeNow -
5       atomTime[id]);
  atomTime[id] = timeNow;
}
```

and the entire system can be updated by

```
UpdateSystem () {
  int n;
  for (n = 1; n <= nAtom; n ++) UpdateAtom (n);
}
```

The only other kind of event included in this version of the program simply outputs the measured energy and momentum of the system, as well as a report on the number of events that have occurred so far. The output function is

```
PrintSummary (FILE *fp) {
  fprintf (fp, "%.2f %.10g %.10g %.3f %.3f\n",
    timeNow, collCount, crossCount, vSum, sKinEnergy);
}
```

Additional quantities that are set in SetParams are

```
poolSize = eventMult * nAtom;
for (k = 1; k <= NDIM; k ++) cells[k] = region[k];
```

The variable poolSize determines how much space will be allocated for the event calendar. The size of the cell array assumes that the smallest possible cell is wanted (recall that the atoms have unit diameter). Memory allocation in `AllocArrays` includes, in addition to the usual coordinates and velocities,

```
   atomTime = AllocVecR (nAtom);
   cellList = AllocVecI (nAtom + cells[1] * cells[2] *
     cells[3]);
   inCell = AllocMatI (NDIM, nAtom);
 5 tree = AllocMatI (9, poolSize);
   treeTime = AllocVecR (poolSize);
```

The final two arrays appear only in the event-processing functions described in the next section. The initial state is defined in the same way as for soft spheres (Chapter 3).

12.2.5 Properties

Equilibrium and transport properties for models based on impulsive interactions can be defined by analogy with the continuous case. The only difference occurs in those quantities that depend directly on the interactions, such as the pressure: the virial expression must be replaced [erp77] by its impulsive limit, namely, a sum over the collisions occurring during the measurement period t_m:

$$PV = \tfrac{1}{3}\left[\left\langle\sum_i v_i^2\right\rangle + \frac{1}{t_m}\sum_c r_{i(c)j(c)} \cdot \Delta v_{i(c)}\right] \tag{12.7}$$

where $i(c)$ and $j(c)$ are the atoms involved in a particular collision c; the separation $r_{i(c)j(c)}$ allows for periodic wraparound, so that $|r_{i(c)j(c)}| = \sigma$. Transport properties follow a similar approach [ald70a].

Computation of the RDF is the same as for soft spheres (Chapter 4); the only difference is that, instead of the measurement being performed at fixed multiples of the timestep, a new class of measurement event is required. New variables, in addition to those needed for the RDF computation itself, are

```
   real intervalRdf, nextRdfTime;
```

the input data includes

```
   RNAME (intervalRdf),
```

and the initialization (in `SetupJob`):

```
   nextRdfTime = 0.;
   ScheduleEvent (0, ATOM_LIMIT + 8, nextRdfTime);
```

The event processing (in `SingleEvent`) now includes a test for the new event type:

```
  } else if (evIdB == ATOM_LIMIT + 8) {
    UpdateSystem ();
    EvalRdf ();
    nextRdfTime = nextRdfTime + intervalRdf;
5   ScheduleEvent (0, ATOM_LIMIT + 8, nextRdfTime);
  } else ...
```

Another measurement that will be made here is the distribution of path lengths between collisions; this can only be studied in the hard-sphere framework – for continuous potentials the notion of a collision event is not precisely defined. The average of the distribution is just the familiar mean free path. The new variables required for this analysis are

```
real **rCol, *histFreePath, rangeFreePath;
int countFreePath, limitFreePath, sizeHistFreePath;
```

input data items:

```
INAME (limitFreePath),
RNAME (rangeFreePath),
INAME (sizeHistFreePath),
```

additional array allocation:

```
rCol = AllocMatR (NDIM, nAtom);
histFreePath = AllocVecR (sizeHistFreePath);
```

and initialization:

```
InitFreePath ();
```

After processing each collision, a call is made to the function that calculates the path lengths, allowing for periodic boundaries:

```
EvalFreePath () {
  real dr, rr;
  int j, k, n;
  for (n = evIdA; n <= evIdB; n += evIdB - evIdA) {
5   countFreePath = countFreePath + 1;
    if (countFreePath == 1) {
      for (j = 1; j <= sizeHistFreePath; j ++)
        histFreePath[j] = 0.;
    }
10  rr = 0.;
    for (k = 1; k <= NDIM; k ++) {
      dr = fabs (r[k][n] - rCol[k][n]);
      if (dr > regionH[k]) dr = region[k] - dr;
      rr = rr + Sqr (dr);
15    rCol[k][n] = r[k][n];
    }
    j = (int) (sqrt (rr) / rangeFreePath) + 1;
```

```
          if (j > sizeHistFreePath) j = sizeHistFreePath;
          histFreePath[j] = histFreePath[j] + 1.;
20        if (countFreePath == limitFreePath) {
            for (j = 1; j <= sizeHistFreePath; j ++)
              histFreePath[j] = histFreePath[j] / countFreePath;
            PrintFreePath (stdout);
            countFreePath = 0;
25    } }
      }
```

The output function (not shown) just prints the path distribution histogram; the initialization function is

```
    InitFreePath () {
      int k, n;
      countFreePath = 0;
      for (n = 1; n <= nAtom; n ++) {
5       for (k = 1; k <= NDIM; k ++) rCol[k][n] = r[k][n];
      }
    }
```

12.3 Event management

12.3.1 Calendar design

We have already alluded to the central role played by the event calendar
listing the collisions and cell crossings, as well as events corresponding
to measurements of various kinds conducted at fixed time intervals. For
a large system the calendar will hold a great deal of information, and it
is imperative that it be managed in an efficient way. Efficiency focuses
principally on execution time, but space requirements are also important.

The scheme we describe here [rap80] is based on a binary tree data
structure [knu68, knu73]. The binary tree is a generalization of the
linked list to the case where each node (or list member) has pointers
to two successors rather than just one; the analogy with an inverted
tree is obvious, hence the name. Other data structures could serve the
purpose, but the binary tree is relatively straightforward to implement.
More significantly, its performance in situations relevant to MD can be
analyzed theoretically and, to within a constant factor, can be shown to
be optimal.

Each scheduled event is represented by a node in the tree. The
information contained within the node identifies the time at which the
event is scheduled to occur and the event details: if the event is a collision,
then the atoms involved are specified; if it is a cell crossing, then the

atom is specified, together with an indication of which cell boundary is crossed; for other event types, typically measurement events, the details are given by a suitable numerical code.

The operations that are performed on the tree data include retrieving the earliest event, adding a new event, and deleting an existing event. After retrieving an event the node containing its description must be deleted from the tree, and whenever a collision occurs all other event nodes involving either of the participants must also be deleted. A 'pool' of spare nodes exists from which withdrawals are made when events are added to the calendar, and to which the nodes are returned once no longer needed; this pool must never be allowed to become empty.

Three pointers are used to link event nodes into the tree; these point to the left and right descendant nodes and to the parent node. The time ordering is such that all the left-hand descendants are events scheduled to occur before the event at the current node, while those on the right are due to occur after it. The pointer to the parent is not essential but its presence simplifies algorithms for navigating the tree. The actual tree representation of a given set of events is far from unique and depends on the order in which the event nodes are added.

To support rapid deletion of related event nodes, the storage for each event also provides the pointers needed for linking the node into two circular lists [knu68], one list for each of the atoms involved if the event is a collision, two distinct lists for the same atom if the event is a cell crossing (the explanation follows); the pointers are unused in other cases. The reason for two linked lists per atom is again one of convenience: for a given atom j there is one list joining all collision nodes in which j appears as the first partner in the pair, and another for those in which j is the second partner. The pointers belonging to the cell-crossing node associated with atom j (there is always exactly one such node) are used to access these two lists. To improve performance further, the circular lists are doubly linked, each having pointers that traverse the list in both directions.

12.3.2 Theoretical performance

The amount of work required to perform certain elementary operations on the data in the binary tree can be estimated theoretically [knu73]; these operations lie at the heart of the calendar management functions to be described shortly. Here we summarize the relevant results.

Consider a binary tree with N nodes. If the tree is balanced, in the sense that all paths from the root (the node from which all others descend) to the nodes at the ends of all the branches are essentially the same length, then it can be shown that an average of $1.39 \ln N$ nodes must be tested to find the correct insertion point for a new node (assuming that the value determining the node position – the scheduled time in the case of MD – is randomly chosen). This represents the optimal value. In circumstances more relevant to MD, namely that the entire tree is constructed from a series of events whose scheduled times are (from the tree's point of view) randomly distributed, the average number of tests increases to $2 \ln N$, a value still not too far from optimal. Measurements using actual MD simulations confirm this result [rap80] and lay to rest any concern that the tree might degenerate into a near-linear list (for which average insertion time is proportional to N) over the duration of the run.

Another theoretical result deals with the average number of cycles in a search loop required to relink the neighbors of a particular node after that node is deleted. While this could also have shown a certain amount of N-dependence, in actual fact the value is a constant less than 0.5, and is thus completely independent of the tree size.

12.3.3 *Program details*

We have already introduced the pointers associated with the event nodes: there are three pointers for linking each node into the tree, and every node (except for measurement events) also belongs to two circular lists, each of which is doubly linked. Thus the total number of pointers per node is seven, and rather than allocating a separate array for each kind of pointer we pack them into the two-dimensional array `tree` defined earlier; we also include the two values specifying what the event actually is. (The node values are not stored consecutively because of the way C organizes arrays; an alternative is to pack the values into consecutive elements of a one-dimensional array.) For convenience the following definitions are used:

```
  #define treeLeft    tree[1]
  #define treeRight   tree[2]
  #define treeUp      tree[3]
  #define treeCircAL  tree[4]
5 #define treeCircBL  tree[5]
  #define treeCircAR  tree[6]
  #define treeCircBR  tree[7]
  #define treeIdA     tree[8]
  #define treeIdB     tree[9]
```

Each set of nine values defines a node. The first three values are used as pointers for traversing the tree. The next four are associated with the two circular lists to which each collision and cell-crossing event node belongs; the two lists are denoted A and B, and the two pointers for each list L and R. The final two values describe the event, as explained previously. There is one other quantity associated with each node, treeTime, the scheduled time of the event; because this is a real rather than an integer quantity it is stored separately.

The tree fluctuates in size over the course of the simulation as nodes are added and removed. A pool of spare nodes is provided to accommodate these size variations. The first node of tree is not used to hold events but serves as a fixed root from which the rest of the tree grows; one of its pointers is used to access the pool, the nodes of which are joined into a linked list (using the pointer treeCircAR). The nodes corresponding to cell crossings also occupy reserved locations (the N_a nodes immediately following the root) since there is always one such event scheduled per atom; these nodes also serve as anchors for the circular lists associated with each atom. The remaining nodes are dynamically assigned to collisions and other events as necessary. Figure 12.1 shows an example of a very small event tree.

The three functions that schedule an event, determine the next event, and delete an event follow. The list and tree manipulations are entirely standard [knu73] and, with just a little effort, it should be possible to follow the logic of the algorithms.

The first of these functions inserts an event node at the correct location in the tree, and then links it into the two circular lists; the node itself is taken from the pool, with a check being made to ensure that the pool is not empty. The tests (here and subsequently) to determine event type make provision for event types that are not included in the present case study (in particular, collisions with impenetrable boundary walls, which, from the point of view of event management, combine features of both collisions and cell crossings – see Section 12.5).

```
    ScheduleEvent (int idA, int idB, real tEvent) {
      int id, idNew, more;
      id = 1;
      if (idB <= ATOM_LIMIT || idB > ATOM_LIMIT + NDIM * 2 &&
5       idB <= ATOM_LIMIT + 100) {
        if (treeIdA[1] == 0) ErrExit ("empty event pool");
        idNew = treeIdA[1];
        treeIdA[1] = treeCircAR[treeIdA[1]];
      } else idNew = idA + 1;
```

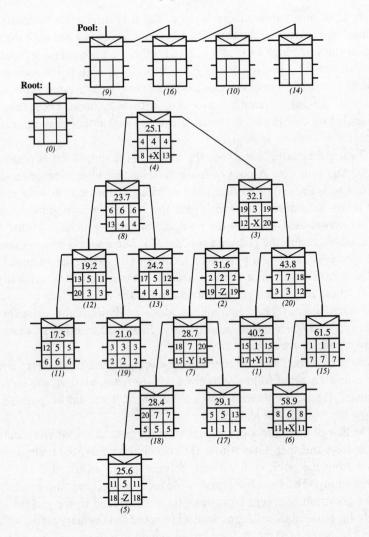

Fig. 12.1 A hand-crafted miniature event tree (the real one is much larger) – the tree links are shown, but for clarity the circular lists are omitted and the four pointer values are shown instead (on either side of the node). Each node includes the event time, the atom(s) involved, and the cell face crossed (if relevant); the value beneath the node is its 'address' in the tree.

```
10    if (treeRight[id] == 0) treeRight[id] = idNew;
      else {
        more = 1;    id = treeRight[id];
        while (more) {
```

```
         if (tEvent <= treeTime[id]) {
15           if (treeLeft[id] > 0) id = treeLeft[id];
             else {
               more = 0;     treeLeft[id] = idNew;
             }
         } else {
20           if (treeRight[id] > 0) id = treeRight[id];
             else {
               more = 0;     treeRight[id] = idNew;
             }
     } } } }
     if (idB <= ATOM_LIMIT) {
25     treeCircAR[idNew] = treeCircAR[idA + 1];
       treeCircAL[idNew] = idA + 1;
       treeCircAL[treeCircAR[idA + 1]] = idNew;
       treeCircAR[idA + 1] = idNew;
       treeCircBR[idNew] = treeCircBR[idB + 1];
30     treeCircBL[idNew] = idB + 1;
       treeCircBL[treeCircBR[idB + 1]] = idNew;
       treeCircBR[idB + 1] = idNew;
     }
     treeTime[idNew] = tEvent;
35   treeIdA[idNew] = idA;     treeIdB[idNew] = idB;
     treeLeft[idNew] = treeRight[idNew] = 0;
     treeUp[idNew] = id;
   }
```

We mentioned earlier that the average number of times the loop in this function is executed grows logarithmically with tree size.

The second function determines the next event about to occur, removes it from the tree, and if it is a collision event removes all other scheduled events involving the affected atoms from the tree by traversing all four circular lists to which the atoms belong; nodes removed are returned to the pool:

```
   NextEvent () {
     int id, idAx, idBx, idd, idNow, idtx;
     idNow = treeRight[1];
     while (treeLeft[idNow] > 0) idNow = treeLeft[idNow];
5    timeNow = treeTime[idNow];
     evIdA = treeIdA[idNow];     evIdB = treeIdB[idNow];
     if (evIdB <= ATOM_LIMIT + NDIM * 2) {
       if (evIdA < evIdB) {
         idAx = evIdA + 1;     idBx = evIdB + 1;
10     } else {
         idAx = evIdB + 1;     idBx = evIdA + 1;
       }
       idtx = idBx - idAx;
```

```
        if (evIdB > ATOM_LIMIT) idBx = idAx;
15      for (id = idAx; id <= idBx; id += idtx) {
          DeleteEvent (id);
          for (idd = treeCircAL[id]; idd != id;
              idd = treeCircAL[idd]) {
            treeCircBR[treeCircBL[idd]] = treeCircBR[idd];
20          treeCircBL[treeCircBR[idd]] = treeCircBL[idd];
            DeleteEvent (idd);
          }
          treeCircAR[treeCircAL[id]] = treeIdA[1];
          treeIdA[1] = treeCircAR[id];
25        treeCircAL[id] = treeCircAR[id] = id;
          for (idd = treeCircBL[id]; idd != id;
              idd = treeCircBL[idd]) {
            treeCircAR[treeCircAL[idd]] = treeCircAR[idd];
            treeCircAL[treeCircAR[idd]] = treeCircAL[idd];
30          DeleteEvent (idd);
            treeCircAR[idd] = treeIdA[1];    treeIdA[1] = idd;
          }
          treeCircBL[id] = treeCircBR[id] = id;
        }
35    } else {
        DeleteEvent (idNow);
        if (evIdB <= ATOM_LIMIT + 100) {
          treeCircAR[idNow] = treeIdA[1];
          treeIdA[1] = idNow;
40  } }
    }
```

The last of these three functions rearranges the tree pointers following
the removal of a node. All possible eventualities are handled and, as
indicated previously, the average number of times the short loop in this
function is executed is less than unity:

```
    DeleteEvent (int id) {
      int idp, idq, idr;
      idr = treeRight[id];
      if (idr == 0) idq = treeLeft[id];
5     else {
        if (treeLeft[id] == 0) idq = idr;
        else {
          if (treeLeft[idr] == 0) idq = idr;
          else {
10          idq = treeLeft[idr];
            while (treeLeft[idq] > 0) {
              idr = idq;    idq = treeLeft[idr];
            }
```

```
           treeLeft[idr] = treeRight[idq];
15         treeUp[treeRight[idq]] = idr;
           treeRight[idq] = treeRight[id];
           treeUp[treeRight[id]] = idq;
        }
        treeUp[treeLeft[id]] = idq;
20      treeLeft[idq] = treeLeft[id];
   } }
   idp = treeUp[id];    treeUp[idq] = idp;
   if (treeRight[idp] != id) treeLeft[idp] = idq;
   else treeRight[idp] = idq;
25 }
```

One further function is required for initializing the data structures. All
nodes are placed in the pool, and all circular lists are constructed so that
each contains just the one node associated with the cell crossing for the
particular atom:

```
InitEventList () {
   int id;
   treeLeft[1] = treeRight[1] = 0;
   treeIdA[1] = nAtom + 2;
5  for (id = treeIdA[1]; id <= poolSize - 1; id ++)
      treeCircAR[id] = id + 1;
   treeCircAR[poolSize] = 0;
   for (id = 2; id <= nAtom + 1; id ++) {
      treeCircAL[id] = treeCircBL[id] = id;
10     treeCircAR[id] = treeCircBR[id] = id;
   }
}
```

12.4 Results

12.4.1 Efficiency

Comparisons between the computational efficiency of hard- and soft-
sphere systems require some way of quantifying equivalent amounts of
computation. One could, for example, measure the computational effort
needed to determine pressure to a given degree of accuracy, although we
will not do this here. The results will certainly depend on the density, with
soft spheres having the advantage at high density (where the collision
rate is high), and the hard spheres at low density (where the mean free
path is relatively long). The whole question is usually irrelevant, however,
because the nature of the problem tends to dictate which approach is
required.

Actual timing measurements for hard-sphere systems at $\rho = 0.8$ are
shown in Table 12.1. The results are based on runs of $1.5 - 4.5 \times 10^5$

Table 12.1. *Hard-sphere timing estimates and analysis* [a]

N_a	τ	N_a/N_c	τ'	$\tau'/\log N_a$
256	410	1.19	345	62
864	368	0.86	428	63
2048	483	0.93	519	68
4000	497	0.81	613	73

[a] See text for details.

collisions; at this density there are approximately 0.2 cell crossings per collision. In order to compare the different measurements the results must be adjusted to allow for variations in mean cell occupancy, and to simplify the discussion we will assume that all (!) the computational effort goes into collision scheduling. Then if τ is the measured processing time per collision (in μs) and N_c the number of cells, $\tau' = \tau/(N_a/N_c)$ is proportional to the average time needed for scheduling one collision. Theory predicts that this time grows logarithmically with tree (and hence system) size, so we examine the ratio $\tau'/\log N_a$. The results are included in Table 12.1, where the ratio is seen to increase slowly with N_a; more careful study would be needed to determine whether or not this is due to overly simplistic analysis.

12.4.2 Properties

The RDF is obtained from a run at $\rho = 0.8$, $T = 1$, with additional input data

```
initUcell        8 8 8
eventCountLimit  9999999
eventMult        4
intervalRdf      0.25
limitRdf         100
rangeRdf         4.0
sizeHistRdf      200
```

The initial state is an FCC lattice, so that $N_a = 2048$; the conditions are the same as those used in Chapter 4 for measuring the soft-sphere RDF.

With these input values a set of RDF results is produced roughly every 6×10^5 collisions. The first set is discarded, the second is shown in Figure 12.2. The soft-sphere RDF is included for comparison; apart from the sharper first peak the two curves are practically identical.

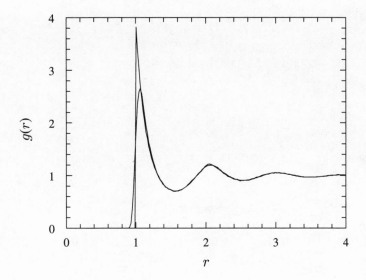

Fig. 12.2 RDFs for hard- and soft-sphere fluids.

The free-path distribution is studied during runs with $N_a = 500$ that include the input data

```
limitFreePath    20000
rangeFreePath    0.2
sizeHistFreePath 100
```

with ρ ranging from 0.025 to 0.2. The results from the first 40 000 collisions are discarded, and those for the next 20 000 are shown in Figure 12.3. The distributions all decrease with path length l (ignoring noise); if each is scaled by the mean free path at the corresponding ρ, all should collapse onto a single curve [ein68].

12.5 Generalizations

12.5.1 Outline

In this section we deal briefly with several useful extensions of the hard-sphere approach: the addition of a gravitational field, the use of hard-wall boundaries, the construction of polymer chains, and the way in which both rotational motion and inelasticity can be added (the latter intended for use in a macroscopic context only). In the first three examples (the first two of these are required for a case study in Chapter 13) the additions involve procedural details, so the program modifications are shown; in

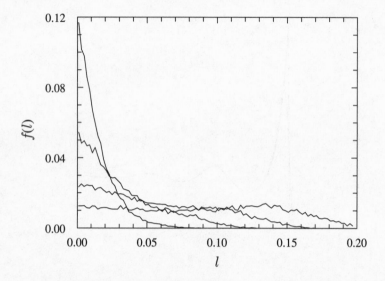

Fig. 12.3 Free-path distributions for hard spheres at $\rho = 0.025, 0.05, 0.1,$ and 0.2.

the final example the changes are in the collision dynamics and so we concentrate on the mathematical details.

12.5.2 Motion in a gravitational field

Surprisingly enough, extending the hard-sphere method to include the effect of a uniform gravitational field is a relatively simple exercise. The most important fact is that the collision prediction process is not affected at all because the same uniform acceleration is experienced by all atoms, and so the computation for determining the existence of a collision and when it occurs is independent of the field. Between collisions, atoms follow parabolic rather than linear trajectories.

It is only the prediction of cell crossings that calls for special attention because parabolic motion means, for example, that it is possible for an atom to leave a cell through the face of entry. The changes to PredictEvent, where the gravitational field (forceField) is assumed to act in a negative direction, are as follows; the algebraic problem solved here is locating the intersection of a parabola with a straight line:

```
real ... h1, h2;
...
for (k = 1; k <= NDIM; k ++) {
  if (forceField[k] != 0.) {
```

```
5      h1 = rv[k][na] * rv[k][na] - 2. * forceField[k] *
          (r[k][na] + regionH[k] - (inCell[k][na] - 1) *
          region[k] / cells[k]);
       if (rv[k][na] > 0.) {
          h2 = h1 + 2. * forceField[k] * region[k] / cells[k];
10        if (h2 > 0.) {
             tm[k] = - (rv[k][na] - sqrt (h2)) / forceField[k];
             signDir[k] = 0;
          } else {
             tm[k] = - (rv[k][na] + sqrt (h1)) / forceField[k];
15           signDir[k] = 1;
          }
       } else {
          tm[k] = - (rv[k][na] + sqrt (h1)) / forceField[k];
          signDir[k] = 1;
20     }
    } else {
       ... (same as without field) ...
    }
 }
```

The position updating function UpdateAtom must be modified to allow for parabolic trajectories, with the velocity also being updated:

```
   real tInt;
   tInt = timeNow - atomTime[id];
   ...
   r[k][id] = r[k][id] + (rv[k][id] + 0.5 * forceField[k] *
5     tInt) * tInt;
   rv[k][id] = rv[k][id] + forceField[k] * tInt;
```

Because local time variables are associated with each atom a similar modification is required in PredictEvent for the collision prediction:

```
   dr[k] = r[k][na] - (r[k][n] + (rv[k][n] +
      0.5 * forceField[k] * tInt) * tInt) - shift[k];
   dv[k] = rv[k][na] - (rv[k][n] + forceField[k] * tInt);
```

When checking for energy conservation, allowance should be made for the potential energy contribution.

12.5.3 Hard-wall boundaries

For some kinds of simulation the periodic boundaries are replaced by hard walls. These walls can be smooth, in which case collisions with the walls are energy-conserving specular collisions, but in most situations the walls will be rough, so that after undergoing a collision an atom loses all memory of its prior velocity. Simply randomizing the velocity direction (and possibly magnitude) will achieve this effect, as in Chapter 7, but here we will demonstrate an alternative approach in which the random element is absent. Each wall is divided into a series of narrow strips with width typically equal to the atom diameter; when an atom collides with the wall the outcome alternates between specular collisions and velocity reversals, depending on the strip – a kind of corrugation effect. If the wall is also attached to a constant-temperature heat bath the velocity magnitude can be changed to the required value (as in Chapter 7). Combinations of periodic, rough, and smooth walls can be used, although this option will not be shown here.

Modifications to the program to incorporate these effects are as follows. New variables specifying the width of the wall corrugations, whether each pair of opposite walls act as thermal reservoirs, and if so their temperatures, are

```
real roughWid[NDIM + 1], wallTemp[3];
int thermalWall[NDIM + 1];
```

and the necessary additional input data:

```
INAME (thermalWall),
RNAME (wallTemp),
```

Collisions with the hard walls require a new form of processing, so an additional class of cell-boundary event is introduced; in SingleEvent all cell events are now identified by the modified test:

```
if (evIdB <= ATOM_LIMIT + NDIM * 2 ||
    evIdB > ATOM_LIMIT + 100)
```

The usual cell events are covered by the latter part of the test, while the former – with evIdB greater than ATOM_LIMIT – accounts for the wall collisions.

The function ProcessCellCrossing is modified to handle wall collisions as well as regular cell crossings:

```
   real vFac, vv;
   int ... j, kk;
   ...
   j = evIdB - ATOM_LIMIT;
 5 if (j > 100) {
     k = j - 100;
     ... (same as before) ...
   } else {
     k = (j + 1) / 2;    cellRange[j] = 0;
10   rv[k][evIdA] = - rv[k][evIdA];
```

```
     for (kk = 1; kk <= NDIM; kk ++) {
       if (kk != k && (int) ((r[kk][evIdA] + regionH[kk]) /
           roughWid[kk]) % 2 == 0)
           rv[kk][evIdA] = - rv[kk][evIdA];
15   }
     if (thermalWall[k]) {
       vv = 0.;
       for (kk = 1; kk <= NDIM; kk ++)
           vv = vv + Sqr (rv[kk][evIdA]);
20     vFac = sqrt (NDIM * wallTemp[2 - j % 2] / vv);
       for (kk = 1; kk <= NDIM; kk ++)
           rv[kk][evIdA] = rv[kk][evIdA] * vFac;
   } }
```

In ProcessCollision the reference to periodic boundaries:

```
   if (fabs (dr[k]) > regionH[k]) dr[k] = ...
```

must be removed.

In PredictEvent changes are required in determining the identity of the cell event about to be scheduled and the range of cells that must be examined:

```
   int cellRangeT[2 * NDIM + 1] ...
   ...
   evCode = 100 + k;
   if (inCell[k][na] == 1 && signDir[k] == 1)
5    evCode = 2 * k - 1;
   else if (inCell[k][na] == cells[k] && signDir[k] == 0)
       evCode = 2 * k;
   ScheduleEvent ...
   for (k = 1; k <= 2 * NDIM; k ++)
10   cellRangeT[k] = cellRange[k];
   for (k = 1; k <= NDIM; k ++) {
     if (inCell[k][na] + cellRangeT[2 * k - 1] < 1)
         cellRangeT[2 * k - 1] = 0;
     if (inCell[k][na] + cellRangeT[2 * k] > cells[k])
15       cellRangeT[2 * k] = 0;
   }
   for (iZ = cellRangeT[5]; iZ <= cellRangeT[6]; iZ ++) ...
```

The code for dealing with periodic wraparound, namely,

```
   if (jZ == 0) {
     ...
   } else if (jZ > cells[3]) {
     ...
5  }
```

(etc.) is no longer required, and cellRangeT replaces cellRange for the three nested cell-loop limits.

As formulated above, an atom is deemed to collide with the wall when its center reaches the wall position. While not affecting the results here, it is more sensible (and necessary if atoms of mixed sizes are involved) if the collision occurs when the wall–atom distance equals the atom radius. In the present case the only changes needed are to enlarge the region slightly and shift the initial coordinates; in SetParams add

```
for (k = 1; k <= NDIM; k ++) region[k] = region[k] + 1.;
```

and in InitCoords (Chapter 3):

```
gap[k] = (region[k] - 1.) / initUcell[k];
...
c[1] = (nX - 0.5) * gap[1] - regionH[1] + 0.5;
... (same for other coordinates) ...
```

The width of the wall 'corrugations' is also specified in SetParams:

```
for (k = 1; k <= NDIM; k ++)
    roughWid[k] = region[k] / (int) region[k];
```

12.5.4 Hard-sphere polymer chains

Flexible polymer chains (reminiscent of a bead necklace) can be constructed by placing each pair of chain neighbors in a potential well with infinitely high walls and width corresponding to the maximum bond elongation [rap79]. While this kind of model lacks the refinement of the approach described in Chapter 10, it allows polymer studies to benefit from the advantages of event-driven MD.

We assume that the atoms still have unit diameter, and define the maximum bond elongation to be bondStretch. Cell sizes will now be determined by this value, so that the collision event occurring when a bond becomes fully stretched will also be detected by the usual scan over cells. The initial state must be constructed correctly, and we will also assume that all chain atoms are numbered consecutively. The only additional change needed is in PredictEvent to ensure that when predicting collisions for chain neighbors both normal and stretched-bond collisions are examined, and the appropriate choice made:

```
    int ... collCode;
    real .. rr;
    ...
    for (n = cellList[n]; n > 0; n = cellList[n]) {
5     if (n != na && n != nb && (nb >= 0 || n < na)) {
        tInt = ...
        for (k = ...
```

```
         ...
         collCode = 0;
10       if (abs (n - na) > 1 ||
             (n - 1) / chainLen != (na - 1) / chainLen) {
           b = ...
           if (b < 0.) {
             vv = ... ;    d = ... ;
15           if (d >= 0.) collCode = 1;
           }
         } else if (nb <= ATOM_LIMIT + 2 * NDIM) {
           collCode = 2;
           b = ... ;    vv = ... ;
20         rr = Sqr (dr[1]) + Sqr (dr[2]) + Sqr (dr[3]);
           if (b < 0.) {
             d = Sqr (b) - vv * (rr - 1.);
             if (d >= 0.) collCode = 1;
           } }
25       if (collCode > 0) {
           if (collCode == 1) t = - (sqrt (d) + b) / vv;
           else {
             d = Sqr (b) - vv * (rr - Sqr (bondStretch));
             t = (sqrt (d) - b) / vv;
30         }
           ScheduleEvent ...
     } } }
```

The fact that chain neighbors are never more than one cell apart removes the need to search for new collisions of this type after a cell crossing. The role of collCode is to distinguish between the two types of collision event that are now possible; the formula used for the stretched-bond collision is a particular instance of (12.4).

12.5.5 Rotation and inelasticity

Rotational motion is another feature that is readily added to the hard-sphere model [ber77]. The surfaces of the spheres are assumed to be rough, so that when a collision occurs not only is there a change in translational motion but the spins of the spheres also change.

It is a simple exercise in kinematics to show that the relative velocity at the point of impact (we now include the possibility that the spheres have different sizes and masses) is

$$\boldsymbol{g} = \boldsymbol{v}_{ij} - (\sigma_i \omega_i + \sigma_j \omega_j) \times \boldsymbol{r}_{ij} / \bar{\sigma} \tag{12.8}$$

where $\bar{\sigma} = (\sigma_i + \sigma_j)/2$. If the impulse is $\boldsymbol{b}$, the velocity and angular velocity changes required to conserve linear and angular momentum are given by

$$m_i \Delta v_i = -m_j \Delta v_j = b \tag{12.9}$$

$$m_i \sigma_i \Delta \omega_i = m_j \sigma_j \Delta \omega_j = -(2/\kappa \bar{\sigma}) r_{ij} \times b \tag{12.10}$$

where κ is the numerical factor in the moment of inertia, $I = \kappa m \sigma^2/4$; solid spheres have $\kappa = 2/5$, whereas for solid disks $\kappa = 1/2$.

The impulse can be expressed in terms of the components of g parallel and perpendicular to r_{ij}:

$$b = \bar{m}\left[g'^{\parallel} - g^{\parallel} + \kappa_1\left(g'^{\perp} - g^{\perp}\right)\right] \tag{12.11}$$

where the primes denote values after the collision, $\kappa_1 = \kappa/(1 + \kappa)$, and the reduced mass $\bar{m}$ is $m_i m_j/(m_i + m_j)$. The change in kinetic energy can also be written in terms of these components:

$$\Delta E_k = \tfrac{1}{2}\bar{m}\left[g'^{\parallel^2} - g^{\parallel^2} + \kappa_1\left(g'^{\perp^2} - g^{\perp^2}\right)\right] \tag{12.12}$$

Since $\Delta E_k = 0$ for an elastic collision, $g'^{\parallel} = -g^{\parallel}$, and $g'^{\perp} = \pm g^{\perp}$, with the negative solution being applicable for rough surfaces (the positive solution is for smooth surfaces, in which case $r_{ij} \times b = 0$ and the spins do not change).

Another feature that can be incorporated into the hard-sphere model is inelasticity. While not relevant in the atomic context, the same general MD approach can be used for modeling granular materials, and here energy dissipation by means of highly inelastic collisions is a key element in the dynamics. For inelastic collisions a coefficient of restitution ϵ is introduced, with $0 \le \epsilon \le 1$, and the relative velocities before and after collision are related by $g' = -\epsilon g$ (assuming, for convenience, that the same ϵ applies to both components). Expressions for b and ΔE_k follow immediately from (12.11) and (12.12).

12.6 Further work

12.1 Compare the thermodynamic properties of square-well and LJ fluids.

12.2 Study the equation of state of the hard-sphere fluid [erp84].

12.3 Examine trajectory sensitivity to small perturbations; here, unlike the corresponding soft-sphere treatment of this problem (Chapter 3), there is no numerical integration error.

12.4 Construct polymer chains using linked hard spheres and compare the effectiveness of the approach with the soft-sphere chains described in Chapter 9.

12.5 Measure the diffusion coefficient directly and by using the velocity autocorrelation function [erp85].

12.6 Implement rotational motion and examine the effect on diffusion.

12.7 Consider the inelastic particles as a model for granular materials [hon92]; most of the algorithmic details needed for studying granular flow problems [jae92] have been provided here (a continuous potential for granular material modeling appears in [tho91]).

12.8 Nonspherical particles can be handled by checking for overlaps and then using an iterative method to determine the instant of collision [kus76, reb77, all89]; explore this approach.

13

Time-dependent phenomena

13.1 Introduction

The simulations described so far have all involved systems that are either in equilibrium or are in some time-independent stationary state; while individual results are subject to fluctuation, it is the well-defined averages over sufficiently long time intervals that are of interest. In this chapter we extend the MD approach to a class of problem in which the behavior is not only time-dependent but the properties themselves are also spatially dependent in ways that cannot always be predicted. The analysis of the behavior of such systems cannot be carried out following the methods described earlier, which generally involve the evaluation of system-wide averages or correlations, and one is therefore compelled to resort to graphical methods. Here we focus on MD applications in fluid dynamics, a subject in which atomic matter is conventionally replaced by a continuous medium for practical purposes: recovering the atomic basis is part of trying to understand more complex fluid behavior of the type studied in rheology. For more on the microscopic approach to hydrodynamics see [mar92].

13.2 Open systems

Most current MD applications deal with closed systems; this implies either total isolation from the outside world or coupling to the environment in a way described by one of the ensembles of statistical mechanics. The coupling can occur with the aid of a thermostat for instance (Chapter 6), in which case the equilibrium properties are those of the canonical ensemble. Extending MD to open systems, where coupling to the external world is of a more general kind, introduces new problems, some of which will be encountered here. Not only are open systems out of thermodynamic equilibrium, but in many cases they are also spatially inhomogeneous and time-dependent. In some situations the presence of physical walls – as opposed to periodic boundaries – is essential to obtain the desired behavior.

Two examples of open systems will be treated here, both from the realm of fluid dynamics. One of the problems involves the flow of a fluid past a rigid obstacle, the other is a study of convective flow driven by a temperature gradient. These represent examples of attempts to bridge the gap between the atomistic picture of the microscopic world, so ably captured by MD, and the more conventional world of fluid dynamics. These examples also benefit from the fact that the phenomena are primarily two-dimensional; three-dimensional simulations of the required size are still beyond reach.

The problem of flow past a rigid obstacle [tri88] is one of the more extensively studied problems in fluid dynamics. Although the existence of atoms is generally irrelevant, and is ignored by the continuum fluid-dynamical equations, in order to learn what the molecular constituents of a fluid really do while the fluid is flowing around the obstacle and exhibiting a range of quite complex behavior it is necessary to return to the roots, and this implies MD simulation. In a wide variety of situations the entire description of a flow experiment can be condensed into a single dimensionless quantity, the Reynolds number $Re = dv/v$, where d is a characteristic length scale, here the diameter of a cylindrical obstacle oriented perpendicular to the flow, v is the flow speed, and v ($= \eta/\rho$) the kinematic viscosity. Irrespective of the type of fluid (liquid or gas), the obstacle size and the flow speed, the flow patterns depend only on Re; this kind of scaling behavior is known as dynamic similarity. At small Re values the flow is laminar, but the flow itself can be either stationary or time-dependent, with a pair of fixed eddies or a highly structured set of moving vortices forming in the wake of the obstacle.

Another flow problem of considerable importance is thermal convection [tri88]: a horizontal layer of fluid is heated from below, and the resulting interplay between the upward flow produced by heating and the downward flow due to gravity leads to the formation of structured flow patterns in the shape of rolls and various other forms, either stationary or time-dependent. Here it is the Rayleigh number $Ra = \alpha g d^3 \Delta T / v\kappa$ that determines the behavior, where α and κ ($= \lambda/\rho C_P$) are the thermal expansion coefficient and thermal diffusivity of the fluid, d is the layer height, and ΔT the temperature difference. (To complete the story, a second dimensionless quantity, the Prandtl number $Pr = \nu/\kappa$, is involved; this partly determines the nature of the convective motion.)

In setting up MD simulations of these systems it is important to ensure parameter combinations that produce the correct values of the dimensionless numbers; if they are too small nothing interesting should be expected, since in each case there exist threshold values for the onset of the instability responsible for the flow patterns. Even if the threshold is exceeded there is no guarantee that a microscopically small MD system will resemble its macroscopic counterpart: there must exist a minimum region size – measurable in atomic diameters or mean free paths (whichever is larger) – below which the characteristic fluid flow patterns cannot develop (the edge length of a square region containing 10^5 atoms at liquid density is only a mere 1000 Å). In addition to the size requirements, the duration of the simulation must be long enough to allow observation of any time-dependent behavior. The pessimist will also point out the use of highly exaggerated shear rates or temperature gradients that must be many orders of magnitude larger than their real-world counterparts in order to overcome the inherent thermal fluctuations and compensate for the small system size, and question whether the concept of dynamic similarity has not been stretched a little too far.

13.3 Thermal convection

13.3.1 Program details

The boundaries used in the study of thermal convection are all rigid (the side walls could also be periodic); when atoms collide with the top and bottom boundaries they are assigned a new velocity with magnitude determined by the wall temperature and, due to the wall roughness, an effectively random direction. The simulation uses the two-dimensional version of the hard-sphere/disk program, supplemented by four rough hard boundaries, top and bottom walls that act as thermal reservoirs, and a downward gravitational field. All the necessary implementation details are to be found in Chapter 12.

The flow is analyzed using the grid method of Chapter 7. Data collection uses the same function GridAverage; two minor changes that are required are the removal of the z-velocity components, and the use of the y-coordinate (rather than z) to determine grid cell membership. The grid measurements are recorded at fixed time intervals; to do this an additional event category similar to those used previously for other measurements is introduced.

Whenever the specified number of grid samples have been collected the results are appended to a file that stores the grid measurements (or 'snapshots') for the entire run. This is the task of the function PutGridAverage shown below. Data is output in binary form, after scaling the values so that they can be stored as short (16-bit) integers in order to save disk space. Enough information accompanies the grid data to permit reconstruction of the original values (with reduced precision). The program parameter NHIST has the value 4, with the four quantities treated in GridAverage being density, velocity squared, and the two velocity components; each quantity is computed for all atoms in every cell at a given instant, and the cell results are averaged over time:

```
   PutGridAverage () {
     real histMax[NHIST + 1], w;
     int blockSize, dataFormat, fOk, hSize, j, n;
     short *hI;
5    dataFormat = 1;
     hSize = sizeHistGrid[1] * sizeHistGrid[2];
     for (j = 1; j <= NHIST; j ++) {
       histMax[j] = 0.;
       for (n = 1; n <= hSize; n ++) {
10       w = fabs (histGrid[j][n]);
         if (histMax[j] < w) histMax[j] = w;
       }
       if (histMax[j] == 0.) histMax[j] = 1.;
     }
15   fOk = 1;
     blockSize = (NHIST + NDIM + 1) * sizeof (real) +
       (NDIM + 4) * sizeof (int) + NHIST * hSize *
       sizeof (short);
     if ((fp = fopen (fileName[FL_SNAP], "a"))) {
20     fwrite (&blockSize, sizeof (int), 1, fp);
       fwrite (&dataFormat, sizeof (int), 1, fp);
       fwrite (&runId, sizeof (int), 1, fp);
       fwrite (&timeNow, sizeof (real), 1, fp);
       fwrite (&region[1], sizeof (real), NDIM, fp);
25     fwrite (&snapNumber, sizeof (int), 1, fp);
       fwrite (&sizeHistGrid[1], sizeof (int), NDIM, fp);
       fwrite (&histMax[1], sizeof (real), NHIST, fp);
       hI = AllocVecS (hSize);
       for (j = 1; j <= NHIST; j ++) {
30       for (n = 1; n <= hSize; n ++)
```

```
          hI[n] = 32767. * histGrid[j][n] / histMax[j];
          fwrite (&hI[1], sizeof (short), hSize, fp);
        }
        FreeVecS (hI);
35      if (ferror (fp)) fOk = 0;
        fclose (fp);
      } else fOk = 0;
      if (! fOk) ErrExit ("write snap data");
    }
```

The variable `dataFormat` is added to allow a common analysis program
to distinguish between different kinds of stored information, although
it is not used here. The serial number of each set of data is stored
in `snapNumber`. For a discussion of other aspects of file usage see the
Appendix.

An analysis program (not shown in full here) would read this file, one
snapshot at a time, using the function below that is the complement of
`PutGridAverage`. For the initial call the variable `blockNum` is given the
value -1 (as in `GetConfig` shown in Chapter 4), and this ensures that
the necessary storage is allocated:

```
    GetGridAverage () {
      int fOk, hSize, j, n;
      short *hI;
      fOk = 1;
5     if (blockNum == -1) {
        if (! (fp = fopen (fileName[FL_SNAP], "r"))) fOk = 0;
      } else {
        fseek (fp, blockNum * blockSize, 0);
        blockNum = blockNum + 1;
10    }
      if (fOk) {
        fread (&blockSize, sizeof (int), 1, fp);
        if (feof (fp)) {
          more = 0;
15        return;
        }
        fread (&dataFormat, sizeof (int), 1, fp);
        ... (more data written by PutGridAverage) ...
        if (blockNum == -1) {
20        AllocArrays ();
          blockNum = 1;
        }
        hSize = sizeHistGrid[1] * sizeHistGrid[2];
        hI = AllocVecS (hSize);
25      for (j = 1; j <= NHIST; j ++) {
```

```
      fread (&hI[1], sizeof (short), hSize, fp);
      for (n = 1; n <= hSize; n ++)
         histGrid[j][n] = hI[n] * histMax[j] / 32767.;
   }
30   FreeVecS (hI);
      if (ferror (fp)) fOk = 0;
   }
   if (! fOk) ErrExit ("read snap data");
}
```

The size of the grid cells used for the analysis and the number of samples that contribute to a single time-averaged snapshot have yet to be specified. Ideally, both should allow the smallest spatial structures and the most rapid changes in the flow to be resolved. Opposing this goal is the sampling issue, since the smaller the number of atoms participating in the average for a single cell the larger the fluctuation. Compromise is necessary, and a typical (although very much a problem-dependent) tradeoff for a square system with $N_a = 10^5$ might involve a 50×50 grid, with measurements collected at time intervals of between 0.1 and 1 (there is no benefit in having them too closely spaced) and a snapshot after every 100−500 measurements.

There is no shortage of methods for displaying the results of the flow analysis. Here we only use arrow plots to show the grid-averaged flow direction (another method appears in the next case study). Contour plots of local scalar properties such as density and temperature are also readily produced. Successive plots can be combined to produce an animated sequence that will clearly reveal any time-dependent behavior. We avoid discussing the details of how to produce different kinds of graphic output, since standard software packages are usually available for this kind of work.

13.3.2 Results

The results shown here are from a hard-disk run using the following set of input data:

```
initUcell      200 100
density        0.4
temperature    1.0
intervalSum    1.0
forceField     0. -0.15
thermalWall    0 1
wallTemp       20.0 1.0
gridTime       1.0
```

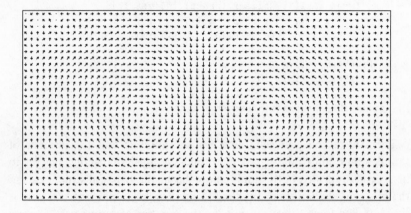

Fig. 13.1 Arrow plot showing convective flow at $t = 2000$.

```
limitGrid        100
sizeHistGrid     60 30
```

The system contains a total of $N_a = 20\,000$ disks, the walls are hard and rough, the lower boundary is hot and the upper cold, and the gravitational field acts in the downward direction. There is no temperature gradient in the initial state.

In Figure 13.1 we show the convective flow that has become firmly established by $t = 2000$ (corresponding to approximately 131 million collisions). Temperature is subject to both vertical and horizontal variation, as can be seen in Figure 13.2; the simple vertical profile used in Chapter 7 would have concealed this information. Density (not shown) is found to be practically constant, except at the cold wall. The value of *Ra* can only be estimated very roughly [puh89, rap91a], but it turns out to be consistent with what is expected from continuum fluid dynamics based on the observed behavior.

In simulations of this type there is normally an initial transient phase during which the system gradually evolves into a final state appropriate to the choice of parameters [rap92]. Identifying the point at which the system may be said to have reached this final state is not always obvious. A steady state is easily detected, but slow periodic oscillations or completely random behavior are less readily identifiable as 'final states'. There is also the possibility of long-lived metastable states preceding the true final state. In short, there are few general rules.

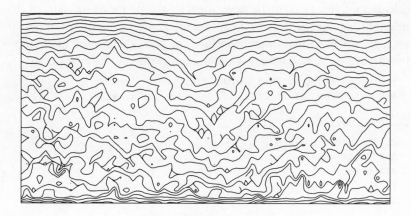

Fig. 13.2 Temperature contour plot.

13.4 Obstructed flow

13.4.1 Boundaries and driving force

There are two kinds of boundary that must be considered in this problem – the region boundaries and the obstacle itself. Region boundaries can be made periodic so that the flow recirculates, provided all memory of nonuniform flow is erased by (for example) periodically randomizing the velocities of all atoms located very close to either end of the system. The obstacle boundary should be rough to ensure that the adjacent fluid layer is at rest, corresponding to the experimental nonslip boundary. There are various ways of accomplishing this; a particularly simple approach for two-dimensional flow around a circular obstacle is to represent the obstacle as a ring of fixed atoms identical to those in the fluid.

The flow is driven by superimposing a velocity bias in the flow direction when randomizing the velocities; a constant external field could have been used instead, exactly as in the case of Poiseuille flow (Chapter 7). Heat will be generated as a result of the sheared flow near the obstacle, but now there are no walls that can be used to remove excess heat; a simple method of overcoming this problem is to cool the atoms while randomizing the velocities.

13.4.2 Program details

The program used in this case study is an extension of the basic soft-disk MD program using either cells or neighbor lists to assist the interaction

calculations (because of the flow, some of the performance gain of the
neighbor-list method is lost). Leapfrog integration is used. Features that
must be added to the program include the obstacle and the mechanism
for driving the flow. The new variables appearing in this problem are

```
real obsPos[3], bdyStripWidth, flowSpeed, obsSize;
int *atomType, nFixedAtom, nFreeAtom, stepDrive;
```

the input data are

```
  RNAME (bdyStripWidth),
  RNAME (flowSpeed),
  RNAME (obsPos),
  RNAME (obsSize),
5 INAME (stepDrive),
```

and an extra array is needed (`AllocArrays`):

```
atomType = AllocVecI (nAtom);
```

Initialization of the run begins by determining the precise number
of atoms in the system, given that the circular region occupied by the
obstacle is devoid of atoms. The fluid atoms initially occupy the sites
of a square lattice; the atoms used to form the obstacle boundary are
evenly spaced around the perimeter with separation $\approx r_c$. The obstacle
size and position are specified relative to the region size:

```
  EvalAtomCount () {
    real c[3], gap[3];
    int k, nX, nY;
    nFixedAtom = pi * obsSize * region[2] / rCut;
5   for (k = 1; k <= 2; k ++)
      gap[k] = region[k] / initUcell[k];
    nFreeAtom = 0;
    for (nY = 1; nY <= initUcell[2]; nY ++) {
      c[2] = (nY - 0.5) * gap[2] - regionH[2];
10    for (nX = 1; nX <= initUcell[1]; nX ++) {
        c[1] = (nX - 0.5) * gap[1] - regionH[1];
        if (OutsideObs (c[1], c[2]))
          nFreeAtom = nFreeAtom + 1;
    } }
15  nAtom = nFixedAtom + nFreeAtom;
  }

  int OutsideObs (real x, real y) {
    int outside;
    outside = (Sqr (x - obsPos[1] * region[1]) +
      Sqr (y - obsPos[2] * region[2]) >
5     Sqr (obsSize * regionH[2] + rCut));
```

```
      return (outside);
    }
```

(This test is easily modified for obstacles with other shapes.) The array
atomType that is used to distinguish obstacle boundary atoms from those
in the fluid is then filled:

```
  SetAtomType () {
    int n;
    for (n = 1; n <= nFixedAtom; n ++) atomType[n] = 1;
    for (n = nFixedAtom + 1; n <= nAtom; n ++)
5       atomType[n] = 0;
  }
```

The initial state can now be prepared; flow will be in the $+x$-direction:

```
  InitCoords () {
    real c[3], gap[3], ang;
    int k, n, nX, nY;
    for (k = 1; k <= 2; k ++)
5       gap[k] = region[k] / initUcell[k];
    for (n = 1; n <= nFixedAtom; n ++) {
      ang = 2. * pi * (n - 1) / nFixedAtom;
      r[1][n] = obsPos[1] * region[1] + obsSize *
         regionH[2] * cos (ang);
10     r[2][n] = obsPos[2] * region[2] + obsSize *
         regionH[2] * sin (ang);
    }
    n = nFixedAtom;
    for (nY = 1; nY <= initUcell[2]; nY ++) {
15    c[2] = ...
      for (nX = 1; nX <= initUcell[1]; nX ++) {
        c[1] = ...
        if (OutsideObs (c[1], c[2])) {
          n = n + 1;
20        for (k = 1; k <= 2; k ++) r[k][n] = c[k];
    } } }
  }

  InitVels () {
    real vsum[3], ang;
    int k, n;
    for (k = 1; k <= 2; k ++) vsum[k] = 0.;
5   for (n = 1; n <= nAtom; n ++) {
      if (atomType[n] == 0) {
        ... (normal treatment) ...
      } else {
        rv[1][n] = 0.;    rv[2][n] = 0.;
10  } }
```

```
      for (n = 1; n <= nAtom; n ++) {
        if (atomType[n] == 0) {
          for (k = 1; k <= 2; k ++)
            rv[k][n] = rv[k][n] - vsum[k] / nFreeAtom;
15        rv[1][n] = rv[1][n] + flowSpeed;
      } }
    }
```

The threefold task of maintaining fluid motion, removing excess heat, and erasing the flow pattern is achieved by the following call from SingleStep:

```
    if (stepCount % stepDrive == 0) DriveFlow ();
```

and the function is

```
    DriveFlow () {
      real ang;
      int n;
      for (n = 1; n <= nAtom; n ++) {
5       if (atomType[n] == 0 && fabs (r[1][n]) > regionH[1] -
            bdyStripWidth) {
          ang = 2. * pi * RandR (&randSeed);
          rv[1][n] = vMag * cos (ang) + flowSpeed;
          rv[2][n] = vMag * sin (ang);
10      } }
      }
```

The following test must be added to LeapfrogStep to prevent the atoms belonging to the obstacle boundary from moving:

```
    if (atomType[n] != 0) continue;
```

Data collection is again based on the grid method, with the fixed atoms forming the obstacle boundary excluded by testing atomType.

Another way of examining flow is to plot the streamlines. This amounts to evaluating the stream function at each grid point and then constructing a contour plot. The stream function [tri88] is defined as the line integral

$$\psi(\boldsymbol{r}) = \int \rho(v_y \, dl_x - v_x \, dl_y) \tag{13.1}$$

along an arbitrary path from the origin to the point $\boldsymbol{r}$. The following approach to computing the stream function can be included in an analysis or display program:

```
    ComputeStreamFun () {
      real sFirst, wX, wY;
      int ix, iy, n;
      wX = region[1] / sizeHistGrid[1];
```

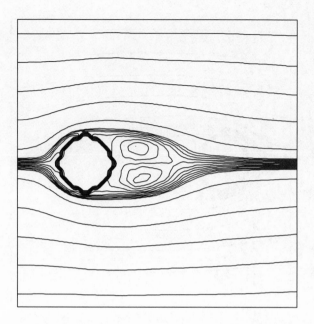

Fig. 13.3 Streamline plot of the vortices that form during the transient phase.

```
5    wY = region[2] / sizeHistGrid[2];
     sFirst = 0.;    n = 0;
     for (iy = 1; iy <= sizeHistGrid[2]; iy ++) {
       for (ix = 1; ix <= sizeHistGrid[1]; ix ++) {
         n = n + 1;
10       if (ix == 1) {
           sFirst = sFirst - histGrid[1][n] *
             histGrid[3][n] * wY;
           streamFun[n] = sFirst;
         } else {
15         streamFun[n] = streamFun[n - 1] +
             histGrid[1][n] * histGrid[4][n] * wX;
     } } }
   }
```

13.4.3 Results

The results demonstrated here use a soft-disk simulation with the following input data:

```
initUcell       250 250
```

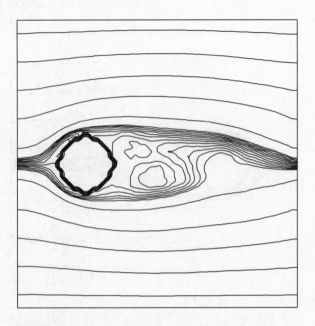

Fig. 13.4 An example of the oscillating flow pattern.

```
bdyStripWidth    3.0
flowSpeed        2.0
obsPos           -0.25 0.
obsSize          0.2
stepDrive        40
limitGrid        200
sizeHistGrid     60 60
stepGrid         20
```

together with $\rho = 0.8$, $T = 1$, and $\Delta t = 0.005$. There are $N_a = 62\,500$ atoms – large systems are required in order to allow the flow patterns to develop properly [rap87].

Two pictures of the flow streamlines are shown here. Figure 13.3 illustrates the transient phase – here at $t = 80$ – in which a pair of counter-rotating eddies have formed at the downstream edge of the obstacle. The flow eventually develops oscillatory behavior (actually corresponding to vortex shedding, but the present system is too small to see this properly); Figure 13.4 shows a snapshot at $t = 320$ that is typical of what happens.

13.5 Further work

13.1 The thermal convection problem can also be treated using a soft-disk fluid (an extension of the problem in Chapter 7); explore.

13.2 Other examples of more complex flow problems that call for more detailed analysis include the pipe flow of immiscible fluids [kop89, tho89], stick-slip flow [tho91], and polymer flow [edb87, rap94]. These are examples of fluids that cannot be described adequately at the continuum level and there is considerable scope for research using the MD approach.

14

Algorithms for supercomputers

14.1 Introduction

The previous chapters concentrated on translating physical problems into practical simulations. Computational efficiency, beyond the use of cells and neighbor lists, received little attention. For 'conventional' computers there is not a great deal more that the average user can do in this respect, assuming that a reasonably effective programming style has been adopted. This attitude is no longer adequate when modern multiprocessor and vector machines are to serve as platforms for large-scale simulation.

In this chapter we focus on ways of adapting the basic MD approach to take advantage of advanced computer architectures; since enhanced performance comes not only from a faster processor clock cycle, but also from a number of fundamental changes in the way computers process data, this is a subject that cannot be ignored. The subject is also a relatively complex one, and at best only peripheral to the interests of the practicing simulator. We will therefore not delve too deeply into the issues involved, but will merely focus on two examples; one involves message-passing parallelism, the other demonstrates how to rearrange data to achieve effective vector processing – both approaches have proved valuable in large-scale MD simulation. The vectorizeable program can be run on a non-vector machine in its present form (although this would only be done for development purposes), and the parallel program can also be run on a single-processor system if software for emulating message-passing is available.

14.2 The quest for performance

It comes as no surprise to learn that spreading a computational task over several processors is a way to complete the job sooner. Multipro-

330

cessor systems benefit from an economy of scale, and high-performance computers now almost always consist of at least a few processing units, if not more; the number of processors per machine begins at two and extends to over a thousand, and there is no limit in sight. The ideal kind of problem for such a machine (assuming that each processor has its own private memory) is one which can be partitioned into a number of smaller computations that are carried out in parallel on all the processors without too much data having to be shared between them. Molecular dynamics systems with short-range forces fall comfortably into this category.

The other major means for extracting higher performance from computer hardware is to resort to vector processing. This entails pipelining the computations in assembly-line fashion. The constraint placed on a computation if it is to be effectively vectorized is that data items should be organized into relatively long vectors in such a way that all items can be processed independently without fear that the processing of one item will affect a later one. This condition is not always easily satisfied.

Needless to say, both these architectural features are increasingly likely to be encountered. The price of utilizing them effectively is increased algorithm and software complexity. Unlike 'simple' computers, where optimizing compilers can take the average program and massage it to give reasonable performance, the needs of vector and distributed processing cannot always be resolved in this way, because it is not always obvious – assuming that it is possible at all – how to automate the process of optimally mapping an intrinsically serial computation onto the more 'complex' computer architectures. In addition to actually producing a working vectorized and/or distributed program, the efficiency of the end result must be considered: a parallel computation with large interprocessor communication overheads is doomed to failure, as is a vectorized computation that either uses the vector capability inefficiently because the vectors are too short, or spends a disproportionate amount of time rearranging data into vectorizeable form.

Beyond these two most visible features of modern computers there are a number of more subtle and not so well-documented processor characteristics that can have a significant impact on performance; these features are equally important in high-performance workstations. Just to name some of them: primary and secondary caches, address-space mapping, and memory interleaving. An algorithm that in some way conflicts with certain engineering design assumptions, for example in its pattern of memory accesses, can experience a drastic performance drop. Beyond drawing the attention of the reader to the existence of such potential pitfalls, there is little more that can be said without addressing each computer model individually. So after the extra work to develop software that appears to be effectively vectorized or distributed, there is still the task of establishing that the hardware is being used in at least a reasonably effective manner.

14.3 Techniques for distributed processing

14.3.1 Living with multiprocessor computers

The taxonomy of multiprocessing is far from simple. Among the factors to be taken into account are whether individual processors all carry out the same operation during each cycle, or whether they are able to act independently; whether each processor has its own private memory, or all share a common memory, or both; whether processors communicate with one another by passing messages across a communication network, or through common memory; the nature and topology of the communication network, if any. Some of these features can influence the way in which software ought to be organized.

We will avoid becoming involved in these issues here by assuming a particular generic architecture, one that is in fact widespread because it is the simplest to implement. The assumption is that there are several independent processors, each with private memory, communicating over a network – the so-called message-passing approach. Systems of this type can be built by simply linking workstations using standard network hardware; the performance may be less than satisfactory, however, but the same concept is often found embedded in customized hardware designs.

In order for the message-passing scheme to work efficiently it is vital that the communication overheads are kept low compared with the time a processor spends computing; the ideal application consists of a lot of calculation with small amounts of data being transferred from time to time. The communication overhead is composed of two parts, the time to initiate a message transfer, typically a constant value, and a transfer time that is roughly proportional to the message length. There is also the issue of load balancing; obviously if all processors can be kept busy doing useful computing the overall system utilization is optimal, but if some processors have more work to do than others overall effectiveness is reduced.

14.3.2 Algorithm organization

There are different ways to partition an MD computation among multiple processors. The act of partitioning can focus on the computations, on the atoms involved, or on the simulation region. While all three elements are part of every scheme, the emphasis differs; this has a considerable impact on the memory and communication requirements of each method.

If it is just the computations that are partitioned, then all information about the system resides in the memory of each processor, but each only carries out the interaction computations for certain atom pairs. The information about the forces on each atom is then combined. This approach is extremely wasteful in terms of memory and is best suited to small computations only (or shared-memory computers).

The second partitioning scheme is based on the atoms themselves, and assigns each atom to a particular processor for the duration of the simulation, irrespective of its spatial location [rai89]. While conceptually simple, large amounts of communication are required to handle interactions between atoms assigned to different processors. For long-range interactions this may not be a problem, but for the short-range case the third choice turns out to be far more efficient.

The third scheme subdivides space and assigns each processor a particular subregion [rap91b]. All the atoms that are in a given subregion at some moment in time reside in the processor responsible, and when an atom moves between subregions all the associated variables are explicitly transferred from one processor to another. Thus there is economy insofar as memory is concerned, and also in the communication required to allow atoms to transfer between processors, since comparatively few atoms make such a move during a single timestep. More importantly, assuming that there are some 10^4 or more atoms per subregion (in three dimensions) and a soft-sphere potential with very short range, most of the interactions will occur among atoms in the subregion, and relatively few between atoms in adjacent subregions – see Figure 14.1. In order to accommodate the latter, copies of the coordinates of atoms close to any subregion boundary are transferred to the processor handling the adjacent subregion prior to the interaction computation. This transfer also involves only a small fraction of the atoms.

It is this third scheme that will be described here. The only requirement is that communication be reasonably efficient, with the associated system overheads – both hardware and software – kept to a low level in comparison with the amount of computation involved. Under these

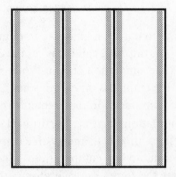

Fig. 14.1 A simulation region divided into slices assigned to different processors; each shaded region contains atoms that can interact with atoms in the adjacent processor.

circumstances both computation speed and memory requirements scale in the expected linear way with the numbers of atoms and processors.

14.4 Distributed MD simulation

14.4.1 Overview

The program described here is based on the three-dimensional soft-sphere computation described in Chapter 3 that uses both cells and neighbor lists. Much of the original program can be used unchanged, but wherever atoms become aware of the existence of subregions the computations must take this into account. While conceptually straightforward, the distributed computation involves numerous details that must be treated with care.

The functions handling communication-related tasks are referenced in a generic form that does not assume any particular message-passing system. The text describes what each of these functions is supposed to do and, in practice, since all such software must provide similar functionality, a simple series of macro substitutions may be all that is required to produce a working program; an actual implementation is shown at the end of the chapter.

In addition to the interaction calculations, integration of the equations of motion, initialization and measurements, each of which is modified to a lesser or greater degree for the distributed implementation, it is also necessary to specify which processor is responsible for each spatial subregion, to identify the atoms participating in each data transfer, and to carry out the transfers. Several kinds of data transfer are used: for interaction calculations it is necessary to copy information about the coordinates of atoms close to subregion boundaries; when atoms move between subregions their entire descriptions are transferred; when making measurements the values computed separately in each processor must be combined to produce an overall result.

14.4.2 Basic computations

The program listing begins with neighbor-list construction, on the assumption that the atoms are already in the correct processors and that copies of the coordinates of atoms in adjacent subregions have been appended to the coordinate arrays. These tasks will be described later. Two new variables play an important role here: nAtomMe is the number of atoms presently in the subregion, and nAtomCopy is the total number of additional atoms from adjacent subregions whose coordinates have been copied to this processor because those atoms are potential interaction candidates. Given the additional information, this function (and those which follow) can be executed independently on each processor. The reader new to distributed processing should bear in mind that all variables are local to each processor; the concept of a global variable does not exist in a message-passing environment. (For brevity we avoid the use of NDIM+1 in the array declarations, although this should not be done for input data.)

```
  BuildNebrList () {
    real cellBase[4], dr[4], invWid[4], rr, rrNebr;
    int c, iof, j1, j2, k, m1, m1X, m1Y, m1Z, m2, m2X, m2Y,
      m2Z, n, offset, tOffset,
5     iofX[] = {0,0,1,1,0,-1,0,1,1,0,-1,-1,-1,0,1},
      iofY[] = {0,0,0,1,1,1,0,0,1,1,1,0,-1,-1,-1},
      iofZ[] = {0,0,0,0,0,0,1,1,1,1,1,1,1,1,1},
      iofTableLen[] = {1,3,1,3,9,3,1,3,1,2,6,2,5,14,4,
      1,3,1}, offsetList[][14] = {{8}, {8,9,10}, {10},
10    {7,8,14}, {6,7,8,9,10,11,12,13,14}, {10,11,12}, {14},
      {12,13,14}, {12}, {3,8}, {3,4,5,8,9,10}, {5,10},
      {2,3,7,8,14}, {1,2,3,4,5,6,7,8,9,10,11,12,13,14},
      {5,10,11,12}, {14}, {12,13,14}, {12}};
    rrNebr = Sqr (rCut + rNebrShell);
15  for (n = nAtomMe + nAtomCopy + 1; n <= nAtomMe +
      nAtomCopy + cells[1] * cells[2] * cells[3]; n ++)
      cellList[n] = 0;
    for (k = 1; k <= NDIM; k ++) {
```

```
           invWid[k] = (cells[k] - 2) / (subRegionHi[k] -
20            subRegionLo[k]);
           cellBase[k] = subRegionLo[k] - 1. / invWid[k];
        }
        for (n = 1; n <= nAtomMe + nAtomCopy; n ++) {
          c = ((int) ((r[3][n] - cellBase[3]) * invWid[3]) *
25           cells[2] + (int) ((r[2][n] - cellBase[2]) *
             invWid[2])) * cells[1] + (int) ((r[1][n] -
             cellBase[1]) * invWid[1]) + nAtomMe + nAtomCopy + 1;
          cellList[n] = cellList[c];    cellList[c] = n;
        }
30      nebrTabLen = 0;
        for (m1Z = 1; m1Z <= cells[3] - 1; m1Z ++) {
          for (m1Y = 1; m1Y <= cells[2]; m1Y ++) {
            for (m1X = 1; m1X <= cells[1]; m1X ++) {
              tOffset = 14;
35            if (m1Z == 1) tOffset = tOffset - 9;
              if (m1Y == 1) tOffset = tOffset - 3;
              else if (m1Y == cells[2]) tOffset = tOffset + 3;
              if (m1X == 1) tOffset = tOffset - 1;
              else if (m1X == cells[1]) tOffset = tOffset + 1;
40            m1 = ((m1Z - 1) * cells[2] + m1Y - 1) * cells[1] +
                 m1X + nAtomMe + nAtomCopy;
              for (offset = 1; offset <=
                 iofTableLen[tOffset - 1]; offset ++) {
                iof = offsetList[tOffset - 1][offset - 1];
45              m2X = m1X + iofX[iof];
                m2Y = m1Y + iofY[iof];
                m2Z = m1Z + iofZ[iof];
                m2 = ((m2Z - 1) * cells[2] + m2Y - 1) *
                   cells[1] + m2X + nAtomMe + nAtomCopy;
50              j1 = cellList[m1];
                while (j1 > 0) {
                  j2 = cellList[m2];
                  while (j2 > 0) {
                    if (m1 != m2 || j2 < j1) {
55                    for (k = 1; k <= NDIM; k ++)
                        dr[k] = r[k][j1] - r[k][j2];
                      rr = ...
                      ... (identical to standard version) ...
        }
```

Several points should be noted. The cell array is defined separately for each subregion, and includes an additional layer of cells that completely surrounds the subregion; it is here that all the atoms copied from adjacent subregions are located. Periodic boundaries are not mentioned at this stage of the computation because, as will be shown later, they are handled during the copying operation. The array variables subRegionLo and subRegionHi contain the spatial limits of the subregion handled by each processor. Because periodic boundaries are handled by other means, the range of adjacent cells scanned during neighbor-list construction depends on the cell location; for cells that are located on a subregion face, edge, or corner, fewer adjacent cells need be examined – the data in offsetList makes provision for all 18 distinct cases and the way it is used (together with iofTableLen) should be apparent from the listing.

Only minor changes are needed in the force calculation. The principal reason for the changes is in order to evaluate accumulated properties such as the potential energy and the virial sum. The force computation does not distinguish between atoms that really belong to the subregion and those that are merely copies from an adjacent subregion, since the force contributions associated with the latter are simply discarded afterwards. To prevent atom pairs that interact across a subregion boundary from contributing more than once, separate energy and virial values are maintained for each atom, and at the end of the computation only those atoms that belong to the subregion are allowed to contribute to the summed values. Once again, no mention of periodic wraparound is needed:

```
ComputeForces () {
   ...
   for (n = 1; n <= nAtomMe + nAtomCopy; n ++) {
     for (k = 1; k <= NDIM; k ++) ra[k][n] = 0.;
5    u[n] = 0.;     vir[n] = 0.;
   }
   for (n = 1; n <= nebrTabLen; n ++) {
     j1 = nebrTab[1][n];     j2 = nebrTab[2][n];
     for (k = 1; k <= NDIM; k ++)
10      dr[k] = r[k][j1] - r[k][j2];
     rr = ...
     if (rr < rrCut) {
       ...
       u[j1] = u[j1] + uVal;     u[j2] = u[j2] + uVal;
15      vir[j1] = vir[j1] + fcVal * rr;
       vir[j2] = vir[j2] + fcVal * rr;
   } }
   uSum = 0.;     virSum = 0.;
   for (n = 1; n <= nAtomMe; n ++) {
20    uSum = uSum + u[n];     virSum = virSum + vir[n];
   }
   uSum = uSum * 0.5;     virSum = virSum * 0.5;
}
```

Integration uses the leapfrog method. The only change to Leapfrog-Step is the use of nAtomMe for the number of atoms to be processed. The atoms whose coordinates were copied from adjacent subregions are excluded from this calculation since they are not among the first nAtomMe atoms stored in the arrays.

In establishing the initial state we encounter communication for the first time, albeit still in a very minor role. Keep in mind that the initialization function is executed concurrently on all processors, but each decides which atoms belong to its subregion. An FCC lattice is used for the initial state, and the coordinate and velocity initializations have been combined into a single function. Since a particular atom is no longer associated with a fixed memory location, unlike the single-processor version of the program, each atom is labeled with a unique identifier atomId; although not really needed in this particular example it is sometimes necessary to be able to distinguish individual atoms (a polymer fluid is one such example). The variable totAtom is used for ensuring the uniqueness of this identifier:

```
     InitState () {
        real c[4], e[4], gap[4], rc[4], vSum[4], vSumL[4];
        int j, k, n, np, nX, nY, nZ, totAtom;
        for (k = 1; k <= NDIM; k ++) {
5         gap[k] = region[k] / initUcell[k];     vSum[k] = 0.;
        }
        totAtom = 0;     nAtomMe = 0;
        for (nZ = 1; nZ <= initUcell[3]; nZ ++) {
        ... (see InitCoords) ...
10          for (j = 1; j <= 4; j ++) {
              for (k = 1; k <= NDIM; k ++) {
                if (j == k || j == 4) rc[k] = c[k];
                else rc[k] = c[k] + 0.5 * gap[k];
              }
15            totAtom = totAtom + 1;
              RandVec3 (e, &randSeed);
              for (k = 1; k <= NDIM; k ++) {
                if (rc[k] < subRegionLo[k] ||
                  rc[k] >= subRegionHi[k]) break;
20            }
              if (k == NDIM + 1) {
                nAtomMe = nAtomMe + 1;
                for (k = 1; k <= NDIM; k ++) {
                  r[k][nAtomMe] = rc[k];
25                rv[k][nAtomMe] = vMag * e[k];
                  vSum[k] = vSum[k] + rv[k][nAtomMe];
                }
```

```
                 atomId[nAtomMe] = totAtom;
       } } } } }
30     if (procMe == 0) {
           for (np = 1; np <= nProcs - 1; np ++) {
             MsgRecv (np, 121);
             MsgRecvInit ();
             MsgUnpackR (&vSumL[1], NDIM);
35           for (k = 1; k <= NDIM; k ++)
                 vSum[k] = vSum[k] + vSumL[k];
           }
           for (k = 1; k <= NDIM; k ++) vSum[k] = vSum[k] / nAtom;
           if (nProcs > 1) {
40           MsgSendInit ();
             MsgPackR (&vSum[1], NDIM);
             MsgBcSend (122);
           }
       } else {
45         MsgSendInit ();
           MsgPackR (&vSum[1], NDIM);
           MsgSend (0, 121);
           MsgBcRecv (122);
           MsgRecvInit ();
50         MsgUnpackR (&vSum[1], NDIM);
       }
       for (n = 1; n <= nAtomMe; n ++) {
           for (k = 1; k <= NDIM; k ++)
               rv[k][n] = rv[k][n] - vSum[k];
55     }
   }
```

The latter part of this function includes several communication operations. A summary of what occurs here is as follows. There is a variable procMe that distinguishes processors from one another; the mission of the processor whose procMe value is zero is to collect the values of the array vSum from all the other processors after each has computed its local values, evaluate the total sums, and compute the values that each processor must subtract from all its atoms' velocities to ensure a zero center of mass velocity, and finally return these values to each of the other processors. The other processors perform a complementary task: they accumulate vSum, send it to processor 'zero', and wait for the necessary values to be returned. In a sense, processor number zero could be regarded as the master processor, and the remainder as slaves, but this detracts from the fact that, with a couple of minor exceptions, all processors perform completely equivalent tasks and are mutually synchronized by the data transfers that occur throughout the calculation.

Much of the communication processing is hidden away from the application, as indeed it should be. It is a pity that even the few details included here have to be mentioned at all, but until some standardized form of programming distributed computers achieves widespread acceptance this situation will persist. The reader might want to read the documentation of some message-passing software package (such as the widely available PVM system [gei94]) in order to associate the functions appearing here with those actually used (an example appears at the end of the chapter); depending on the system a few of the functions shown here may be combined or even redundant.

The function MsgRecv waits for and accepts a message from a specified processor; the numerical value appearing here and in other communication functions is an arbitrary value used by the message software to distinguish between different kinds of message (this feature is not used here), and is especially helpful at the programming stage in ensuring that messages sent and received correspond to one another. The MsgRecvInit function is used to begin the processing of the received message, while MsgUnpackR extracts the specified number of real variables from the message and stores them sequentially in the indicated locations (note the use of the address of the variable and not the value itself); a similar function MsgUnpackI deals with integer variables. MsgSendInit begins the processing involved in constructing a new message for sending, MsgPackR places the specified variable(s) in the message, and MsgSend actually sends the message to the indicated destination processor. There is also a broadcast capability; the function MsgBcSend broadcasts an identical message to all other processors and each processor uses MsgBcRecv to receive this message. Other communication functions will appear in due course as needed.

Another MD function that must retrieve a small amount of information from each processor is EvalProps. Note that several consecutive calls to the message packing and unpacking functions can be used when a number of different data items are combined into a single message. Here again processor zero is the one responsible for gathering and combining the data:

```
     EvalProps () {
       real vs[4], vsL[4], uSumL, v, virSumL, vv, vvMaxL,
         vvSumL;
       int k, n, np;
5      for (k = 1; k <= NDIM; k ++) vs[k] = 0.;
       vvSum = 0.;    vvMax = 0.;
       for (n = 1; n <= nAtomMe; n ++) {
```

```
         vv = 0.;
         for (k = 1; k <= NDIM; k ++) {
10           v = rv[k][n] - 0.5 * ra[k][n] * deltaT;
             vs[k] = vs[k] + v;
             vv = vv + Sqr (v);
         }
         vvSum = vvSum + vv;
15       if (vv > vvMax) vvMax = vv;
       }
       if (procMe == 0) {
         for (np = 1; np <= nProcs - 1; np ++) {
           MsgRecv (np, 161);
20         MsgRecvInit ();
           MsgUnpackR (&vvMaxL, 1);      MsgUnpackR (&vvSumL, 1);
           MsgUnpackR (&vsL[1], NDIM);
           MsgUnpackR (&uSumL, 1);       MsgUnpackR (&virSumL, 1);
           if (vvMaxL > vvMax) vvMax = vvMaxL;
25         vvSum = vvSum + vvSumL;
           for (k = 1; k <= NDIM; k ++) vs[k] = vs[k] + vsL[k];
           uSum = uSum + uSumL;      virSum = virSum + virSumL;
         }
         vSum = 0.;
30       for (k = 1; k <= NDIM; k ++)
             vSum = vSum + fabs (vs[k]);
         kinEnergy = 0.5 * vvSum / nAtom;
         ...
         dispHi = dispHi + sqrt (vvMax) * deltaT;
35       if (dispHi > rNebrShell * 0.5) nebrNow = 1;
       } else {
         MsgSendInit ();
         MsgPackR (&vvMax, 1);      MsgPackR (&vvSum, 1);
         MsgPackR (&vs[1], NDIM);
40       MsgPackR (&uSum, 1);       MsgPackR (&virSum, 1);
         MsgSend (0, 161);
       }
     }
```

Finally, the modified version of the function SingleStep is

```
SingleStep () {
   stepCount = stepCount + 1;
   timeNow = stepCount * deltaT;
   if (nebrNow > 0) DoParlMove ();
5  DoParlCopy ();
   if (nebrNow > 0) {
     nebrNow = 0;
     if (procMe == 0) dispHi = 0.;
     BuildNebrList ();
```

```
10   }
     ComputeForces ();
     LeapfrogStep ();
     EvalProps ();
     if (procMe == 0) {
15     if (nProcs > 1) {
         MsgSendInit ();
         MsgPackI (&nebrNow, 1);
         MsgBcSend (151);
       }
20   } else {
       MsgBcRecv (151);
       MsgRecvInit ();
       MsgUnpackI (&nebrNow, 1);
     }
25   if (procMe == 0) {
       if (stepCount >= stepEquil) {
         ...
     } }
   }
```

Processor zero is responsible for collecting the results and handling the output. The communication operations appearing here provide each processor with the current value of nebrNow, telling it whether a neighbor-list update is due. The functions DoParlMove and DoParlCopy – described below – deal with the interprocessor data transfers needed for interaction calculations and atom movements.

An important detail should be apparent from the way communications in the above functions are organized. Obviously, there is a one-to-one correspondence between messages sent and messages received. Equally significant, however, is the fact that these message transfer operations are used to synchronize the processors. None of the additional capabilities found in distributed software systems for explicit synchronization are required here, such as creating a barrier that no processor can pass until it receives permission. When writing parallel software based on a message-passing paradigm it is important to plan the communications carefully, otherwise deadlock and race conditions can occur that are difficult to diagnose.

14.4.3 Message-passing operations

We now turn to the functions where the majority of the interprocessor communication occurs, namely, the movement of atoms between subregions and the copying of coordinate data prior to the interaction calculations.

The functions responsible for deciding which atoms should be moved and then actually doing the work, including coordinate adjustment for periodic wraparound, are as follows. For brevity, both here and elsewhere, we omit various checks that ensure none of the storage arrays are filled to excess; such safety measures should always be present whenever unpredictable amounts of data are involved. There are six directions (in three dimensions) to be considered, and each is treated in turn; atoms can of course participate in more than one such transfer. The special case that a processor is its own neighbor, which occurs when no subdivision of the region is made in a particular direction, is also taken into account. Once all the moves are complete each processor compresses its own data to eliminate gaps in the arrays. The only new communication function appearing here is MsgSendRecv that both transmits data to a neighboring processor and receives data from the opposite neighbor; in some message-passing systems a function of this kind can be used to achieve overlapped data transfers. The size of the multiprocessor configuration and the way the simulation region is subdivided (for example, into smaller boxes, or into slices spanning the entire region – as in Figure 14.1) are specified by the array procArraySize:

```
   DoParlMove () {
      int dir, j, k, n, nHi, nIn, nLo;
      for (dir = 1; dir <= NDIM; dir ++) {
        nLo = nHi = 0;
5       for (n = 1; n <= nAtomMe; n ++) {
          if (atomId[n] > 0) {
            if (r[dir][n] < subRegionLo[dir]) {
              nLo = nLo + 1;     trPtrLo[dir][nLo] = n;
            } else if (r[dir][n] >= subRegionHi[dir]) {
10            nHi = nHi + 1;     trPtrHi[dir][nHi] = n;
        } } }
        nTrLo[dir] = nLo;    nTrHi[dir] = nHi;
        PackMovedData (dir, -1, trPtrLo[dir], nTrLo[dir]);
        if (procArraySize[dir] > 1) {
15        MsgSendInit ();
          MsgPackI (&nTrLo[dir], 1);
          MsgPackR (&trBuff[1], (2 * NDIM + 1) * nTrLo[dir]);
          MsgSendRecv (procNebrHi[dir], procNebrLo[dir],
            140 + 2 * dir - 1);
20        MsgRecvInit ();
          MsgUnpackI (&nIn, 1);
          MsgUnpackR (&trBuff[1], (2 * NDIM + 1) * nIn);
```

```
       } else nIn = nTrLo[dir];
       UnpackMovedData (nIn);
25     PackMovedData (dir, +1, trPtrHi[dir], nTrHi[dir]);
       if (procArraySize[dir] > 1) {
         MsgSendInit ();
         MsgPackI (&nTrHi[dir], 1);
         MsgPackR (&trBuff[1], (2 * NDIM + 1) * nTrHi[dir]);
30       MsgSendRecv (procNebrLo[dir], procNebrHi[dir],
             140 + 2 * dir);
         MsgRecvInit ();
         MsgUnpackI (&nIn, 1);
         MsgUnpackR (&trBuff[1], (2 * NDIM + 1) * nIn);
35     } else nIn = nTrHi[dir];
       UnpackMovedData (nIn);
     }
     j = 0;
     for (n = 1; n <= nAtomMe; n ++) {
40     if (atomId[n] > 0) {
         j = j + 1;
         for (k = 1; k <= NDIM; k ++) {
           r[k][j] = r[k][n];     rv[k][j] = rv[k][n];
         }
45       atomId[j] = atomId[n];
     } }
     nAtomMe = j;
   }
```

Message packing and unpacking are two-stage processes; the work that is specific to the application data (shown below) is kept separate from the actual filling or emptying of message buffers (by calls to MsgPackR and related functions in DoParlMove above). Periodic boundaries are addressed at this stage:

```
PackMovedData (int dir, int sign, int *trPtr, int nOut) {
  int j, k, m, n;
  for (j = 1; j <= nOut; j ++) {
    n = trPtr[j];     m = (2 * NDIM + 1) * (j - 1);
5   for (k = 1; k <= NDIM; k ++) trBuff[m + k] = r[k][n];
    if (sign == 1) {
      if (procArrayMe[dir] == procArraySize[dir] - 1)
        trBuff[m + dir] = trBuff[m + dir] - region[dir];
    } else {
10    if (procArrayMe[dir] == 0)
        trBuff[m + dir] = trBuff[m + dir] + region[dir];
    }
    for (k = 1; k <= NDIM; k ++)
      trBuff[m + NDIM + k] = rv[k][n];
```

```
15      trBuff[m + 2 * NDIM + 1] = atomId[n];    atomId[n] = 0;
     }
  }

  UnpackMovedData (int nIn) {
    int j, k, m;
    for (j = 1; j <= nIn; j ++) {
      m = (2 * NDIM + 1) * (j - 1);
5     for (k = 1; k <= NDIM; k ++) {
        r[k][nAtomMe + j] = trBuff[m + k];
        rv[k][nAtomMe + j] = trBuff[m + NDIM + k];
      }
      atomId[nAtomMe + j] = trBuff[m + 2 * NDIM + 1];
10  }
    nAtomMe = nAtomMe + nIn;
  }
```

The functions for copying the coordinates of atoms close to subregion
boundaries to adjacent processors (prior to the interaction calculations)
are very similar; the set of atoms involved is updated only when the
neighbor list is about to be rebuilt:

```
DoParlCopy () {
  int dir, j, m, n, nHi, nIn, nLo;
  nAtomCopy = 0;
  for (dir = 1; dir <= NDIM; dir ++) {
5   if (nebrNow > 0) {
      nLo = nHi = 0;
      for (n = 1; n <= nAtomMe + nAtomCopy; n ++) {
        if (r[dir][n] < subRegionLo[dir] +
          rCut + rNebrShell) {
10        nLo = nLo + 1;    trPtrLo[dir][nLo] = n;
        } else if (r[dir][n] >= subRegionHi[dir] -
          rCut - rNebrShell) {
          nHi = nHi + 1;    trPtrHi[dir][nHi] = n;
      } }
15    nTrLo[dir] = nLo;    nTrHi[dir] = nHi;
    }
    PackCopiedData (dir, -1, trPtrLo[dir], nTrLo[dir]);
    if (procArraySize[dir] > 1) {
    ... (same as DoParlMove, only NDIM items per atom) ...
20  } else nIn = nTrLo[dir];
    UnpackCopiedData (nIn);
    PackCopiedData (dir, +1, trPtrHi[dir], nTrHi[dir]);
    if (procArraySize[dir] > 1) {
      ... (ditto) ...
25  } else nIn = nTrHi[dir];
```

```
      UnpackCopiedData (nIn);
    }
  }

  PackCopiedData (int dir, int sign, int *trPtr, int nOut) {
    int j, k, m, n;
    for (j = 1; j <= nOut; j ++) {
      n = trPtr[j];     m = NDIM * (j - 1);
5     for (k = 1; k <= NDIM; k ++) trBuff[m + k] = r[k][n];
      if (sign == 1) {
        ... (based on PackMovedData) ...
    } }
  }

  UnpackCopiedData (int nIn) {
    int j, k, m;
    for (j = 1; j <= nIn; j ++) {
      m = NDIM * (j - 1);
5     for (k = 1; k <= NDIM; k ++)
          r[k][nAtomMe + nAtomCopy + j] = trBuff[m + k];
    }
    nAtomCopy = nAtomCopy + nIn;
  }
```

The main program and initialization function for the distributed computation are as follows:

```
  main (int argc, char **argv) {
    ... (get value of procMe by system specific method) ...
    if (procMe == 0) {
      GetNameList (argc, argv);
5     PrintNameList (stdout);
      nProcs = procArraySize[1] * procArraySize[2] *
          procArraySize[3];
    }
    InitParlProcs ();
10  NebrParlProcs ();
    SetParams ();
    SetupJob ();
    moreCycles = 1;
    while (moreCycles) {
15    SingleStep ();
      if (stepCount >= stepLimit) moreCycles = 0;
    }
  }

  SetupJob () {
    AllocArrays ();
```

```
     InitState ();
     stepCount = 0;
5    nebrNow = 1;
     if (procMe == 0) AccumProps (0);
   }
```

We have glossed over two small details that are intimately associated with the message-passing software: how to ensure that all the processors run copies of the same program, and how each processor obtains the correct, and distinct, value of procMe. The former may be automatic, or require some user action either before or while running the program; the latter may be as simple as a function call. Examination of the documentation for the particular software system will resolve these questions; an example appears at the end of the chapter.

New variables introduced in this program are

```
real *trBuff, *u, *vir, subRegionHi[4], subRegionLo[4];
int **trPtrHi, **trPtrLo, *atomId, nTrHi[4], nTrLo[4],
   procArrayMe[4], procArraySize[4], procNebrHi[4],
   procNebrLo[4], nAtomCopy, nAtomMe, nAtomMeMax, nProcs,
5  procMe, trBuffMax;
```

There are new input data items specifying the number of processors and the way the simulation region is subdivided, the maximum number of atoms a processor can hold (including copies), and the size of the buffers used for collecting data to be transferred:

```
INAME (procArraySize),
INAME (nAtomMeMax),
INAME (trBuffMax),
```

In SetParams the subregion limits are established and the cell array size (with an extra cell at each end) is determined by the size of the subregion:

```
   real w;
   int k;
   ...
   nebrTabMax = nebrTabFac * nAtomMeMax;
5  for (k = 1; k <= NDIM; k ++) {
     regionH[k] = 0.5 * region[k];
     w = region[k] / procArraySize[k];
     subRegionLo[k] = - regionH[k] + procArrayMe[k] * w;
     subRegionHi[k] = subRegionLo[k] + w;
10   cells[k] = (int) (w / (rCut + rNebrShell)) + 2;
   }
```

Memory allocation differs from the standard case:

```
   AllocArrays () {
     r = AllocMatR (NDIM, nAtomMeMax);
     rv = AllocMatR (NDIM, nAtomMeMax);
     ra = AllocMatR (NDIM, nAtomMeMax);
5    u = AllocVecR (nAtomMeMax);
     vir = AllocVecR (nAtomMeMax);
     atomId = AllocVecI (nAtomMeMax);
     cellList = AllocVecI (nAtomMeMax + cells[1] * cells[2] *
       cells[3]);
10   nebrTab = AllocMatI (2, nebrTabMax);
     trPtrHi = AllocMatI (NDIM, trBuffMax);
     trPtrLo = AllocMatI (NDIM, trBuffMax);
     trBuff = AllocVecR (NDIM * trBuffMax);
   }
```

Finally, each processor discovers who its neighbors are, based on the value of ProcMe:

```
   NebrParlProcs () {
     int tNum[4], j, k;
     nProcs = procArraySize[1] * procArraySize[2] *
       procArraySize[3];
5    procArrayMe[1] = procMe % procArraySize[1];
     procArrayMe[2] = (procMe / procArraySize[1])
       procArraySize[2];
     procArrayMe[3] = procMe / (procArraySize[1] *
       procArraySize[2]);
10   for (k = 1; k <= NDIM; k ++) {
       for (j = 1; j <= NDIM; j ++) tNum[j] = procArrayMe[j];
       tNum[k] = (procArrayMe[k] + procArraySize[k] - 1)
         procArraySize[k];
       procNebrLo[k] = (tNum[3] * procArraySize[2] +
15       tNum[2]) * procArraySize[1] + tNum[1];
       tNum[k] = (procArrayMe[k] + 1) % procArraySize[k];
       procNebrHi[k] = (tNum[3] * procArraySize[2] +
         tNum[2]) * procArraySize[1] + tNum[1];
     }
20 }
```

and processor zero, the only one with access to the input data file, distributes its contents to all the other processors:

```
   InitParlProcs () {
     if (procMe == 0) {
     if (nProcs > 1) {
       MsgSendInit ();
5      MsgPackI (&procArraySize[1], NDIM);
       MsgPackI (&initUcell[1], NDIM);
```

```
         MsgPackR (&density, 1);
         MsgPackR (&temperature, 1);
         MsgPackR (&deltaT, 1);
10       MsgPackI (&nAtomMeMax, 1);
         MsgPackI (&nebrTabFac, 1);
         MsgPackR (&rNebrShell, 1);
         MsgPackI (&randSeed, 1);
         MsgPackI (&stepEquil, 1);
15       MsgPackI (&stepLimit, 1);
         MsgPackI (&trBuffMax, 1);
         MsgBcSend (111);
       }
     } else {
20     MsgBcRecv (111);
       MsgRecvInit ();
       MsgUnpackI (&procArraySize[1], NDIM);
       ... (unpack remaining variables) ...
     }
25 }
```

The distributed environment makes its presence felt both during algorithm development and at the programming stage; any features that are added to the simulation must take this extra software overhead into account. A lot of tedious detail must be handled, but hopefully standardized programming languages will one day succeed in eliminating much of this effort.

14.5 Techniques for vector processing

14.5.1 Pipelined computation

When supercomputers first appeared their performance derived from two principal features: a fast processor clock, and the use of pipelined vector processing. With the appearance of comparatively cheap microprocessors and their employment in parallel computers the role of vector processing has temporarily diminished. However, the reader may rest assured that it will return once again to boost the performance of the microprocessors. Some pipelining is already to be found in these chips (as well as multiple instruction units), but the need for application software to take its existence into account is less of an issue than when serious vector processing is crucial to performance. Mainframe supercomputers still rely heavily on vector processing.

We will again concentrate on the soft-sphere simulation; this turns out to be the most difficult to vectorize effectively. For medium- to long-range forces, the number of interacting neighbors of each atom is large, and since a slightly modified form of the all-pairs version of the interaction function can then be used (provided there are no more than a few hundred atoms) the compiler is able to handle the vectorization automatically [bro86]. Likewise for small systems with only short-range forces, where the loss of efficiency due to using the all-pairs method is adequately compensated by the vectorization speedup. But for bigger systems (over a few hundred atoms) with short-range forces, for which cells are essential, and neighbor lists advisable if the memory is available, compilers are unable to rearrange the conventional MD program so that it may be vectorized effectively. The problem must be solved the hard way – by rearranging the computation in a more suitable form [rap91c].

A specific computer architecture will not be invoked here. Instead, the computations will be expressed in a form that any processor with certain common hardware features – to be enumerated later – and an effective compiler will be able to execute relatively efficiently. The approach differs from the previous section where we spelled out the communication functions to be used; a similar approach could have been used here as well, but since the result of reformulating the algorithm is a program that can be vectorized automatically there is little point in doing the work manually (by calling an assortment of vector functions directly).

Effective vector processing requires long sequences of data items (the vectors) that can be processed independently in pipeline fashion. Because of the overhead in filling these pipes initially, there is an advantage to using long vectors, although the minimum useful length depends on the specific machine. Vector computers emphasize floating-point performance, but in recognition of the fact that most problems are not organized in precisely the form needed, the machines are also able to carry out certain kinds of data rearrangement at high speed. More precisely, the ability to speedily gather randomly distributed data items into a single array based on a vector of addresses, as well as the converse scatter operation, exist on current machines; the MD code below relies heavily on such operations.

14.5.2 Layer method

The problem with the cell method, which makes it impossible to vectorize effectively, is its use of linked lists to identify the atoms in each cell. The method we describe here takes the same cell-occupancy data and rearranges it into a form more suitable for vector processing. The approach can also be extended to produce neighbor lists [gre89b] organized in a way that can be automatically vectorized.

In the original cell-based version of ComputeForces in Chapter 3 there are several nested loops; the outermost is over cells, and within this there is a loop over the offsets to neighboring cells; two further inner loops consider all pairs of cell occupants. We will now proceed to rearrange these loops. If the cell occupants are assumed to be ordered in some (arbitrary) way, then the role of the two outermost loops in the revised version is to produce all valid pairings of the *i*th and *j*th occupants of whatever pair of cells happens to be under consideration by the inner loops (the case where both cells are the same is also covered). Within these two outer loops there is a loop over possible relative offsets between cell pairs. Finally, the innermost loop is over all cells, with the second cell of the pair obtained using the known offset. If the number of cells is close to the number of atoms the innermost loop will fulfill the principal requirement of effective vectorization – a large repetition count. Of course this is not the whole story, because it is the details of the processing carried out by the innermost loop that determine performance and, in addition, the work needed to rearrange the data for this computation must be taken into account.

The reordered cell-occupancy data are stored in 'layers' [rap91c], as shown in Figure 14.2. Each layer consists of one storage element per cell, and there are as many layers as the maximum cell occupancy. The first atom in a cell (the order has no special significance) is placed in the first layer, and so on; the first layer will be practically full, but later layers will be less densely occupied. Since the two outer loops scan all pairs of layers, and the number of cycles of the inner loops is independent of cell occupancy, it follows that computation time varies quadratically with the number of layers. Thus the method is most effective for relatively dense systems where cell occupancy does not vary too widely.

The presence of periodic boundaries complicates the algorithm, so they are eliminated by a process of replication (Chapter 3), reminiscent of the copying operation used in the distributed approach. Before beginning the interaction computation (ComputeForces), all atoms close to any boundary are duplicated so that they appear just beyond the opposite boundary. These replicas are used for all interactions that would otherwise extend across one or more boundaries. The cell array must be enlarged to provide a shell of thickness r_c (the interaction cutoff range) surrounding the region.

The replication function follows. Each spatial dimension is treated in turn, and atoms near edges or corners can be replicated more than once.

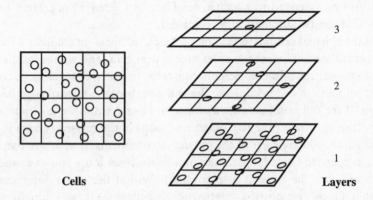

Fig. 14.2 The reorganization of the cell contents into layers for efficient vector processing.

An example of checking for array overflow is included. (To actually vectorize the loops in this function it may be necessary to divide each loop over atoms into two parts; the first loop would identify the atoms to be replicated and store their indices in a separate array, the second would do the replication using this index array.)

```
   ReplicateAtoms () {
     int i, k, kk, n;
     nAtomRep = nAtom;
     for (k = 1; k <= NDIM; k ++) {
5      n = nAtomRep;
       for (i = 1; i <= nAtomRep; i ++) {
         if (r[k][i] < - regionH[k] + rCut) {
           n = n + 1;
           for (kk = 1; kk <= NDIM; kk ++)
10           r[kk][n] = r[kk][i];
           r[k][n] = r[k][n] + region[k];
       } }
       for (i = 1; i <= nAtomRep; i ++) {
         if (r[k][i] >= regionH[k] - rCut) {
15         n = n + 1;
           for (kk = ...
           r[k][n] = r[k][n] - region[k];
       } }
       nAtomRep = n;
20     if (nAtomRep > nAtomMax) ErrExit ("too many atoms");
     }
   }
```

The interactions are computed in several stages. The first determines the cell occupied by each atom. Next the layers are constructed by a method to be discussed below. Then come the multiply-nested loops: the outer two loops select all possible layer pairs, within them is the loop over cell offsets, and inside this are four successive loops where all the interactions are computed. The first of the innermost loops considers all cell pairings between the chosen layers with the specified offset, determines whether a valid pair of atoms is to be found there, and if so adds this information to a list. The second loop processes the listed atom pairs and computes their interactions. The final two loops add these newly computed terms to the accumulated interactions of the respective atoms.

New variables appearing in the program are

```
real **fL, *u, *uL;
int **layerAtom, *atomId, *atomPtr1, *atomPtr2, *inCell,
    *inside, offsetVal[28], bdyMargin, nAtomMax, nAtomRep,
    nCells, nLayerMax;
```

The arrays are allocated by a new version of `AllocArrays`:

```
   AllocArrays () {
     r = AllocMatR (NDIM, nAtomMax);
     rv = AllocMatR (NDIM, nAtom);
     ra = AllocMatR (NDIM, nAtomMax);
5    fL = AllocMatR (NDIM, nCells);
     u = AllocVecR (nAtomMax);
     uL = AllocVecR (nCells);
     layerAtom = AllocMatI (nLayerMax, nCells +
       2 * bdyMargin);
10   atomId = AllocVecI (nAtomMax);
     atomPtr1 = AllocVecI (nCells);
     atomPtr2 = AllocVecI (nCells);
     inCell = AllocVecI (nAtomMax);
     inside = AllocVecI (nCells + 2 * bdyMargin);
15 }
```

The variable `nAtomMax` is the maximum number of atoms, including replicas, for which storage is available, `nLayerMax` is the number of available layers, and `bdyMargin` is a constant used to avoid any problems with negative array indices when shifted layers are paired (see later). The meaning of the other variables will become clear in due course. If more than a few layers are needed the array `layerAtom` will dominate the storage requirements of the program.

The interaction calculation follows:

```
   ComputeForces () {
     real dr[4], invWid[4], regionExH[4], fcVal, rr, rrCut,
       rri, rri3;
     int i, ii, k, layer1, layer2, m1, m2, n, nLayer, nn,
5      nPair, offset, offsetHi, offsetLo;
     for (k = 1; k <= NDIM; k ++) {
       invWid[k] = (cells[k] - 2) / region[k];
       regionExH[k] = regionH[k] + 1. / invWid[k];
     }
10   for (i = 1; i <= nAtomRep; i ++) {
       inCell[i] = ((int) ((r[3][i] + regionExH[3]) *
         invWid[3]) * cells[2] + (int) ((r[2][i] +
         regionExH[2]) * invWid[2])) * cells[1] +
         (int) ((r[1][i] + regionExH[1]) * invWid[1]) + 1;
15     atomId[i] = i;
     }
     nLayer = 0;     nn = nAtomRep;
     while (nn > 0) {
       nLayer = nLayer + 1;
20     if (nLayer > nLayerMax) ErrExit ("too many layers");
       for (n = 1; n <= nCells; n ++)
         layerAtom[nLayer][bdyMargin + n] = 0;
       for (i = 1; i <= nn; i ++)
         layerAtom[nLayer][bdyMargin + inCell[atomId[i]]] =
25           atomId[i];
       for (n = 1; n <= nCells; n ++) {
         i = layerAtom[nLayer][bdyMargin + n];
         if (i != 0) inCell[i] = 0;
       }
30     ii = 0;
       for (i = 1; i <= nn; i ++) {
         if (inCell[atomId[i]] != 0) {
           ii = ii + 1;     atomId[ii] = atomId[i];
       } }
35     nn = ii;
     }
     for (i = 1; i <= nAtomRep; i ++) {
       for (k = 1; k <= NDIM; k ++) ra[k][i] = 0.;
       u[i] = 0.;
40   }
     rrCut = Sqr (rCut);
     for (layer1 = 1; layer1 <= nLayer; layer1 ++) {
       for (layer2 = layer1; layer2 <= nLayer; layer2 ++) {
         if (layer2 == layer1) {
45         offsetLo = 2;     offsetHi = 14;
         } else {
           offsetLo = 1;     offsetHi = 27;
```

```
          }
          for (offset = offsetLo; offset <= offsetHi;
50           offset ++) {
            nPair = 0;    m2 = offsetVal[offset];
            for (m1 = 1; m1 <= nCells; m1 ++) {
              m2 = m2 + 1;
              if (inside[bdyMargin + m1] +
55               inside[bdyMargin + m2] != 0 &&
                 layerAtom[layer1][bdyMargin + m1] != 0 &&
                 layerAtom[layer2][bdyMargin + m2] != 0) {
                nPair = nPair + 1;
                atomPtr1[nPair] =
60                 layerAtom[layer1][bdyMargin + m1];
                atomPtr2[nPair] =
                   layerAtom[layer2][bdyMargin + m2];
            } }
            for (n = 1; n <= nPair; n ++) {
65            for (k = 1; k <= NDIM; k ++)
                 dr[k] = r[k][atomPtr1[n]] - r[k][atomPtr2[n]];
              rr = Sqr (dr[1]) + Sqr (dr[2]) + Sqr (dr[3]);
              if (rr < rrCut) {
                rri = 1. / rr;     rri3 = rri * rri * rri;
70              fcVal = 48. * rri3 * (rri3 - 0.5) * rri;
                for (k = 1; k <= NDIM; k ++)
                   fL[k][n] = fcVal * dr[k];
                uL[n] = 4. * rri3 * (rri3 - 1.) + 1.;
              } else {
75              for (k = 1; k <= NDIM; k ++) fL[k][n] = 0.;
                uL[n] = 0.;
            } }
            for (n = 1; n <= nPair; n ++) {
              for (k = 1; k <= NDIM; k ++)
80               ra[k][atomPtr1[n]] = ra[k][atomPtr1[n]] +
                     fL[k][n];
              u[atomPtr1[n]] = u[atomPtr1[n]] + uL[n];
            }
            for (n = 1; n <= nPair; n ++) {
85            for (k = 1; k <= NDIM; k ++)
                 ra[k][atomPtr2[n]] = ra[k][atomPtr2[n]] -
                     fL[k][n];
              u[atomPtr2[n]] = u[atomPtr2[n]] + uL[n];
      } } } }
90  uSum = 0.;
    for (i = 1; i <= nAtom; i ++) uSum = uSum + u[i];
    uSum = 0.5 * uSum;
    }
```

Each of the innermost loops (disregarding loops over the spatial dimensions that can be unrolled) can be shown to satisfy the basic requirement of vectorization, namely, that each item processed is independent of all others. Supplementary system-specific compiler directives will have to be used to inform the compiler of this fact since it is not at all apparent from the code itself. The other characteristic of the loops is that, at least for the earliest layers, the vectors processed are of length proportional to N_a.

The one slightly subtle detail in this computation is the assignment of atoms to layers. The way this is done is to make several passes over the set of atoms, one pass for each layer, each time assigning whichever atoms happen to be in each cell to the layer being filled. Thus several atoms may be assigned to the same position in a layer (in the array layerAtom), but all atoms except the last one found in each cell will be overwritten. A check is made after each pass to see which atoms were recorded in the layer; these are removed from the set before proceeding to the next layer by zeroing the values of inCell for these atoms and compressing the identities of the remaining atoms held in atomId. This process is repeated until no unassigned atoms remain. Clearly, such a scheme would be wasteful in terms of computation were it not for the fact that it can be vectorized.

Other comments about the above function are the following. The computation of the cell size ignores the outermost cells because these are outside the simulation region. The array inside is used to quickly distinguish cells that are within the simulation region from those adjoining the boundary; this is preferable to a test based on the three indices needed to specify cell position. When a layer is paired with itself, only positive cell offsets need be considered.

Changes to SetParams are

```
nAtomMax = 4 * (initUcell[1] + 2) * (initUcell[2] + 2) *
    (initUcell[3] + 2);
for (k = 1; k <= NDIM; k ++)
    cells[k] = (int) (region[k] / rCut) + 2;
5 nCells = cells[1] * cells[2] * cells[3];
bdyMargin = cells[1] * (cells[2] + 1) + 1;
```

and one additional item of input data is needed:

```
INAME (nLayerMax),
```

The final function listed here initializes the new arrays:

```
SetupLayers () {
  int n, nLayer, nX, nY, nZ;
  for (n = 1; n <= nCells + 2 * bdyMargin; n ++)
```

```
       inside[n] = 0;
5    for (nZ = 2; nZ <= cells[3] - 1; nZ ++) {
       for (nY = 2; nY <= cells[2] - 1; nY ++) {
         for (nX = 2; nX <= cells[1] - 1; nX ++) {
           inside[bdyMargin + ((nZ - 1) * cells[2] +
             nY - 1) * cells[1] + nX] = 1;
10   } } }
     for (nLayer = 1; nLayer <= nLayerMax; nLayer ++) {
       for (n = 1; n <= nCells + 2 * bdyMargin; n ++)
         layerAtom[nLayer][n] = 0;
     }
15   offsetVal[1] = 0;    offsetVal[2] = cells[1];
     offsetVal[3] = cells[1] * cells[2];
     offsetVal[4] = cells[1] * (cells[2] + 1);
     offsetVal[5] = cells[1] * (cells[2] - 1);
     for (n = 6; n <= 10; n ++)
20     offsetVal[n] = offsetVal[n - 5] + 1;
     for (n = 11; n <= 14; n ++)
       offsetVal[n] = - offsetVal[n - 9] + 1;
     for (n = 15; n <= 27; n ++)
       offsetVal[n] = - offsetVal[n - 13];
25 }
```

The rest of the program is unchanged.

14.6 Further work

14.1 Implement the distributed MD computation on your favorite multiprocessor computer and measure the communication overheads.

14.2 Generalize the distributed MD approach to accommodate some of the other kinds of system described in this book.

14.3 Extend the vectorized layer method to include neighbor lists.

14.7 Additional material

In order to complete the distributed MD demonstration we provide the extra information needed to produce a working program for use with PVM [gei94], a widely available software package that supports distributed applications in a message-passing environment. The two missing details are (a) how to actually initialize the computation and let each processor know its identity, and (b) translation of the generic communication functions used in the listing to actual PVM functions.

The following additions to main are needed for initialization; note that the test using procMe is replaced by one using taskIds[0] since procMe is not yet known. We assume that the reader is familiar with the PVM functions used here:

```
   int taskMe;
   taskMe = pvm_mytid ();
   taskIds[0] = pvm_parent ();
   if (taskIds[0] < 0) {
5    GetNameList (argc, argv);
     nProcs = ...
   }
   SetupParlProcs (taskMe);
   ...
10 pvm_exit ();
```

We also require:

```
   #define PROG_NAME "mdparl"
   #define MAX_PROCS 16
   int taskIds[MAX_PROCS];
   char procNames[MAX_PROCS][32];
```

The real work is done by the function SetupParlProcs – remember that each processor executes this function using its own private copy of the data. We assume that the user provides a list of machine 'names' so that PVM knows where the program is to be run. This is just one of a number of ways of handling the initialization task (error conditions are not examined here).

```
   SetupParlProcs (int taskMe) {
     int np, npr;
     if (taskIds[0] < 0) {
       for (np = 0; np < MAX_PROCS; np ++) taskIds[np] = 0;
5      for (np = 0; np < nProcs; np ++)
           scanf ("%s", procNames[np]);
       procMe = 0;    taskIds[0] = taskMe;
       for (np = 1; np < nProcs; np ++) {
         npr = pvm_spawn (PROG_NAME, NULL, PvmTaskHost,
10           procNames[np], 1, &taskIds[np]);
       }
       if (nProcs > 1) {
         MsgSendInit ();    MsgPackI (taskIds, MAX_PROCS);
         for (np = 1; np < nProcs; np ++) MsgSend (np, 101);
15     }
     } else {
       MsgRecv (0, 101);    MsgRecvInit ();
       MsgUnpackI (taskIds, MAX_PROCS);
       for (np = 1; np < MAX_PROCS; np ++) {
20       if (taskMe == taskIds[np]) {
```

```
      procMe = np;      break;
    }
  }
}
25 }
```

The replacement of the generic communication functions by calls to the PVM equivalents is done using the following macro definitions:

```
 #define MsgSendInit()                    \
         pvm_initsend (PvmDataDefault)
 #define MsgRecvInit()
 #define MsgSend(to, id)                  \
5        pvm_send (taskIds[to], id)
 #define MsgRecv(from, id)                \
         pvm_recv (taskIds[from], id)
 #define MsgSendRecv(from, to, id)        \
         pvm_send (taskIds[to], id),     \
10       pvm_recv (taskIds[from], id)
 #define MsgBcSend(id)                    \
         pvm_mcast (&taskIds[1], nProcs - 1, id)
 #define MsgBcRecv(id)                    \
         pvm_recv (taskIds[0], id)
15 #define MsgPackR(v, nv)                 \
         pvm_pkdouble (v, nv, 1)
 #define MsgPackI(v, nv)                  \
         pvm_pkint (v, nv, 1)
 #define MsgUnpackR(v, nv)                \
20       pvm_upkdouble (v, nv, 1)
 #define MsgUnpackI(v, nv)                \
         pvm_upkint (v, nv, 1)
```

This approach is preferable to using actual PVM calls in the body of the program from the point of view of portability.

15

The future

15.1 Role of simulation

Computer simulation in general, and molecular dynamics in particular, represent a new scientific methodology. Rather than adopt the traditional theoretical practice of constructing layer upon layer of assumption and approximation, this modern alternative attacks the original problem in all its detail. Unfortunately, phenomena that are primarily quantum mechanical in nature still present conceptual and technical obstacles, but, insofar as classical problems are concerned, the simulational approach is advancing as rapidly as computer technology permits. For this class of problem the limits of what can be achieved remain well beyond the horizon.

Theoretical breakthroughs involve both new concepts and the mathematical tools with which to develop them. Most of the major theoretical advances of the twentieth century rest upon mathematical foundations developed during the preceding century, if not earlier. Whether still undeveloped mathematical tools and new concepts will ever replace the information presently only obtainable by computer simulation (or whether the simulation is the solution) is something only the future can tell. Whether computer modeling will become an integral part of theoretical science or whether it will continue to exist independently is also a big unknown. After all, theory as we know it has not been around for very long.

To what extent can simulation replace experiment? In the more general sense this is already happening in engineering fields, where models are routinely constructed from well-established foundations. Most MD applications are still at the stage of attempting to solve the inverse problem, working backwards from experimental data to elucidate the invisible microscopic details; this is not done directly, however, but over the course of time the models gradually evolve until they are capable of yielding quantitatively correct behavior. To use MD as a predictive tool demands a high degree of confidence in the model and the methodology; this implies a thorough understanding of the effects of finite system size, limited timescales, numerical integration, and the consequences of using approximate (classical) potential functions.

The systems studied by MD are normally many orders of magnitude smaller than the corresponding systems in nature. With increasing computer power and memory this gap can be reduced, but never eliminated (the largest MD calculations in two dimensions are now approaching the size of mesoscopic systems studied experimentally). The same is true for timescales as well. The great success of the MD approach is due to the fact that these two limitations are in many instances irrelevant: the phenomena MD is used to study can often be exhaustively explored despite these limitations, because the lengths and times involved fall within the regime that MD can handle.

While one of the aims of MD is reproducing experiment, it also offers the opportunity for probing behavior at a much more detailed level than is obtainable by thermodynamic, mechanical, or spectroscopic techniques. While such 'numerical' measurements cannot be confirmed in the laboratory in a direct fashion, they sometimes have implications that can be tested experimentally; if these predictions are substantiated they can lead to improved understanding of the system under study. The wealth of detail potentially available is what makes MD such a useful and important method, the question only being what data to examine, and how to convert it into manageable form. Complete trajectories represent one extreme, thermodynamic averages the other.

15.2 Limits of growth

The MD field is still comparatively young; it has grown but remains tied to the advance of computer technology. Remarkable results have been obtained from what could be thought of as extremely small systems: that a few hundred to a few thousand model atoms not only permit studies of structure and dynamics, but also yield quantitative results in good agreement with experiment, is now a familiar fact of life. Of course life is not always so idyllic, and there are phenomena requiring length and time scales that exceed the capabilities of even the most powerful of computers. But if one is permitted to extrapolate from past rates of computer performance growth, the severity of these limitations should gradually diminish.

MD simulation has been a direct beneficiary of the rapid growth in computing power and the even greater improvement in the cost–performance ratio. Efficient compilers can help to a certain extent, but there is no substitute for a well-tuned algorithm, especially as the system size increases. Even when the performance of individual processors stops growing at the present rate and distributed computing becomes unavoidable, the nature of MD, with its computations often based on highly localized information, make the distributed environment ideal for large-scale problem solving.

It is important to be aware of the way MD computations scale in order to appreciate the kinds of problems that might be approachable in the foreseeable future. The amount of computation grows at least linearly with the number of particles, and so, too, will the processing time (for a fixed number of processing nodes in one's parallel computer). But this is not the whole story because the time over which a simulated system must be observed in order to examine a particular class of phenomenon can also increase with system size. Propagating disturbances such as sound waves cross a system in a time proportional to its linear size L, but any process governed by diffusion requires a time of order L^2. Processes involving, for example, large polymer molecules, occur on timescales that are truly macroscopic, representing an extreme situation beyond the capability of any conceivable computer. Thus the prognosis is mixed.

15.3 Interactive simulation

Computation is no longer merely number crunching. Along with ready access to high-performance computing comes the ability to actually observe the system being simulated. The MD practitioner need no longer be content with graphs of P plotted as a function of T, or some correlation as a function of time, but is now able to observe a system as it freezes, letting the eye capture some of the more subtle cooperative effects as the molecules reorganize. The systems that can be examined in this fashion are already of a size that only a few years ago were considered clients for the supercomputers of the day. Given that the human eye is without peer for many kinds of information reduction it is obvious that the visual approach is an important one – the computational analogy of the optical microscope.

Visualization itself takes many forms. Representing the results of discrete particle simulations offers a choice between direct observation of the particles themselves, or a display, typically involving scalar or vector fields, of suitably averaged quantities such as velocity, vorticity, temperature, concentration, and stress, to name but a few. Data can be represented by means of arrow, contour, and surface plots, as well as in less conventional, but visually rich forms, including extensive use of color and animation.

Once it has become possible to observe the system as it develops, the next step is to introduce a certain amount of interactivity into MD simulation by allowing the user to control both the parameters of the simulation and the way the results are displayed. Visualizing in real-time can prove invaluable while developing models and computational techniques, in debugging, demonstration, comparison of models, and choosing parameters. There are few general rules for adding interactivity to a calculation since the approach is at least partly determined by the software environment; the one critical requirement is responsiveness – if the user fails to receive rapid feedback, interactivity is of little value.

15.4 Coda

In retrospect one can do no better than to borrow from Anatole France: *'Si les plats que je vous offre sont mal préparés, c'est moins la faute de mon cuisinier que celle de la chimie, qui est encore dans l'enfance.'* Quoted apologetically in a text on quantum mechanics [mes64], it is no less appropriate for this volume of recipes despite the entirely classical foundations.

Appendices

A1 Allocating arrays

One particularly useful feature of the C language is the ability to allocate storage for arrays dynamically, with the amount of storage being determined at run time. All arrays used in the programs whose sizes depend on the input parameters are allocated dynamically (Fortran translations may be forced to take a less desirable route). As a consequence there are no preset size limits built into the C implementations of the programs.

The functions for real vectors and matrices are shown below; the corresponding integer and short-integer versions (`AllocVecI`, `AllocVecS`, etc.) are similar; these functions hide both the details of setting the correct offsets (in the MD programs almost all array indices begin at one, rather than at the customary zero used in C) and the way in which arrays are actually represented:

```
  real *AllocVecR (int size) {
    real *v;
    v = (real *) malloc (size * sizeof (real));
    return (v - 1);
5 }

  FreeVecR (real *v) {
    free (v + 1);
  }

  real **AllocMatR (int size1, int size2) {
    real **v;
    int k;
    v = (real **) malloc (size1 * sizeof (real *));
5   v[0] = (real *) malloc (size1 * size2 *
        sizeof (real)) - 1;
    for (k = 1; k < size1; k ++) v[k] = v[k - 1] + size2;
```

364

```
    return (v - 1);
}

FreeMatR (real **v) {
    free ((v + 1)[0] + 1);
    free (v + 1);
}
```

If reducing computation time is a serious issue, then there may be more to memory usage than is apparent here, and the concerned user will have to take into account various hardware architectural features when going about the task. Considerations such as localizing memory access, the actual number of array dimensions, minimizing address computation, and avoiding cache conflicts, are typical examples of the factors influencing efficiency. Disregarding such matters can sometimes lead to computation times several times (even an order of magnitude) larger than is really necessary. Determining whether near-optimal performance has been achieved is not always a simple process.

A2 Organizing input data

Very little attention has been paid to the subject of input data for the programs beyond simply listing the necessary items. The reason we were able to do this is because all a program requires is a file containing a list of variable names and their input values; the processing, completeness checks, and even an annotated printout of the values, are all handled transparently. While such a service is standard in Fortran – the 'namelist' feature – there is no similar facility in C. We have therefore had to roll our own; here it is for the curious, starting with the data structure and macro definitions that must be included in any program using the feature:

```
    typedef struct {
      char *vName;
      void *vPtr;
      enum {NI, NR} vType;
5     int vLen, vStatus;
    } NameList;
    #define INAME(x) {#x, &x, NI, sizeof (x) / sizeof (int)}
    #define RNAME(x) {#x, &x, NR, sizeof (x) / sizeof (real)}
```

The way this is used is typically

```
    NameList nameList[] = {
      INAME (intVariable),
```

```
     RNAME (realVector),
  };
```

and the data input function `GetNameList` called from main employs this information to process the input file (the return code can be used to check for errors). The name of the input file is taken from the variable `progId`, to which a two digit run identifier is appended if entered on the command line. Given some familiarity with standard file and string processing functions, the function listed below should be comprehensible. The restrictions on allowed C syntax that applied to the MD programs are now relaxed:

```
   int GetNameList (int argc, char **argv) {
     int id, j, jMin, k, match, ok;
     char buff[80], *token;
     strcpy (buff, progId);
5    if (argc == 2) {
       id = atoi (argv[1]);    buff[2] = id / 10 + '0';
       buff[3] = id % 10 + '0';    buff[4] = '\0';
     }
     strcat (buff, ".data");
10   if (! (fp = fopen (buff, "r"))) return (0);
     for (k = 0; k < sizeof (nameList) / sizeof (NameList);
        k ++) nameList[k].vStatus = 0;
     ok = 1;
     while (1) {
15     fgets (buff, 80, fp);
       if (feof (fp)) break;
       if (! (token = strtok (buff, " \t\n"))) break;
       match = 0;
       for (k = 0; k < sizeof (nameList) / sizeof (NameList);
20        k ++) {
         if (strcmp (token, nameList[k].vName) == 0) {
           match = 1;
           if (nameList[k].vStatus == 0) {
             nameList[k].vStatus = 1;
25           jMin = (nameList[k].vLen > 1) ? 1 : 0;
             for (j = jMin; j <= nameList[k].vLen - 1;
                j ++) {
               if (token = strtok (NULL, ", \t\n")) {
                 switch (nameList[k].vType) {
30                 case NI:
                     *((int *) (nameList[k].vPtr) + j) =
                         atol (token);
                     break;
                   case NR:
```

```
35              *((real *) (nameList[k].vPtr) + j) =
                    atof (token);
                break;
            }
        } else {
40          nameList[k].vStatus = 2;    ok = 0;
        } }
        if (token = strtok (NULL, ", \t\n")) {
            nameList[k].vStatus = 3;    ok = 0;
        }
45          break;
        } else {
            nameList[k].vStatus = 4;    ok = 0;
    } } }
    if (! match) ok = 0;
50  }
    fclose (fp);
    return (ok);
}
```

The function for printing an annotated record of the input data is

```
PrintNameList (FILE *fp) {
    int j, jMin, k;
    fprintf (fp, "NameList -- data\n");
    for (k = 0; k < sizeof (nameList) / sizeof (NameList);
5       k ++) {
        fprintf (fp, "%s\t", nameList[k].vName);
        if (strlen (nameList[k].vName) < 8) fprintf (fp, "\t");
        if (nameList[k].vStatus > 0) {
            jMin = (nameList[k].vLen > 1) ? 1 : 0;
10          for (j = jMin; j <= nameList[k].vLen - 1; j ++) {
                switch (nameList[k].vType) {
                case NI:
                    fprintf (fp, "%d ",
                        *((int *) (nameList[k].vPtr) + j));
15                  break;
                case NR:
                    fprintf (fp, "%#g ",
                        *((real *) (nameList[k].vPtr) + j));
                    break;
20      } } }
        switch (nameList[k].vStatus) {
        case 0: fprintf (fp, "** no data");  break;
        case 1: break;
        case 2: fprintf (fp, "** missing data");  break;
25      case 3: fprintf (fp, "** extra data");  break;
        case 4: fprintf (fp, "** multiply defined");  break;
```

```
      }
      fprintf (fp, "\n");
   }
30  fprintf (fp, "----\n");
   }
```

A3 Managing extensive computations

Computations short enough to run to completion without fear of interruption are generally inadequate for serious work. A production program must be equipped with the means to save its present state on disk and be able to resume computation from this saved state – the checkpoint/restart mechanism. Since there is no generally available procedure for this task, responsibility falls on the user.

Everything that the program is unable to reconstruct quickly from available information must be included in the checkpoint file. We will assume that the program has access to the original input data, so that this need not be included (a minor detail). For added security, two copies of the checkpoint file will be maintained, and they will be updated alternately; thus if the job aborts while writing the file (the usual reason being a lack of file space) the previous copy will still be available. The newer version can be identified in various ways, one of which is to have yet another file just for this purpose (if file modification times are accessible they could be used instead).

Various other files are also introduced at this point. One file is used to log all program output, others might be used to store various kinds of data – here, for example, grid averages (snapshots) and sets of atomic configurations – for later analysis or graphics work. Here is a sample set of files that might be used; the xxnn prefix in each file name is replaced by the two-letter program mnemonic progId and a two-digit serial number runId identifying the run:

```
   char *fileName[] = {
      "xxnnlog.data",
      "xxnnchecka.data",
      "xxnncheckb.data",
5     "xxnncklast.data",
      "xxnnsnap.data",
      "xxnnconfig.data",
   };
   #define FL_LOG      0
10 #define FL_CHECKA   1
   #define FL_CHECKB   2
```

```
#define FL_CKLAST  3
#define FL_SNAP    4
#define FL_CONFIG  5
```

All necessary files are created at the beginning of what is generally a series of continued runs by the following function. The variable doCheckpoint indicates whether checkpointing is activated, newRun is used to determine whether this is the first run of the series, and several other variables included with the input data – recordLog, recordSnap, recordConfig – specify which of these files are actually required:

```
   SetupFiles () {
     int k;
     for (k = 0; k < 6; k ++) {
       fileName[k][0] = progId[0];
5      fileName[k][1] = progId[1];
       fileName[k][2] = runId / 10 + '0';
       fileName[k][3] = runId % 10 + '0';
     }
     if (! doCheckpoint) {
10     newRun = 1;
     } else if ((fp = fopen (fileName[FL_CKLAST], "r"))) {
       newRun = 0;
       fclose (fp);
     } else {
15     newRun = 1;
       fp = fopen (fileName[FL_CHECKA], "w");    fclose (fp);
       fp = fopen (fileName[FL_CHECKB], "w");    fclose (fp);
       fp = fopen (fileName[FL_CKLAST], "w");
       fputc ('0' + FL_CHECKA, fp);     fclose (fp);
20   }
     if (newRun && recordLog) {
       fp = fopen (fileName[FL_LOG], "w");    fclose (fp);
     }
     if (newRun && recordSnap) {
25     fp = fopen (fileName[FL_SNAP], "w");    fclose (fp);
     }
     if (newRun && recordConfig) {
       fp = fopen (fileName[FL_CONFIG], "w");    fclose (fp);
     }
30 }
```

The functions used to access checkpoint files are as follows:

```
   PutCheckpoint () {
     int fOk, fVal;
     fOk = 0;
     if (fp = fopen (fileName[FL_CKLAST], "r+")) {
```

```
5        fVal = FL_CHECKA + FL_CHECKB - (fgetc (fp) - '0');
         rewind (fp);
         fputc ('0' + fVal, fp);
         fclose (fp);
         fOk = 1;
10   }
     if (fOk && (fp = fopen (fileName[fVal], "w"))) {
       WriteCheckpointData (fp);
       if (feof (fp) || ferror (fp)) fOk = 0;
       fclose (fp);
15   } else fOk = 0;
     if (! fOk) ErrExit ("write checkpoint data");
   }

   GetCheckpoint () {
     int fOk, fVal;
     fOk = 0;
     if (fp = fopen (fileName[FL_CKLAST], "r")) {
5      fVal = fgetc (fp) - '0';
       fclose (fp);
       fOk = 1;
     }
     if (fOk && (fp = fopen (fileName[fVal], "r"))) {
10     ReadCheckpointData (fp);
       if (ferror (fp)) fOk = 0;
       fclose (fp);
     } else fOk = 0;
     if (! fOk) ErrExit ("read checkpoint data");
15 }
```

The details of the actual data to be read/written are specific to the problem. Here, as an example, is WriteCheckpointData for a basic soft-sphere simulation using the leapfrog method. The data is output in binary form. Note that it is the address of each variable which must be passed as the argument (the 1-based indexing is also taken into account here):

```
WriteCheckpointData (FILE *fp) {
  fwrite (&timeNow, sizeof (real), 1, fp);
  fwrite (&stepCount, sizeof (int), 1, fp);
  fwrite (&r[1][1], sizeof (real), NDIM * nAtom, fp);
5  fwrite (&rv[1][1], sizeof (real), NDIM * nAtom, fp);
  fwrite (&sKinEnergy, sizeof (real), 1, fp);
  ...
}
```

The function ReadCheckpointData is similar; merely replace fwrite by fread.

To use these capabilities main requires the additions

```
SetupFiles ();
...
while (moreCycles) {
   ...
5  if (doCheckpoint && stepCount % stepCheckpoint == 0)
      PutCheckpoint ();
}
if (doCheckpoint) PutCheckpoint ();
```

and SetupJob must include

```
   if (newRun) {
     printf ("new run\n");
     InitCoords ();
     ...
5    stepCount = 0;
     AccumProps (0);
     if (doCheckpoint) {
       PutCheckpoint ();
       PutCheckpoint ();
10   }
   } else {
     printf ("continued run\n");
     GetCheckpoint ();
   }
```

The variable moreCycles determines when a run should terminate. Orderly termination can occur as a result of the number of timesteps reaching a limit, the execution time exceeding some preset value, or some other reason. It is possible (in some environments) to send a signal to the program manually telling it to stop (after writing its checkpoint file); in Unix making provision for such a signal requires inserting the call

```
SetupInterrupt ();
```

near the start of main and adding the functions

```
void ProcInterrupt () {
   moreCycles = 0;
}

SetupInterrupt () {
   signal (SIGUSR1, ProcInterrupt);
}
```

There is a user command for sending any desired signal (here the signal used is SIGUSR1) to an executing job.

Multiple runs with minor variations of the input data are most readily handled using shell scripts. For example, create a data file called `ss-org.data` containing all the input data except for the values that change between runs. Then if the aim is to cover a series of temperature values the following shell script will handle the job, leaving all the output in the file `ss-out`:

```
#!/bin/sh/
for temp in 0.4 0.6 0.8
do
cp ss-org.data ss01.data
cat >> ss01.data << EOD
temperature $temp
EOD
mdsoft 1 >> ss-out
done
```

This requires just a minimal acquaintance with Unix shell programming.

A4 Utility functions

A4.1 Random number generation

Random numbers generally play only a minor part in MD simulations and the need to ensure 'high quality' values is far less important than in Monte Carlo work. Molecular chaos will tend to eradicate all but the most egregious irregularities in whatever random numbers are generated. We have therefore (out of habit) used a simple (but adequate) method for producing uniformly distributed values in $(0, 1)$, the origins of which have long since been lost!

```
  #define IMUL    314159269
  #define IADD    453806245
  #define MASK    2147483647
  #define SCALE   0.4656612873e-9
5 real RandR (int *seed) {
    *seed = (*seed * IMUL + IADD) & MASK;
    return (*seed * SCALE);
  }
```

The variable seed is supplied by the user at each call and updated by the function. The reader who cares for a better documented procedure should turn to the literature [knu69, jam90, pre92], or check what is available on the local computer.

Random vectors are also required. The following function produces unit vectors in three dimensions with uniformly distributed random orientation by using a rejection method [mar72]:

```
   RandVec3 (real *p, int *pSeed) {
     real x, y, s;
     s = 2.;
     while (s > 1.) {
5      x = 2. * RandR (pSeed) - 1.;
       y = 2. * RandR (pSeed) - 1.;
       s = x * x + y * y;
     }
     p[3] = 1. - 2. * s;    s = 2. * sqrt (1. - s);
10   p[1] = s * x;    p[2] = s * y;
   }
```

The average success rate at each attempt is $\pi/4$ (76%).

A4.2 Mathematical and other functions

Several standard computational functions are required by a few of the case studies. They are all to be found in [pre92], as well as in various widely available mathematical libraries. In the interest of completeness, we include listings of the specially-written versions of these functions used in the case studies.

The function needed to solve a set of linear equations efficiently (Chapter 10) – based on the Crout version of the LU method – is as follows [ral78]. The matrix a is stored by columns as a one-dimensional array; the vector on the right-hand side of the equation is stored in x and is overwritten by the solution:

```
   #define N_MAX 100
   SolveLineq (real *a, real *x, int n) {
     real vMaxI[N_MAX + 1], v, vMax;
     int ptrMax[N_MAX + 1], i, ij, ik, ir, j, k, kj, kk, kr,
5      r, rj, rk;
     if (n > N_MAX) ErrExit ("matrix too large");
     for (i = 1; i <= n; i ++) {
       vMax = 0.;    ij = i;
       for (j = 1; j <= n; j ++) {
10       v = fabs (a[ij]);    if (v > vMax) vMax = v;
         ij = ij + n;
       }
       vMaxI[i] = 1. / vMax;
     }
15   for (r = 1; r <= n; r ++) {
```

```
       vMax = 0.;    ir = r * (n + 1) - n;
       for (i = r; i <= n; i ++) {
         v = a[ir];    ik = i;    kr = (r - 1) * n + 1;
         for (k = 1; k <= r - 1; k ++) {
20         v = v - a[ik] * a[kr];
           ik = ik + n;    kr = kr + 1;
         }
         a[ir] = v;    v = fabs (a[ir]) * vMaxI[i];
         if (v > vMax) {
25         vMax = v;    ptrMax[r] = i;
         }
         ir = ir + 1;
       }
       if (r != ptrMax[r]) {
30       kk = 0;
         for (k = 1; k <= n ; k ++) {
           v = a[kk + r];    a[kk + r] = a[kk + ptrMax[r]];
           a[kk + ptrMax[r]] = v;    kk = kk + n;
         }
35       vMaxI[ptrMax[r]] = vMaxI[r];
       }
       rj = r * (n + 1);
       for (j = r + 1; j <= n; j ++) {
         v = a[rj];
40       rk = r;    kj = (j - 1) * n + 1;
         for (k = 1; k <= r - 1; k ++) {
           v = v - a[rk] * a[kj];
           rk = rk + n;    kj = kj + 1;
         }
45       a[rj] = v / a[rk];    rj = rj + n;
       }
     }
     for (i = 1; i <= n; i ++) {
       v = x[ptrMax[i]];    x[ptrMax[i]] = x[i];    ij = i;
50     for (j = 1; j <= i - 1; j ++) {
         v = v - a[ij] * x[j];    ij = ij + n;
       }
       x[i] = v / a[ij];
     }
55   for (i = n - 1; i >= 1; i --) {
       v = x[i];    ij = i * (n + 1);
       for (j = i + 1; j <= n; j ++) {
         v = v - a[ij] * x[j];    ij = ij + n;
       }
60     x[i] = v;
     }
   }
```

The function for fast (in-place) Fourier transformation of complex data arrays of size 2^n (Chapter 5) is [hig76]:

```
    FtFastComplex (real *a, int size, int iSign) {
      real pi, theta, tI, tR, w, wI, wIo, wR, wRo;
      int i, j, m, mMax;
      pi = 4. * atan (1.);
 5    j = 1;
      for (i = 1; i <= 2 * size; i += 2) {
        if (i < j) {
          tR = a[j];      tI = a[j + 1];
          a[j] = a[i];    a[j + 1] = a[i + 1];
10        a[i] = tR;      a[i + 1] = tI;
        }
        m = size;
        while (m >= 2 && j > m) {
          j = j - m;      m = m / 2;
15      }
        j = j + m;
      }
      mMax = 2;
      while (mMax < size * 2) {
20      theta = iSign * pi * 2. / mMax;
        wRo = cos (theta) - 1.;     wIo = sin (theta);
        wR = 1.;    wI = 0.;
        for (m = 1; m <= mMax; m += 2) {
          for (i = m; i <= 2 * size; i += 2 * mMax) {
25          j = i + mMax;
            tR = wR * a[j] - wI * a[j + 1];
            tI = wR * a[j + 1] + wI * a[j];
            a[j] = a[i] - tR;       a[j + 1] = a[i + 1] - tI;
            a[i] = a[i] + tR;       a[i + 1] = a[i + 1] + tI;
30        }
          tR = wR;
          wR = wR * wRo - wI * wIo + wR;
          wI = wI * wRo + tR * wIo + wI;
        }
35      mMax = mMax * 2;
      }
      if (iSign < 0) {
        w = 1. / size;
        for (m = 1; m <= size; m ++) a[m] = a[m] * w;
40    }
    }
```

A sorting function is required (Chapter 4). This version shows the Heapsort method [knu73]; the data are not rearranged but the correctly (ascending) ordered indices are returned:

```
Sort (real *a, int *seq, int n) {
  real aa;
  int i, j, k, m, s;
  for (j = 1; j <= n; j ++) seq[j] = j;
5 k = n / 2 + 1;    m = n;
  while (1) {
    if (k > 1) {
      k = k - 1;    s = seq[k];    aa = a[s];
    } else {
10     s = seq[m];    aa = a[s];
      seq[m] = seq[1];    m = m - 1;
      if (m == 1) {
        seq[1] = s;
        return;
15     }
    }
    i = k;    j = 2 * k;
    while (j <= m) {
      if (j < m && a[seq[j]] < a[seq[j + 1]]) j = j + 1;
20     if (aa < a[seq[j]]) {
        seq[i] = seq[j];    i = j;    j = 2 * j;
      } else break;
    }
    seq[i] = s;
25 }
}
```

A function for integration (Chapter 5), using the very simple trape-zoidal rule (assuming unit spacing between the points), is

```
real Integrate (real *f, int nf) {
  real s;
  int i;
  s = 0.5 * (f[1] + f[nf]);
5 for (i = 2; i <= nf - 1; i ++) s = s + f[i];
  return (s);
}
```

The solution of a cubic equation (Chapter 9) – assuming that all roots are real and that the cubic term has a coefficient of unity – is given by

```
SolveCubic (real a1, real a2, real a3, real *g) {
  real q1, q2, t;
  q1 = sqrt (a1 * a1 - 3. * a2) / 3.;
  q2 = (2. * a1 * a1 * a1 - 9. * a1 * a2 + 27. * a3) / 54.;
5 t = acos (q2 / (q1 * q1 * q1));
  g[1] = - 2. * q1 * cos (t / 3.) - a1 / 3.;
```

```
  g[2] = - 2. * q1 * cos ((t + 2. * pi) / 3.) - a1 / 3.;
  g[3] = - 2. * q1 * cos ((t + 4. * pi) / 3.) - a1 / 3.;
}
```

with the roots being returned in the array g.

A few other very simple functions are used by the programs, some of which may appear in various libraries under different names:

```
real DotProd3 (real *x, real *y) {
  return (x[1] * y[1] + x[2] * y[2] + x[3] * y[3]);
}

real Max3R (real x1, real x2, real x3) {
  if (x1 > x2 && x1 > x3) return (x1);
  else if (x2 > x1 && x2 > x3) return (x2);
  else return (x3);
}

real Min3R (real x1, real x2, real x3) {
  ... (opposite of Max3R) ...
}

int Nint (real x) {
  if (x < 0.) return (- (int) (0.5 - x));
  return ((int) (0.5 + x));
}

real SignR (real x, real y) {
  if (y >= 0.) return (x);
  else return (- x);
}

real Sqr (real x) {
  return (x * x);
}
```

Implementation of these functions as macros is also possible, and sometimes advisable for performance reasons; for example,

```
#define SignR(x,y) (((y) >= 0) ? (x) : (- (x)))
#define Sqr(x) ((x) * (x))
```

with enough enveloping parentheses to ensure the desired effect.

Error reporting is done with a minimum of fuss; all errors terminate the job:

```
ErrExit (char *s) {
  printf ("Error: %s\n", s);
  exit (0);
}
```

A5 Header files

All the MD programs share a common header file that includes various definitions, all the details needed for the file operations mentioned previously, function prototypes (these are essential for functions which return non-integer results and optional – though recommended – in other cases), and statements to include several standard Unix/C header files (the actual list may vary slightly). The header file begins with

```
  #include <math.h>
  #include <stdio.h>
  #include <stdlib.h>
  #include <string.h>
5 #include <signal.h>
  typedef double real;
  FILE *fp;
```

followed by the definitions – listed earlier – required to support the namelist feature. The header file also includes function prototypes and any other necessary definitions; a minimal list would typically be

```
  int *AllocVecI (int);
  int **AllocMatI (int, int);
  real *AllocVecR (int);
  real **AllocMatR (int, int);
5 real RandR (int *);
  #define NDIM 3
```

For added flexibility conditional compilation can be used; this will allow several different forms of a program to share the same source code (typically several related case studies, or the two- and three-dimensional versions of a program). Since the program may be divided among several source files the definitions used for this purpose should also appear in the header file.

When compiling the programs it is usually necessary to specify that the C mathematical function library be accessed at link/load time. It is also advisable to request at least a moderate level of compiler optimization.

A6 Variables

An alphabetical listing of the global variables used in the MD programs follows; we omit variables connected with file handling or used in the separate analysis programs (as well as local variables):

> aaDistSq – average bond length squared
> aCon... – numerical constants

aDiffuse – angular diffusion measurement
aDiffuseAv/Org – angular diffusion mean and origin
alpha – α
atomId – atom labels
atomPtr1/2 – atom pointers
atomTime – when atom coordinates last updated
atomType – used to identify fixed and mobile atoms
bdyMargin – array offset (layers)
bdyStripWidth – size of region where flow speed adjusted
bondAngle/Len – polymer chain bond angle and length
bondLim – limit of bond stretching
boundarySlide – offset across sliding boundary
boundPairEng – energy threshold for bound molecules
cAtom1/2 – atoms involved in constraint
cDistSq – constrained distance squared
cellList – pointers used by cell method
cellRange – range of cells to be examined after collision
cells – size of cell array
cfOrg/Val – used for space–time correlations
chainLen – number of monomers in chain
cMat – constraint matrix **M**
coll/crossCount – collision and cell event counts
constraintDevA/L – measured constraint deviations
constraintPrec – desired constraint accuracy
cosV/sinV – tabulated trigonometric functions
count... – counters used in measurements
curBondLenSq – current bond length squared
cVec – constraint vectors
cvMat – **L**
cycleR/V – iterations to restore constraints
deltaT – Δt
density – ρ
diffuseAcfInt – integrated autocorrelation function
dihedAngCorr/Org – used in measuring dihedral angle correlations
dilateRate/1/2 – current and previous γ-values
dipoleInt – μ^2
dispHi – accumulated maximum displacement (for neighbor list)
dvirSum1/2 – used in computing dilateRate
eeDistSq – $\langle R^2 \rangle$
enAtom – e_j
eventCount – current event number
eventCountLimit – run length
eventMult – used to determine poolSize
evIdA/B – event identifiers
extPressure – applied P
f/uL – force and energy for each atom (layers)
flowSpeed – nominal flow speed
forceField – external field
fSite – force acting on interaction site
fSpaceLimit – n_c
gMomRatio1/2 – $\langle g_2/g_1 \rangle$, $\langle g_3/g_1 \rangle$

gravField – force driving flow
g1Sum/g2Sum – G_1, G_2
heatForce – fictitious force for thermal conductivity measurement
hFunction – $h(t)$ for H-function study
histGrid – accumulated coarse-grained measurements
hist... – histograms used for measuring distributions
inCell – cell occupied by atom
index... – number of time-dependent measurements in set
initUcell/chain – size of unit-cell array for initial state
inPoly – chain membership
inside – distinguishes between boundary and interior cells (layers)
interval... – time between measurements
kinEnergy – E_k
latticeCorr – $|\rho(\boldsymbol{k})|$
layerAtom – storage for layer contents
limit... – number of measurements needed for result
massS/V – M_s, M_v
maxEdgeCells – maximum size of cell array
max/minPairEng – range of pair energies to be studied
molSite – interaction site coordinates
momInertia – I, or I_x, etc.
moreCycles – flag controlling program execution
nAtom – N_a
nAtomCopy – number of copied atoms
nAtomMax – maximum number of atoms (layers)
nAtomMe – actual number of atoms in processor
nAtomMeMax – maximum number of atoms per processor
nAtomRep – number of replica atoms (layers)
nBuff... – number of buffers used for data collection
nCellEdge – number of cells in each direction
nCells – number of cells
nChain – number of chains
nConstraint – number of constraints
nDihedAng – number of dihedral angles in polymer chain
NDIM – dimensionality of simulation
nebrNow – neighbor-list refresh due
nebrTab – storage for neighbor list
nebrTabFac – used to determine nebrTabMax
nebrTabLen – neighbor-list length
nebrTabMax – maximum neighbor-list length
nebrTabPtr – pointers used with neighbor list
next...Time – next measurement event time
nFixed/FreeAtom – numbers of fixed and mobile atoms
nFunCorr – number of k-values used for correlation functions
nLayerMax – maximum number of layers
nMol – N_m
nMolBonds – number of bonds formed by molecule
nPressCycle – pressure correction cycle count
nProcs – number of processors
nTrHi/Lo – number of atoms requiring transfer
nVal... – number of values in time-dependent measurement

`obsPos/Size` – position and size of flow obstacle
`offsetVal` – cell offsets (layers)
`pertTrajDev` – velocity perturbation
`pi` – π
`poolSize` – size of event pool
`potEnergy` – E_p
`pressure` – P
`pressureTensorXZ` – P_{xz}
`procArrayMe` – location in distributed-processor array
`procArraySize` – dimensions of distributed-processor array
`procMe` – identifies processor
`procNebrHi/Lo` – neighboring processors
`profileT/V` – temperature and velocity profiles
`progId` – program identification label
`q,qv,qa,qa1/2,qo,qvo` – quaternion components, derivatives, etc.
`r,rv,ra,ra1/2,ro,rvo` – positions, velocities, accelerations, etc.
`radGyrSq` – $\langle S^2 \rangle$
`randSeed` – seed for random number generator
`rangeRdf` – RDF limit
`range...` – upper limit of distribution
`rCol` – coordinates of latest collision
`rCut` – r_c
`rDiffuse` – atom displacements
`rDiffuseAv` – averages of `rDiffuse`
`rDiffuseOrg` – origins for diffusion measurement
`rDiffuseTrue` – 'true' coordinates
`region` – size of simulation region edges
`regionH` – half of `region`
`regionVol` – volume of simulation region
`rfAtom` – $\sum_i r_{xij} f_{yij}$
`rMat` – rotation matrix **A**
`rNebrShell` – r_n
`roughWid` – wall corrugation width
`rSite` – interaction-site coordinates
`runId` – job identification number
`rvAcf` – velocity autocorrelation values
`rvAcfAv` – averages of `rvAcf`
`rvAcfOrg` – origins for velocity autocorrelation measurements
`s,sv,sa,sa1/2,so,svo` – linear molecule s, derivatives, etc.
`sdSum` – value of order parameter M
`sEnergyTrans` – kinetic energy transferred across system
`shearRate` – γ
`shearVisc` – η
`sInitKinEnergy` – accumulated E_k-sum for setting initial T
`sitesMol` – number of interaction sites in molecule
`sizeHistGrid` – size of `histGrid`
`sizeHist...` – size of histograms used for measuring distributions
`sKinEnergy/ssKinEnergy` – accumulated E_k- and E_k^2-sums
`snapNumber` – serial number of snapshot data
`sOrder/ssOrder` – accumulated M- and M^2-sums
`spacetimeCorr` – space–time correlation values

spacetimeCorrAv – averages of spacetimeCorr
sPressure/ssPressure – accumulated P- and P^2-sums
sShearVisc/ssShearVisc – accumulated η- and η^2-sums
stepCount – current timestep
stepLimit – total run length
step... – timesteps between measurements or other activities
sThermalCond/ssThermalCond – accumulated λ- and λ^2-sums
sTotEnergy/ssTotEnergy – accumulated E- and E^2-sums
subRegionHi/Lo – subregion limits (distributed processing)
temperature – T
thermAcf – heat current autocorrelation values
thermAcfAv – averages of thermAcf
thermAcfInt – see diffuseAcfInt
thermAcfOrg – origins for thermAcf measurements
thermalCond – λ
thermalWall – specifies wall attached to heat bath
timeNow – current time
tolPressure – tolerance for pressure correction
torq – N
totEnergy – E
trBuff – buffer for interprocessor transfers
trBuffMax – size of trBuff
tree – pointers and data for event simulation
treeTime – scheduled event time
trPtrHi/Lo – pointers to atoms requiring transfer
typeSite – molecular interaction-site type
typeSiteRdf – molecular RDF site type
u – interaction energy per atom
uSum – total interaction energy
valTrajDev – trajectory deviation
varL,varLv,varLv1/2,varLo – region size variables
varS,varSv/a,varSa1/2,varS/Svo – temperature feedback variables
varV,varVv/a,varVa1/2,varV/Vvo – volume feedback variables
vir – virial contribution per atom
virSum – virial sum
visc... – see therm...
vMag – initial velocity value
vSum – $\sum v_i$
vv...Sum... – $\sum v_i^2$, etc.
vVec – used in solving constraint equations
vvMax – $\max_i v_i^2$
wallTemp/Hi/Lo – thermal wall temperatures

Bibliography

[abr86] Abraham, F. F., *Adv. Phys.* **35** (1986) 1.

[abr89] Abraham, F. F., Rudge, W. E., and Plischke, M., *Phys. Rev. Lett.* **62** (1989) 1757.

[ada76] Adams, D. J. and McDonald, I., *Mol. Phys.* **32** (1976) 931.

[ada80] Adams, D. J., in Ceperley, D., ed., *The Problem of Long-Range Forces in the Computer Simulation of Condensed Media*, Lawrence Berkeley Lab. Rept. LBL-10634, 1980, p. 13.

[ald57] Alder, B. J. and Wainwright, T. E., *J. Chem. Phys.* **27** (1957) 1208.

[ald58] Alder, B. J. and Wainwright, T. E., in Prigogine, I., ed., *Transport Processes in Statistical Mechanics*, Interscience Publishers, New York, 1958, p. 97.

[ald59] Alder, B. J. and Wainwright, T. E., *J. Chem. Phys.* **31** (1959) 459.

[ald62] Alder, B. J. and Wainwright, T. E., *Phys. Rev.* **127** (1962) 359.

[ald67] Alder, B. J. and Wainwright, T. E., *Phys. Rev. Lett.* **18** (1967) 988.

[ald70a] Alder, B. J., Gass, D. M., and Wainwright, T. E., *J. Chem. Phys.* **53** (1970) 3813.

[ald70b] Alder, B. J. and Wainwright, T. E., *Phys. Rev. A* **1** (1970) 18.

[all83] Alley, W. E., Alder, B. J., and Yip, S., *Phys. Rev. A* **27** (1983) 3174.

[all87] Allen, M. P. and Tildesley, D. J., *Computer Simulation of Liquids*, Oxford University Press, Oxford, 1987.

[all89] Allen, M. P., Frenkel, D., and Talbot, J., *Comp. Phys. Repts.* **9** (1989) 301.

[all93a] Allen, M. P. and Masters, A. J., *Mol. Phys.* **79** (1993) 435.

[all93b] Allen, M. P. and Tildesley, D. J., eds., *Computer Simulation in Chemical Physics*, Kluwer Academic Publishers, Dordrecht, 1993.

[and80] Anderson, H. C., *J. Chem. Phys.* **72** (1980) 2384.

[and83] Anderson, H. C., *J. Comp. Phys.* **52** (1983) 24.

[ash75] Ashurst, W. T. and Hoover, W. G., *Phys. Rev. A* **11** (1975) 658.

[bar71] Barker, J. A., Fisher, R. A., and Watts, R. O., *Mol. Phys.* **21** (1971) 657.

[bar88] Barrat, J.-L., Hansen, J. P., and Pastore, G., *Mol. Phys.* **63** (1988) 747.

[bee66] Beeler Jr, J. R., in Meeron, E., ed., *Physics of Many-Particle Systems: Methods and Problems*, Gordon and Breach, New York, 1966, p. 1.

[bee76] Beeman, D., *J. Comp. Phys.* **20** (1976) 130.

[ber77] Berne, B. J., *J. Chem. Phys.* **66** (1977) 2821.

[ber86a] Berendsen, H. J. C., in [cic86a], p. 496.

[ber86b] Berendsen, H. J. C. and van Gunsteren, W. F., in [cic86a], p. 43.

[ber86c] Berne, B. J. and Thirumalai, D., *Ann. Rev. Phys. Chem.* **37** (1986) 401.

[bil94] Billeter, S. R., King, P. M., and van Gunsteren, W. F., *J. Chem. Phys.* **100** (1994) 6692.

[bin92] Binder, K., ed., *Monte Carlo Methods in Condensed Matter Physics*, Springer, Berlin, 1992.

[bin95] Binder, K., ed., *Monte Carlo and Molecular Dynamics Simulations in Polymer Science*, Oxford University Press, Oxford, 1995.

[bir94] Bird, G. A., *Molecular Gas Dynamics and the Direct Simulation of Gas Flows*, Oxford University Press, Oxford, 1994.

[boo91] Boon, J.-P. and Yip, S., *Molecular Hydrodynamics*, Dover, New York, 1991.

[bro78] Brostow, W., Dussault, J.-P., and Fox, B. L., *J. Comp. Phys.* **29** (1978) 81.

[bro84] Brown, D. and Clark, J. H. R., *Mol. Phys.* **51** (1984) 1243.

[bro86] Brode, S. and Ahlrichs, R., *Comp. Phys. Comm.* **42** (1986) 51.

[bro88] Brooks, III, C. L., Karplus, M., and Pettit, B. M., *Proteins: A Theoretical Perspective of Dynamics, Structure, and Thermodynamics*, Wiley, New York, 1988.

[bro90a] Brooks, III, C. L., in [cat90], p. 289.

[bro90b] Brown, D. and Clark, J. H. R., *J. Chem. Phys.* **92** (1990) 3062.

[cap81] Cape, J. N., Finney, J. L., and Woodcock, L. V., *J. Chem. Phys.* **75** (1981) 2366.

[cat90] Catlow, C. R. A., Parker, S. C., and Allen, M. P., eds., *Computer Modelling of Fluids Polymers and Solids*, Kluwer Academic Publishers, Dordrecht, 1990.

[cic82] Ciccotti, G., Ferrario, M., and Ryckaert, J.-P., *Mol. Phys.* **47** (1982) 1253.

[cic86a] Ciccotti, G. and Hoover, W. G., eds., *Molecular Dynamics Simulation of Statistical Mechanical Systems*, North-Holland, Amsterdam, 1986.

[cic86b] Ciccotti, G. and Ryckaert, J.-P., *Comp. Phys. Repts.* **4** (1986) 345.

[cic87] Ciccotti, G., Frenkel, D., and McDonald, I. R., eds., *Simulation of Liquids and Solids. Molecular Dynamics and Monte Carlo Methods in Statistical Mechanics*, North-Holland, Amsterdam, 1987.

[cla90] Clark, J. H. R., in [cat90], p. 203.

[cor60] Corben, H. C. and Stehle, P., *Classical Mechanics*, Wiley, NY, 2nd edition, 1960.

[del51] de Laplace P. S., *A Philosophical Essay on Probabilities* (transl.), Dover, New York, 1951.

[del80] de Leeuw, S. W., Perram, J. W., and Smith, E. R., *Proc. R. Soc. Lond. A* **373** (1980) 27.

[del86] de Leeuw, S. W., Perram, J. W., and Smith, E. R., *Ann. Rev. Phys. Chem.* **37** (1986) 245.

[des88] de Schepper, I. M., Cohen, E. G. D., Bruin, C., van Rijs, J. C., Montfrooij, W., and de Graaf, L. A., *Phys. Rev. A* **38** (1988) 271.

[doo91] Doolen, G. D., ed., *Lattice Gas Methods for PDEs*, North-Holland, Amsterdam, 1991.

[dun92] Dunn, J. H., Lambrakos, S. G., Moore, P. G., and Nagumo, M., *J. Comp. Phys.* **100** (1992) 17.

[dun93] Dünweg, B. and Kremer, K., *J. Chem. Phys.* **99** (1993) 6983.

[edb86] Edberg, R., Evans, D. J., and Morriss, G. P., *J. Chem. Phys.* **84** (1986) 6933.

[edb87] Edberg, R., Morriss, G. P., and Evans, D. J., *J. Chem. Phys.* **86** (1987) 4555.

[ein68] Einwohner, T. and Alder, B. J., *J. Chem. Phys.* **49** (1968) 1458.

[erm80] Ermak, D. L. and Buckholz, H., *J. Comp. Phys.* **35** (1980) 169.

[erp77] Erpenbeck, J. J. and Wood, W. W., in Berne, B. J., ed., *Modern Theoretical Chemistry*, Plenum, New York, 1977, vol. 6B, p. 1.

[erp84] Erpenbeck, J. J. and Wood, W. W., *J. Stat. Phys.* **35** (1984) 321.

[erp85] Erpenbeck, J. J. and Wood, W. W., *Phys. Rev. A* **32** (1985) 412.

[erp88] Erpenbeck, J. J., *Phys. Rev. A* **38** (1988) 6255.

[eva77a] Evans, D. J., *Mol. Phys.* **34** (1977) 317.

[eva77b] Evans, D. J. and Murad, S., *Mol. Phys.* **34** (1977) 327.

[eva82] Evans, D. J., *Phys. Lett.* **91A** (1982) 45.

[eva83a] Evans, D. J., *J. Chem. Phys.* **78** (1983) 3297.

[eva83b] Evans, D. J., Hoover, W. G., Failor, B. H., Moran, B., and Ladd, A. J. C., *Phys. Rev. A* **28** (1983) 1016.

[eva84] Evans, D. J. and Morriss, G. P., *Comp. Phys. Repts.* **1** (1984) 297.

[eva86] Evans, D. J. and Morriss, G. P., *Phys. Rev. Lett.* **56** (1986) 2172.

[eva90] Evans, D. J. and Morriss, G. P., *Statistical Mechanics of Nonequilibrium Liquids*, Academic Press, London, 1990.

[fer91] Ferrario, M., Ciccotti, G., Holian, B. L., and Ryckaert, J.-P., *Phys. Rev. A* **44** (1991) 6936.

[fey63] Feynman, R. P., Leighton, R. B., and Sands, M., *The Feynman Lectures on Physics*, vol. 1, Addison-Wesley, Reading, 1963.

[fin79] Finney, J. L., *J. Comp. Phys.* **32** (1979) 137.

[fin93] Fincham, D., *Mol. Simulation* **11** (1993) 79.

[fly89] Flyvberg, H. and Petersen, H. G., *J. Chem. Phys.* **91** (1989) 461.

[fri75] Friedman, H. L., *Mol. Phys.* **29** (1975) 1533.

[gay81] Gay, J. G. and Berne, B. J., *J. Chem. Phys.* **74** (1981) 3316.

[gea71] Gear, C. W., *Numerical Initial Value Problems in Ordinary Differential Equations*, Prentice-Hall, Englewood Cliffs, NJ, 1971.

[gei79] Geiger, A., Stillinger, F. H., and Rahman, A., *J. Chem. Phys.* **70** (1979) 4185.

[gei94] Geist, A., Beguelin, A., Dongarra, J., Jiang, W., Manchek, R., and Sunderam, V., *PVM3 User's Guide and Reference Manual*, Oak Ridge National Laboratory Technical Report ORNL/TM-12187, 1994.

[gib60] Gibson, J. B., Goland, A. N., Milgram, M., and Vineyard, G. H., *Phys. Rev.* **120** (1960) 1229.

[gil83] Gillan, M. J. and Dixon, M., *J. Phys. C* **16** (1983) 869.

[gil90] Gillan, M. J., in [cat90], p. 155.

[gol80] Goldstein, H., *Classical Mechanics*, Addison-Wesley, Reading, MA, 2nd edition, 1980.

[gra84] Gray, C. G. and Gubbins, K. E., *Theory of Molecular Fluids*, vol. 1, Clarendon Press, Oxford, 1984.

[gre89a] Greengard, L. and Rokhlin, V., *Chem. Scripta* **29A** (1989) 139.

[gre89b] Grest, G. S., Dünweg, B., and Kremer, K., *Comp. Phys. Comm.* **55** (1989) 269.

[gre94] Grest, G. S., *Macromolecules* **27** (1994) 3493.

[han86a] Hansen, J.-P., in [cic86a], p. 89.

[han86b] Hansen, J.-P. and McDonald, I. R., *Theory of Simple Liquids*, Academic Press, London, 2nd edition, 1986.

[han94] Hansen, D. P. and Evans, D. J., *Mol. Phys.* **81** (1994) 767.

[hel60] Helfand, E., *Phys. Rev.* **119** (1960) 1.

[hel79] Helfand, E., *J. Chem. Phys.* **71** (1979) 5000.

[hey89] Heyes, D. M. and Melrose, J. R., *Mol. Phys.* **68** (1989) 359.

[hig76] Higgins, R. J., *Am. J. Phys.* **44** (1976) 772.

[hir54] Hirschfelder, J. O., Curtis, C. F., and Bird, R. B., *Molecular Theory of Gases and Liquids*, Wiley, New York, 1954.

[hoc74] Hockney, R. W., Goel, S. P., and Eastwood, J. W., *J. Comp. Phys.* **14** (1974) 148.

[hon92] Hong, D. C. and McLennan, J. A., *Physica* **187** (1992) 159.

[hoo82] Hoover, W. G., Ladd, A. J. C., and Moran, B., *Phys. Rev. Lett.* **48** (1982) 1818.

[hoo85] Hoover, W. G., *Phys. Rev. A* **31** (1985) 1695.

[hoo91] Hoover, W. G., *Computational Statistical Mechanics*, Elsevier, Amsterdam, 1991.

[hsu79] Hsu, C. S. and Rahman, A., *J. Chem. Phys.* **71** (1979) 4974.

[jae92] Jaeger, H. M. and Nagel, S. R., *Science* **255** (1992) 1523.

[jam90] James, F., *Comp. Phys. Comm.* **60** (1990) 329.

[jor83] Jorgensen, W. L., Chandrasekhar, J., Madura, J. D., Impey, R. W., and Klein, M. L., *J. Chem. Phys.* **79** (1983) 926.

[kle86] Klein, M. L., in [cic86a], p. 424.

[knu68] Knuth, D. E., *Fundamental Algorithms*, vol. 1 of *The Art of Computer Programming*, Addison-Wesley, Reading, MA, 1968.

[knu69] Knuth, D. E., *Seminumerical Algorithms*, vol. 2 of *The Art of Computer Programming*, Addison-Wesley, Reading, MA, 1969.

[knu73] Knuth, D. E., *Sorting and Searching*, vol. 3 of *The Art of Computer Programming*, Addison-Wesley, Reading, MA, 1973.

[kol92] Kolafa, J. and Perram, J. W., *Mol. Simulation* **9** (1992) 351.

[kop89] Koplik, J., Banavar, J. R., and Willemsen, J. F., *Phys. Fluids A* **1** (1989) 781.

[kre88] Kremer, K. and Binder, K., *Comp. Phys. Repts.* **7** (1988) 259.

[kre92] Kremer, K. and Grest, G. S., *J. Comp. Soc. Faraday Trans.* **88** (1992) 1707.

[kus76] Kushick, J. and Berne, B. J., *J. Chem. Phys.* **64** (1976) 1362.

[kus90] Kusalik, P. G., *J. Chem. Phys.* **93** (1990) 3520.

[lan59] Landau, L. D. and Lifshitz, E. M., *Fluid Mechanics*, Pergamon Press, Oxford, 1959.

[leb67] Lebowitz, J. L., Percus, J. K., and Verlet, L., *Phys. Rev.* **153** (1967) 250.

[lee72] Lees, A. W. and Edwards, S. F., *J. Phys. C* **5** (1972) 1921.

[lev73] Levesque, D., Verlet, L., and Kürkijarvi, J., *Phys. Rev. A* **7** (1973) 1690.

[lev84] Levesque, D., Weis, J. J., and Hansen, J.-P., in Binder, K., ed., *Applications of the Monte Carlo Method in Statistical Physics*, Springer, Berlin, 1984, p. 37.

[lev87] Levesque, D. and Verlet, L., *Mol. Phys.* **61** (1987) 143.

[lev92] Levesque, D. and Weis, J. J., in [bin92], p. 121.

[lev93] Levesque, D. and Verlet, L., *J. Stat. Phys.* **72** (1993) 519.

[lie92] Liem, S. Y., Brown, D., and Clarke, J. H. R., *Phys. Rev. A* **45** (1992) 3706.

[loo92] Loose, W. and Ciccotti, G., *Phys. Rev. A* **45** (1992) 3859.

[mai81] Maitland, G. C., Rigby, M., Smith, E. B., and Wakeham, W. A., *Intermolecular Forces*, Clarendon Press, Oxford, 1981.

[mar72] Marsaglia, G., *Ann. Math. Stat.* **43** (1972) 645.

[mar92] Mareschal, M. and Holian, B. L., eds., *Microscopic Simulation of Complex Hydrodynamic Phenomena*, Plenum Press, New York, 1992.

[mck92] McKechnie, J. I., Brown, D., and Clarke, J. H. R., *Macromolecules* **25** (1992) 1562.

[mcq76] McQuarrie, D. A., *Statistical Mechanics*, Harper and Row, New York, 1976.

[med90] Medvedev, N. N., Geiger, A., and Brostow, W., *J. Chem. Phys.* **93** (1990) 8337.

[mes64] Messiah, A., *Quantum Mechanics*, North-Holland, Amsterdam, 1964.

[mor85] Morriss, G. P. and Evans, D. J., *Mol. Phys.* **54** (1985) 629.

[mor91] Morriss, G. P. and Evans, D. J., *Comp. Phys. Comm.* **62** (1991) 267.

[neu83] Neumann, M., *Mol. Phys.* **50** (1983) 841.

[nic79] Nicolas, J. J., Gubbins, K. E., Streett, W. B., and Tildesley, D. J., *Mol. Phys.* **37** (1979) 1429.

[nos83] Nosé, S. and Klein, M. L., *Mol. Phys.* **50** (1983) 1055.

[nos84a] Nosé, S., *Mol. Phys.* **52** (1984) 255.

[nos84b] Nosé, S., *J. Chem. Phys.* **81** (1984) 511.

[orb67] Orban, J. and Bellemans, A., *Phys. Lett. A* **24** (1967) 620.

[par80] Parrinello, M. and Rahman, A., *Phys. Rev. Lett.* **45** (1980) 1196.

[par81] Parrinello, M. and Rahman, A., *J. Appl. Phys.* **52** (1981) 7182.

[pay93] Payne, V. A., Forsyth, M., Kolafa, J., Ratner, M. A., and de Leeuw, S. W., *J. Phys. Chem.* **97** (1993) 10478.

[pea79] Pear, M. R. and Weiner, J. H., *J. Chem. Phys.* **71** (1979) 212.

[per88] Perram, J. W., Petersen, H. G., and de Leeuw, S. W., *Mol. Phys.* **65** (1988) 875.

[pie92] Pierleoni, C. and Ryckaert, J.-P., *J. Chem. Phys.* **96** (1992) 8539.

[pol80] Pollock, E. L. and Alder, B. J., *Physica* **102A** (1980) 1.

[pow79] Powles, J. G., Evans, W. A. B., McGrath, E., Gubbins, K. E., and Murad, S., *Mol. Phys.* **38** (1979) 893.

[pre92] Press, W. H., Teukolsky, S. A., Vetterling, W. T., and Flannery, B. R., *Numerical Recipes in C: The Art of Scientific Computing*, Cambridge University Press, Cambridge, 2nd edition, 1992.

[pri84] Price, S. L., Stone, A. J., and Alderton, M., *Mol. Phys.* **52** (1984) 987.

[puh89] Puhl, A., Mansour, M. M., and Mareschal, M., *Phys. Rev. A* **40** (1989) 1999.

[que73] Quentrec, B. and Brot, C., *J. Comp. Phys.* **13** (1973) 430.

[rah64] Rahman, A., *Phys. Rev.* **136A** (1964) 405.

[rah71] Rahman, A. and Stillinger, F. H., *J. Chem. Phys.* **55** (1971) 3336.

[rah73] Rahman, A. and Stillinger, F. H., *J. Am. Chem. Soc.* **95** (1973) 7943.

[rai89] Raine, A. R. C., Fincham, D., and Smith, W., *Comp. Phys. Comm.* **55** (1989) 13.

[ral78] Ralston, A. and Rabinowitz, P., *A First Course in Numerical Analysis*, McGraw-Hill, New York, 2nd edition, 1978.

[rap79] Rapaport, D. C., *J. Chem. Phys.* **71** (1979) 3299.

[rap80] Rapaport, D. C., *J. Comp. Phys.* **34** (1980) 184.

[rap83] Rapaport, D. C., *Mol. Phys.* **48** (1983) 23.

[rap85] Rapaport, D. C., *J. Comp. Phys.* **60** (1985) 306.

[rap87]   Rapaport, D. C., *Phys. Rev. A* **36** (1987) 3288.
[rap91a]   Rapaport, D. C., *Phys. Rev. A* **43** (1991) 7046.
[rap91b]   Rapaport, D. C., *Comp. Phys. Comm.* **62** (1991) 217.
[rap91c]   Rapaport, D. C., *Comp. Phys. Comm.* **62** (1991) 198.
[rap92]   Rapaport, D. C., *Phys. Rev. A* **46** (1992) R6150.
[rap94]   Rapaport, D. C., *Europhys. Lett.* **26** (1994) 401.
[ray72]   Ray, J. R., *Am. J. Phys.* **40** (1972) 179.
[ray91]   Ray, J. R. and Graben, H. W., *Phys. Rev. A* **44** (1991) 6905.
[reb77]   Rebertus, D. W. and Sando, K. M., *J. Chem. Phys.* **67** (1977) 2585.
[rem90]   Remler, D. K. and Madden, P. A., *Mol. Phys.* **70** (1990) 921.
[ryc77]   Ryckaert, J.-P., Ciccotti, G., and Berendsen, H. J. C., *J. Comp. Phys.* **23** (1977) 327.
[ryc78]   Ryckaert, J.-P. and Bellemans, A., *Faraday Disc. Chem. Soc.* **66** (1978) 95.
[ryc90]   Ryckaert, J.-P., in [cat90], p. 189.
[sar93]   Sarman, S. and Evans, D. J., *J. Chem. Phys.* **99** (1993) 620.
[sch73]   Schofield, P., *Comp. Phys. Comm.* **5** (1973) 17.
[sch85]   Schoen, M. and Hoheisel, C., *Mol. Phys.* **56** (1985) 653.
[sch86]   Schoen, M., Vogelsang, R., and Hoheisel, C., *Mol. Phys.* **57** (1986) 445.
[sch92]   Schmidt, K. E. and Ceperley, D., in [bin92], p. 205.
[smi92]   Smith, W. and Rapaport, D. C., *Mol. Simulation* **9** (1992) 25.
[smi94]   Smith, P. E. and van Gunsteren, W. F., *J. Chem. Phys.* **100** (1994) 3169.
[spr88]   Sprik, M. and Klein, M. L., *J. Chem. Phys.* **89** (1988) 7556.
[spr91]   Sprik, M., *J. Chem. Phys.* **95** (1991) 6762.
[sta92]   Stauffer, D. and Aharony, A., *Introduction to Percolation Theory*, Taylor and Francis, London, 2nd edition, 1992.
[sti72]   Stillinger, F. H. and Rahman, A., *J. Chem. Phys.* **57** (1972) 1281.
[sti74]   Stillinger, F. H. and Rahman, A., *J. Chem. Phys.* **60** (1974) 1545.
[sti85]   Stillinger, F. H. and Weber, T. A., *Phys. Rev. B* **31** (1985) 5262.
[str78]   Streett, W. B., Tildesley, D. J., and Saville, G., *Mol. Phys.* **35** (1978) 639.
[tan83]   Tanemura, M., Ogawa, T., and Ogita, N., *J. Comp. Phys.* **51** (1983) 191.
[ten82]   Tenenbaum, A., Ciccotti, G., and Gallico, R., *Phys. Rev. A* **25** (1982) 2778.
[tho89]   Thompson, P. A. and Robbins, M. O., *Phys. Rev. Lett.* **63** (1989) 766.
[tho90]   Thompson, P. A. and Robbins, M. O., *Phys. Rev. A* **41** (1990) 6830.
[tho91]   Thompson, P. A. and Grest, G. S., *Phys. Rev. Lett.* **67** (1991) 1751.
[tox88]   Toxvaerd, S., *J. Chem. Phys.* **89** (1988) 3808.
[tri88]   Tritton, D. J., *Physical Fluid Dynamics*, Oxford University Press, Oxford, 2nd edition, 1988.
[tro84]   Trozzi, C. and Ciccotti, G., *Phys. Rev. A* **29** (1984) 916.
[tuc94]   Tuckerman, M. E. and Parrinello, M., *J. Chem. Phys.* **101** (1994) 1302.
[van82]   van Gunsteren, W. F. and Karplus, M., *Macromolecules* **15** (1982) 1528.
[ver67]   Verlet, L., *Phys. Rev.* **159** (1967) 98.
[ver68]   Verlet, L., *Phys. Rev.* **165** (1968) 201.
[vog84]   Vogelsang, R. and Hoheisel, C., *Mol. Phys.* **53** (1984) 1355.
[vog85]   Vogelsang, R. and Hoheisel, C., *Mol. Phys.* **55** (1985) 1339.
[vog87]   Vogelsang, R., Hoheisel, C., and Ciccotti, G., *J. Chem. Phys.* **86** (1987) 6371.
[vog88]   Vogelsang, R. and Hoheisel, C., *Phys. Rev. A* **38** (1988) 6296.
[whi94]   White, C. A. and Head-Gordon, M., *J. Chem. Phys.* **101** (1994) 6593.

[wig60] Wigner, E. P., *Comm. Pure App. Math.* **13** (1960) 1.

[woo76] Wood, W. W. and Erpenbeck, J. J., *Ann. Rev. Phys. Chem.* **27** (1976) 319.

[zim72] Ziman, J. M., *Principles of the Theory of Solids*, Cambridge University Press, Cambridge, 2nd edition, 1972.

Function index

AccumBondAngDistn, 256
AccumDiffusion, 121, 215
AccumDihedAngDistn, 254
AccumProps, 30, 178, 185, 274
AccumSpacetimeCorr, 133
AccumVacf, 123, 127
AddBondedPair, 96
AdjustDipole, 273
AdjustPressure, 164
AdjustQuat, 199
AdjustTemp, 65, 160, 165, 203
AllocArrays, 29, 36, 49, 53, 73, 88, 112, 122,
    125, 126, 134, 152, 174, 209, 212, 215,
    218, 228, 241, 256, 257, 272, 281, 296,
    297, 324, 347, 353
AllocMatI, 364
AllocMatR, 364
AllocVecI, 364
AllocVecR, 364
AllocVecS, 364
AnalClusterSize, 98
AnalVorPoly, 94
AnlzConstraintDevs, 241, 242
ApplyBarostat, 163
ApplyBoundaryCond, 26, 154, 172, 177,
    184
ApplyThermostat, 159, 163, 185, 202
AssignToChain, 228

BisectPlane, 106
BuildClusters, 96
BuildConstraintMatrix, 239
BuildNebrList, 54, 71, 171, 224, 252, 280,
    335
BuildRotMatrix, 198
BuildRotMatrixT, 199

CompressClusters, 97
ComputeAccelsQ, 196
ComputeAngVel, 197

ComputeChainAngleForces, 251
ComputeChainBondForces, 224
ComputeChainTorsionForces, 249
ComputeConstraints, 240
ComputeDerivsPT, 152
ComputeDipoleAccel, 271
ComputeExternalForce, 171
ComputeForces, 24, 49, 54, 125, 126, 162,
    183, 281, 337, 353
ComputeForcesDipoleF, 269
ComputeForcesDipoleR, 268
ComputeSiteForces, 208, 217
ComputeStreamFun, 326
ComputeThermalForce, 188
ComputeTorqs, 197
CorrectorStep, 61
CorrectorStepBox, 164
CorrectorStepF, 201
CorrectorStepPT, 153
CorrectorStepQ, 199
CorrectorStepS, 273

DefineMol, 210, 212
DeleteEvent, 304
DoParlCopy, 345
DoParlMove, 343
DotProd3, 377
DriveFlow, 326

ErrExit, 377
EulerToQuat, 204
EvalAtomCount, 324
EvalChainProps, 229
EvalDiffusion, 120, 214
EvalDihedAngCorr, 257
EvalFreePath, 297
EvalLatticeCorr, 91
EvalPairEng, 218
EvalProfile, 174

EvalProps, 30, 53, 152, 154, 163, 164, 185, 199, 201, 274, 340
EvalRdf, 87, 212, 254, 275
EvalShearVisc, 185
EvalSinCos, 271
EvalSpacetimeCorr, 132
EvalThermalCond, 188
EvalVacf, 123, 126
EvalVelDist, 38

FindDistVerts, 109
FindTestSites, 103
FreeMatR, 365
FreeVecR, 364
FtFastComplex, 375

GenSiteCoords, 198
GetCheckpoint, 370
GetConfig, 102
GetGridAverage, 320
GetNameList, 366
GridAverage, 173

InitAccels, 64
InitAdjustTemp, 66
InitAngAccels, 204, 272
InitAngCoords, 204, 272
InitAngVels, 204, 272
InitBoxVars, 165
InitClusters, 97
InitCoords, 27, 63, 71, 74–77, 226, 227, 253, 312, 325
InitDiffusion, 121
InitEventList, 305
InitFeedbackVars, 154
InitFreePath, 297, 298
InitParlProcs, 348
InitSpacetimeCorr, 134
InitState, 338
InitVacf, 124
InitVels, 27, 64, 71, 185, 325
InitVorPoly, 104
Integrate, 376

LeapfrogStep, 26, 159, 326

main, 21, 99, 112, 143, 289, 346, 358, 371
Max3R, 377
MeasureTrajDev, 72
Min3R, 377

NebrParlProcs, 348
NextEvent, 303
Nint, 377

OutsideObs, 324

PackCopiedData, 346
PackMovedData, 344
PerturbTrajDev, 71
PolyGeometry, 110
PolySize, 111
PredictEvent, 293, 308, 309, 311, 312
PredictorStep, 61
PredictorStepBox, 163
PredictorStepF, 200
PredictorStepPT, 153
PredictorStepQ, 199
PredictorStepS, 273
PrintDiffusion, 121
PrintNameList, 367
PrintProfile, 174
PrintRdf, 88
PrintSpacetimeCorr, 134
PrintSummary, 32, 253, 256, 295
PrintTrajDev, 72
PrintVacf, 124
PrintVelDist, 36, 38
ProcCutEdges, 107
ProcCutFaces, 107
ProcDelVerts, 106
ProcessCellCrossing, 292, 310
ProcessCollision, 291, 311
ProcInterrupt, 371
ProcNewFace, 109
ProcNewVerts, 108
PutCheckpoint, 369
PutConfig, 101
PutGridAverage, 319

RandR, 372
RandVec3, 373
ReadCheckpointData, 370
RemoveOld, 109
ReplicateAtoms, 352
RestoreConstraints, 244

ScaleCoords, 154
ScheduleEvent, 301
SetAtomType, 325
SetCellSize, 112
SetParams, 29, 49, 53, 64, 75, 76, 152, 209, 227, 254, 257, 295, 312, 347, 356
SetupFiles, 369
SetupInterrupt, 371
SetupJob, 23, 36, 53, 66, 88, 122, 124, 134, 152, 165, 174, 203, 219, 229, 253, 258, 272, 290, 296, 346, 371
SetupLayers, 356
SetupParlProcs, 358
SignR, 377
SingleEvent, 290, 296, 310
SingleStep, 22, 36, 54, 62, 66, 72, 88, 93, 122, 124, 134, 151, 164, 174, 184, 210,

SingleStep (*cont.*)
    217, 229, 252, 253, 255, 258, 273, 326,
    341
SolveCubic, 376
SolveLineq, 373
Sort, 376
Sqr, 377
StartRun, 294
SubdivCells, 112

UnpackCopiedData, 346

UnpackMovedData, 345
UpdateAtom, 295, 309
UpdateCellSize, 154
UpdateSystem, 295

WriteCheckpointData, 370

ZeroDiffusion, 121, 215
ZeroSpacetimeCorr, 134
ZeroVacf, 124, 127

Subject index

adjustment
 constraint, 241
 dipole length, 273
 pressure, 164
 temperature, 65–6, 160, 165, 203, 254,
  273
 velocity, 39, 65, 147
alkane, 234
 barrier crossing, 262
 bond-angle distribution, 256, 259–60
 bond-angle potential, 249, 251
 bond-torsion potential, 249
 dihedral-angle autocorrelation function,
  256, 258, 261
 dihedral-angle distribution, 254, 259–60
 radial distribution function, 254, 259–60
angular momentum conservation, 199, 314
argon, 16, 191
atom labels, 338
autocorrelation function, 180, 189
 heat flux, 118
 integrated, 141
 large-time behavior, 142
 negative, 136
 pressure tensor, 117
 reproducibility, 136
 truncated, 139, 141
 velocity, 116, 122, 135

binary fluid, 142
biopolymer, 211, 223, 262
block averaging, 81–2, 84–5, 186, 232, 277
Boltzmann H-function, 37–8, 41
boundary
 nonslip, 323
 obstacle, 323, 326
 periodic, 17, 26, 41, 48, 51, 55, 62, 93,
  105, 116–18, 151, 169–70, 187, 199,
  264, 288, 291, 293, 297, 323, 336,
  344, 351: ambiguities, 18;

determines region shape, 18;
 interaction range limited, 18; sliding,
 182–4; wraparound effect, 17, 130;
wall, 16, 169, 175, 301, 317: atomic
 structure, 170, 190; collision event,
 310, 312; constant-temperature, 190,
 310, 318, 322; corrugated, 310, 312;
 fixed, 169; rough, 170, 309, 318, 322;
 sliding, 181; smooth, 309; stochastic,
 170–2; temperature, 170, 177
bulk viscosity, 117, 142
butane, 258

case study format, 9
cell
 size, 49
 subdivision, 47, 284, 288
cell-crossing
 event, 288–9, 292
 examination, 293
clusters
 analysis, 95–100
 construction, 96
 geometrical properties, 98
 membership, 95
 size distribution, 98
 soft-sphere results, 99, 100
 spatial properties, 98
collision
 determining next, 288
 event, 286, 289, 291, 296, 313
 examination, 293
 generalized, 287
 impulse, 285, 313
 rules, 286
 velocity change, 286–7
communication
 broadcast, 340
 copying atoms, 345
 functions, 334, 340, 343, 359

393

communication (*cont.*)
  message buffer, 344
  message packing and unpacking, 340, 344
  moving atoms, 343
  overhead, 331–2, 357
  PVM software, 340, 359
compilation
  conditional, 378
  optimization, 378
computational geometry, 93
computer architecture
  'simple', 331
  complex, 331
  distributed memory, 331
  effect on programming, 26, 56, 365
  message-passing, 330, 332
  processor characteristics, 331
  taxonomy, 332
  vector processor, 330–1
computer simulation
  Brownian dynamics, 7
  cellular automata, 3, 7
  educational role, 4
  event-driven, 286
  experimental basis, 3
  interactive, 4, 41, 363
  large-scale, 330
  lattice–Boltzmann method, 3
  Monte Carlo method, 3, 6, 63, 79, 232
  numerical experiment, 3, 361
  quantum methods, 3
  relation to theory, 2
computer, in scientific research, 2
constraint
  adjustment, 253
  bond-angle, 235, 238
  bond-length, 43, 235, 238
  constant-pressure, 160
  constant-temperature, 147, 158, 182, 201
  deviations, 241, 258
  effect on temperature, 236, 254
  equivalent force, 44, 237
  holonomic, 44, 234
  in statistical mechanics, 235
  indexing, 237
  justification, 234
  matrix construction, 239, 262
  matrix solution, 237
  nonholonomic, 158
  penalty function, 237
  planarity, 235, 262
  preservation, 5, 253
  relaxation, 236–7, 242, 245
  'shake' method, 242, 244, 261
  structural, 4, 234–45
  tolerance, 243

continuum limit, 130
convergence factor, 265
coordinate update, 288, 295
correlation function
  current, 131: longitudinal, 140; transverse, 140
  density, 131
  space–time, 128–35, 141–2
  three-body, 79, 86, 100
  time-dependent, 114
  van Hove, 128
critical phenomena, 222
cubic equation, 229, 376
current
  longitudinal, 130
  transverse, 130

data
  averaging and filtration, 78
  input, 22, 31, 365: from file, 32; use of namelist, 31, 365
  interprocessor transfers, 334
  organization of analysis, 142
  output, 32
data file
  atomic configuration, 93, 98, 101, 368
  binary, 101, 319
  checkpoint, 368–9
  grid snapshot, 319–20, 368
  input, 348
  naming convention, 32, 366, 368
  output, 32, 142, 368
data structure
  binary tree, 298: balanced, 299; navigation, 299; pointers, 299; theoretical performance, 299, 300
  circular list, 299
  linked list, 48, 96, 104, 292
dielectric constant, 211, 221, 276, 284
diffusion coefficient, 115–16, 216
  convergence, 142
  direct measurement, 119–22
  Einstein expression, 116
  from velocity autocorrelation function, 122–5
  Green–Kubo expression, 116
dilation equation, 162–3
dimensionless number, 317–18
dimensionless units, *see* reduced units
dipole
  autocorrelation function, 214–16
  direction, 214, 266
  fluid, 264
  orientational order: long-range, 274, 276; short-range, 274, 276–7
distributed computation, 331, 334, 362
  performance, 334

distribution
  free-path, 297, 307–8
  hydrogen-bond, 219–20
  Maxwell, 34, 37, 64
  pair-energy, 217–19
  velocity, 34–7, 41
driving force, 170–1, 188
dynamic similarity, 317–18
dynamics
  constant-pressure, 160–6, 220
  constant-temperature, 158–60, 283
  non-Newtonian, 158
  nonequilibrium, 168
  rigid-body, 191–2
  unphysical, 146, 169

economy of scale, 331
energy
  conservation, 19, 20, 33, 44, 46, 57, 67,
      199, 309
  dissipation, 314
  drift, 39, 258
  kinetic, 15, 199, 201
  potential, 15
  units, 16
ensemble
  average, 6
  canonical, 6, 83, 146, 148, 158, 316
  isothermal–isobaric, 146, 149
  microcanonical, 5, 6, 80, 83, 146
equation of state, 83, 314
equations of motion, 14
  constrained, 158
  dipole vector, 268
  first-order, 200, 268
  integration, 56
  linear molecule, 200
  non-Newtonian, 5
  numerical solution, 5
  quaternion, 196, 202
  rigid-body, 194
  rotational, 268
  second-order, 200, 268
  time-reversible, 71
equilibrium
  characterizing, 34
  thermal, 2, 64
ergodic hypothesis, 6, 78
error
  analysis, 80
  correlated samples, 80
  numerical integration, 19, 39, 65, 80, 285,
      314
  reporting, 377
  statistical, 80
Euler
  angles, 191–3: ambiguity, 193

equations, 4, 195
event
  cell-crossing, 311
  collision, 286
  measurement, 295–6, 319
  prediction, 293, 312
event calendar, 288, 298
  binary tree organization, 298
  circular lists, 299
  event addition, 299, 301
  event deletion, 299, 304
  example, 302
  getting next event, 291, 299, 303
  initialization, 294, 305
  modification, 289
  node contents, 298, 301
  node pool, 299, 301
  pointers, 300
  representation, 300
  size, 295
Ewald method, 263
  charges, 264–6
  computational effort, 266
  dipoles, 266–71
  numerical accuracy, 266, 275
  surrounding medium, 265, 267, 277, 284
  symmetry, 266, 269
exactly soluble problems, 2
experimental design, 80
extended system
  variable region shape, 157
  variable timescale, 147
  variable volume, 149–56

fast Fourier transform, 143, 375
feedback, 146, 166
  equation, 152
  mechanism, 148, 150
  pressure control, 149–56, 167
  temperature control, 147–9
FFT, *see* fast Fourier transform
Fick's law, 115
fictitious field, 187
finite-size effects, 8, 18, 80, 91, 169, 318, 361
flow
  analysis program, 320
  convection rolls, 322
  Couette, 169, 181, 189
  driven, 323
  eddies, 317, 328
  laminar, 317
  nonslip, 170
  nonuniformity erased, 323
  onset of instability, 318
  oscillatory, 328
  past obstacle, 317, 323–8
  Poiseuille, 169–70, 178, 323

flow (*cont.*)
  polymer, 329
  recirculating, 323
  sheared, 175, 181
  time-dependent, 317
  viscous, 170
fluctuation
  dipole long-range order, 274, 277–8
  kinetic energy, 155
  region size, 155–6
  temperature, 149, 181
fluid dynamics
  continuum picture, 114, 129
  microscopic approach, 316, 329
Fourier transform, 86, 90, 129–31, 135, 142
Fourier's law, 115, 118, 177

Gauss's principle, 158
granular materials, 314–15
graphical methods, 174, 316
  animation, 321
  arrow plot, 321
  contour plot, 321, 326
grid analysis, 131, 318
  sampling, 321

Hamilton equations, 44, 180
Hamilton's principle, 43, 158
Hamiltonian, 44, 149, 151, 179
hard disks, 321
hard nonspherical particles, 315
hard spheres, 285
  in gravitational field, 308–9, 322
  inelastic collisions, 314
  performance, 305–6
  polymer chain, 312
  rotation, 313
  rough, 313
heat bath, 147
heat conduction equation, 170
heat current, 125
hexatic phase, 100
histogram
  bonds per molecule, 218
  dihedral-angle, 254
  energy, 217
  pair-separation, 86
  path-length, 297–8
  velocity, 34
hydrodynamic limit, 128
hydrogen bond
  average number, 220
  definition, 216–17
  distribution, 220
  network, 211, 216, 221

initial coordinates, 63, 74–7

BCC lattice, 75
diamond lattice, 75
FCC lattice, 63
polymer chain, 225
random, 77
simple cubic lattice, 74
specified by unit cells, 27
square lattice, 27
triangular lattice, 76
initial state, 67, 231, 253, 272, 276, 283, 306, 324, 338
  effect on results, 19
  nonlinear molecules, 203
  stationary center of mass, 19, 27, 64
  three-dimensional, 63
  two-dimensional, 27
initial velocities, 27, 64, 185
integration method
  accuracy, 66
  Adams–Bashforth, 59
  Adams–Moulton, 60
  adaptive, 57
  comparison, 62, 66
  high or low order, 5
  isothermal-leapfrog, 159
  leapfrog, 19, 26, 30, 57–8, 66, 123, 132, 220, 338
  Nordsieck, 59
  predictor–corrector, 57–62, 64, 66, 73, 151, 153, 159, 163, 185, 188, 199, 200, 204, 258, 273, 276, 283: coefficients, 74–6
  Runge–Kutta, 57
  storage requirements, 57
  Verlet, 57, 242
integration timestep, 16, 19, 33, 57, 210, 213, 235, 258, 276, 285
interaction
  bond-angle, 251, 279
  bond-torsion, 245–9
  cutoff, 67, 80, 162, 205, 208, 221, 265–6, 268, 279: corrections, 84, 90; range, 13, 351
  dipole, 266–7
  electrostatic, 263
  four-body, 234
  internal, 222
  long-range, 263
  orientation dependence, 284
  step potential, 285
  three-body, 234, 278–81
  velocity-dependent, 62
interaction calculation method
  all-pairs, 14, 24, 47, 51, 53, 205, 208, 264, 350
  cell multipole expansion, 264, 284
  cells, 47–52, 73, 151, 162, 205, 350

comparison, 70
efficiency, 69
layers, 350–7
multiple-timestep, 56, 73
neighbor-list, 52–5, 62, 73, 171, 205, 224, 252, 323, 334, 350, 357: three-body, 280
reaction-field, 264
replication, 55, 351
tabulation, 56, 73
interaction site, 197–8, 200, 205–7, 211, 277
associated charge, 206
location in molecule, 205
types, 208
intermediate scattering function, 129, 139, 141

kinematic viscosity, 317

Lagrange
equations, 43–4, 147, 150, 157, 236, 267
multiplier, 44, 159, 161, 180–1, 201–2, 236, 240, 242
Lagrangian, 43, 147–8, 150, 157, 160
Langevin equation, 7
Laplacian view, 1, 70
layer, *see* interaction calculation method, layers
linear equations, LU method, 240, 373
linear response theory, 115, 169, 179
constant-temperature, 181–2
Liouville
equation, 179
theorem, 180
LJ potential, *see* potential function, Lennard-Jones
local density, 90, 128
fluctuation, 128
local time variable, 288
long-range order, 90–1, 131

main program
basic form, 21
cluster analysis, 99
distributed computation, 346
space–time correlations, 143
step potential, 289
Voronoi analysis, 112
MC, *see* computer simulation, Monte Carlo method
MD, *see* molecular dynamics
mean free path, 129, 297
measurement
autocorrelation functions, 137–8
averages, 20, 30
correlated, 81, 119

diffusion coefficient, 136, 138–9, 315
energy, 20, 30, 82
framework for time-dependent, 119
grid, 172–4: resolution, 175
hard-sphere fluid, 306
long-range order, 90
overlapped samples, 119–20, 122, 131
polymer chain conformation, 231
pressure, 20, 30, 83–4, 166
radial distribution function, 88
reproducibility, 67
shear viscosity, 139, 141
size dependence, 84
soft-disk fluid, 39
soft-sphere fluid, 82
structure, 85
thermal conductivity, 139, 141
thermodynamic, 20, 79
timing, 70, 306
transport coefficients, 141
uncorrelated samples, 79
velocity autocorrelation function, 135–7
melting transition, 99, 167
metastable state, 68, 322
molecular chaos, 70
molecular crystal, 192
molecular dynamics
applications, 7–8
atoms, 12
average, 6
challenges, 9
classical nature, 1, 44, 361
classification of problems, 8
combined techniques, 9
limitations, 223, 317, 361–2
nonequilibrium, 168
thermodynamic connection, 20
units, *see* reduced units
molecule
center of mass separation, 197, 208, 211, 213
flexible, 222
formation, 191
linear, 191–2, 200, 268
monatomic, 191
nonlinear, 192
orientation, 192, 197, 199, 200, 203, 211, 214, 220
partially rigid, 43
reference state, 207
rigid, 191, 221
united-atom approximation, 234
moment of inertia, 207, 213, 276, 314
momentum conservation, 30, 33, 158, 187, 199, 314
multiple-angle recurrence relations, 132, 269

multipole
  expansion, 45
  moment, 207

N-body problem, 1
namelist
  Fortran facility, 31, 365
  input, 31, 365–6
  output, 32, 367
Navier–Stokes equation, 115, 117, 170
neigboring processors, 348
neighbor list
  construction, 54, 280, 335
  refresh, 52–3
  representation, 53, 280
nematic state, 284
networked workstations, 332
Newton's second law, 4, 14, 42–3
Newton–Raphson method, 162, 164
number density, 128
numerical precision, 67, 286

Occam's razor, 2, 4
optimizing compiler, 331

packing of atoms, 91
pair-distribution function, 85
particle current, 129
partition function, 2, 85, 148
partitioning schemes, 333–4
PC method, *see* integration method,
    predictor–corrector
peak
  Brillouin, 130
  Rayleigh, 130
percolation theory, 99, 100
phase-space integral, 6
piston, 150
polymer
  atom assignment, 228
  bead necklace, 312
  bond: interaction, 224; length, 223, 231;
      snapped, 225; stretched, 312–13
  chain, 223
  conformational change, 223
  constraint, 234
  end-to-end distance, 228, 231
  entanglement, 223
  excluded volume, 223
  hard-sphere, 312, 315
  in solution, 223–4, 226, 231–2
  initial state, 225
  internal motion, 234
  labeling elements, 246
  lattice studies, 223, 232
  mass distribution, 228–9, 231
  overlap, 225, 227

radius of gyration, 228, 231
relaxation, 232
reptation, 233
stiffness, 224
topology, 222, 237, 246, 262
polynomial fit, 171, 176
potential function
  argon, 46
  bond-angle, 251
  bond-torsion, 247–8
  Coulomb, 205–6
  effective, 45
  generic, 46
  Lennard-Jones, 5, 13, 15, 46, 206
  liquid silicon, 278–9
  smoothing discontinuity, 46, 73
  soft-sphere, 13, 15, 46, 223–4, 252, 267
  spherical atoms, 12
  square-well, 285, 287
  Stockmayer, 267
  truncated, 46, 99, 264
  various forms, 47
Prandtl number, 318
pressure, 156
pressure tensor, 118, 125, 182
processor
  'zero', 339–40, 342, 348
  deadlock, 342
  synchronization, 339, 342
profile
  temperature, 171, 175–6, 179
  velocity, 171, 175–6, 181
programming
  argument list, 28
  array: allocation, 23, 25, 29, 296, 347,
      364; indexing, 25, 73, 131, 300, 364,
      370
  checkpoint/restart, 24, 368–71
  compiler directives, 356
  efficiency, 26
  error detection, 33, 103
  floating-point precision, 25
  initialization, 22
  job termination, 22, 371
  layout, 21
  loop over timesteps, 22
  loop unrolling, 26, 356
  macros, 334, 359, 377
  nested loops, 52
  obfuscated code, 26, 366
  organization, 21, 23
  planning communications, 342
  real variables, 25
  safety measures, 343
  size assumptions, 29, 364
  software portability, 359
  style, 21

Unix: command line, 22, 32; file
    functions, 101; header file, 378; shell
    script, 372; signals, 371
  variable names, 21
  variables – listing, 378–82
programming language
  C, 21: data structures, 21; flexibility, 21;
    library functions, 32, 377
  familiarity assumed, 21
  Fortran, 21
  translation, 21
protein folding, 223
PVM, *see* communication, PVM software
  working program, 357

quantum mechanics, 5, 44
quaternions, 4, 191, 193–9, 220
  benefits, 194
  normalization, 199
  relation to Euler angles, 193, 204

radial distribution function, 86–90, 100
  computation, 87
  dipole fluid, 276–7
  hard spheres, 296, 306
  silicon, 283
  site–site, 211–14
  soft spheres, 88–9
random
  numbers, 27, 77, 372
  vectors, 64, 203, 373
Rayleigh number, 317
RDF, *see* radial distribution function
reduced units
  alkane, 254
  argon, 16, 142
  Lennard-Jones, 15
  reasons for use, 14
  silicon, 279
  water, 207
region half-length, 29
relativity, 5
Reynolds number, 317
rheology, 114, 223, 316
rotation matrix, 192–3, 198
rotational diffusion, 211, 214

scaled
  coordinates, 149, 151, 154, 157, 160, 166
  time, 147, 150
  velocities, 154
scattering
  neutron, 128
  X-ray, 86, 90
shear viscosity, 117, 171, 175–6, 182, 185–6
  Green–Kubo expression, 117
  measurement, 125–7

Newtonian definition, 115
silicon, tetrahedral structure, 278
solvation, 211
sorting, 103
  Heapsort, 375
space-filling region, 18
  hexagon, 18
  truncated octahedron, 18
specific heat, 83, 99
statistical mechanics, 2, 6, 78, 115, 168,
    316
stream function, 326
streamlines, 326
string phase, 190
structure factor
  dynamic, 129–30, 140–1
  static, 86, 129
subregion, 333–4, 336
  atoms in adjacent, 337
  limits, 337, 347
supercomputer, 330, 349
symbolic software, 248
system
  closed or open, 316
  homogeneous or inhomogeneous, 169
  inhomogeneous, 79
  isotropic, 86, 130–1
  mesoscopic, 361
  replicated, 264
  time-dependent, 79

temperature
  drift, 160, 203, 213, 276
  effect of motion on, 150, 181
  units, 15
thermal
  boundary layer, 178
  conduction, 118, 177
  convection, 118, 317–22, 329
  diffusivity, 317
  expansion coefficient, 317
thermal conductivity, 118, 171, 175–8,
    186–9
  Green–Kubo expression, 118
  measurement, 125–7
thermostat, 158–9, 170, 180–1, 184, 187,
    202, 261, 273, 316
torque, 192, 195, 197, 200, 247
trajectory
  accuracy, 5, 57, 243
  display, 39–41
  linear, 285
  parabolic, 308
  perturbed, 5, 70
  relation to diffusion, 41
  sensitivity, 70–4, 314
transient state, 322

transport coefficient, 114–15, 142, 180
   hard spheres, 296
trapezoidal integration, 124, 376
'true' displacement, 116, 120

van der Waals interaction, 13
vector processing, 349, 356
   gather and scatter operations, 350
   pipelined computation, 331, 349–50
vibration frequency, 235
virial, 20, 49, 83, 211, 261
   step-potential version, 296
visualization, 362
Voronoi polyhedra, 92–5
   analysis, 92, 100
   computational checks, 94, 113

generation, 93, 102–13
geometrical description, 105
properties, 110
soft-sphere results, 94–5

water
   density maximum, 221
   ice-like correlations, 213
   model, 192, 220
   polarizability, 221
   popularity, 206
   TIP4P model, 206–7
waves
   shear, 142
   sound, 142